Computational Photonics

A comprehensive manual on the efficient modelling and analysis of photonic devices through building numerical codes, this book provides graduate students and researchers with the theoretical background and MATLAB programs necessary for them to start their own numerical experiments.

Beginning by summarizing topics in optics and electromagnetism, the book discusses optical planar waveguides, linear optical fibres, the propagation of linear pulses, laser diodes, optical amplifiers, optical receivers, the finite-difference time-domain method, the beam propagation method and some wavelength division devices, solitons, solar cells and metamaterials.

Assuming only a basic knowledge of physics and numerical methods, the book is ideal for engineers, physicists and practicing scientists. It concentrates on the operating principles of optical devices, as well as the models and numerical methods used to describe them.

Marek S. Wartak is a Professor in the Department of Physics and Computer Science, Wilfrid Laurier University, Waterloo, Ontario. He has over 25 years of experience in semiconductor physics, photonics and optoelectronics, analytical methods, modelling and computer-aided design tools.

Computational Photonics
An Introduction with MATLAB

MAREK S. WARTAK
Wilfrid Laurier University

CAMBRIDGE
UNIVERSITY PRESS

University Printing House, Cambridge CB2 8BS, United Kingdom

Published in the United States of America by Cambridge University Press, New York

Cambridge University Press is part of the University of Cambridge.

It furthers the University's mission by disseminating knowledge in the pursuit of
education, learning and research at the highest international levels of excellence.

www.cambridge.org
Information on this title: www.cambridge.org/9781107005525

© M. S. Wartak 2013

This publication is in copyright. Subject to statutory exception
and to the provisions of relevant collective licensing agreements,
no reproduction of any part may take place without the written
permission of Cambridge University Press.

First published 2013

A catalogue record for this publication is available from the British Library

Library of Congress Cataloguing in Publication data
Wartak, Marek S., author.
Computational photonics : an introduction with MATLAB / Marek S. Wartak,
Department of Physics and Computer Science, Wilfrid Laurier University,
Waterloo, Ontario, Canada.
pages cm
Includes bibliographical references and index.
ISBN 978-1-107-00552-5 (hardback)
1. Optoelectronic devices – Mathematical models.   2. Photonics – Mathematics.
3. MATLAB.   I. Title.
TK8304.W37   2012
621.3815'2 – dc23   2012037576

ISBN 978-1-107-00552-5 Hardback

Cambridge University Press has no responsibility for the persistence or accuracy of
URLs for external or third-party internet websites referred to in this publication,
and does not guarantee that any content on such websites is, or will remain, accurate
or appropriate.

Contents

Preface		*page* xi
1 Introduction		1
1.1	What is photonics?	1
1.2	What is computational photonics?	2
1.3	Optical fibre communication	5
1.4	Biological and medical photonics	12
1.5	Photonic sensors	12
1.6	Silicon photonics	13
1.7	Photonic quantum information science	14
	References	14
2 Basic facts about optics		17
2.1	Geometrical optics	17
2.2	Wave optics	21
2.3	Problems	30
	Appendix 2A: MATLAB listings	31
	References	34
3 Basic facts from electromagnetism		35
3.1	Maxwell's equations	35
3.2	Boundary conditions	36
3.3	Wave equation	38
3.4	Time-harmonic fields	39
3.5	Polarized waves	42
3.6	Fresnel coefficients and phases	44
3.7	Polarization by reflection from dielectric surfaces	48
3.8	Antireflection coating	50
3.9	Bragg mirrors	53
3.10	Goos-Hänchen shift	58
3.11	Poynting theorem	59
3.12	Problems	60
3.13	Project	60
	Appendix 3A: MATLAB listings	61
	References	63

4 Slab waveguides — 64
- 4.1 Ray optics of the slab waveguide — 64
- 4.2 Fundamentals of EM theory of dielectric waveguides — 69
- 4.3 Wave equation for a planar wide waveguide — 71
- 4.4 Three-layer symmetrical guiding structure (TE modes) — 72
- 4.5 Modes of the arbitrary three-layer asymmetric planar waveguide in 1D — 75
- 4.6 Multilayer slab waveguides: 1D approach — 79
- 4.7 Examples: 1D approach — 85
- 4.8 Two-dimensional (2D) structures — 88
- 4.9 Problems — 92
- 4.10 Projects — 92
- Appendix 4A: MATLAB listings — 93
- References — 104

5 Linear optical fibre and signal degradation — 106
- 5.1 Geometrical-optics description — 106
- 5.2 Fibre modes in cylindrical coordinates — 110
- 5.3 Dispersion — 123
- 5.4 Pulse dispersion during propagation — 127
- 5.5 Problems — 128
- 5.6 Projects — 129
- Appendix 5A: Some properties of Bessel functions — 129
- Appendix 5B: Characteristic determinant — 130
- Appendix 5C: MATLAB listings — 131
- References — 137

6 Propagation of linear pulses — 138
- 6.1 Basic pulses — 138
- 6.2 Modulation of a semiconductor laser — 143
- 6.3 Simple derivation of the pulse propagation equation in the presence of dispersion — 146
- 6.4 Mathematical theory of linear pulses — 148
- 6.5 Propagation of pulses — 152
- 6.6 Problems — 155
- Appendix 6A: MATLAB listings — 156
- References — 165

7 Optical sources — 167
- 7.1 Overview of lasers — 167
- 7.2 Semiconductor lasers — 172
- 7.3 Rate equations — 181
- 7.4 Analysis based on rate equations — 187
- 7.5 Problems — 196
- 7.6 Project — 196

    Appendix 7A: MATLAB listings — 196
    References — 202

8 Optical amplifiers and EDFA — 204
    8.1 General properties — 205
    8.2 Erbium-doped fibre amplifiers (EDFA) — 209
    8.3 Gain characteristics of erbium-doped fibre amplifiers — 213
    8.4 Problems — 215
    8.5 Projects — 215
    Appendix 8A: MATLAB listings — 215
    References — 222

9 Semiconductor optical amplifiers (SOA) — 223
    9.1 General discussion — 223
    9.2 SOA rate equations for pulse propagation — 228
    9.3 Design of SOA — 231
    9.4 Some applications of SOA — 233
    9.5 Problem — 236
    9.6 Project — 236
    Appendix 9A: MATLAB listings — 236
    References — 238

10 Optical receivers — 240
    10.1 Main characteristics — 241
    10.2 Photodetectors — 242
    10.3 Receiver analysis — 251
    10.4 Modelling of a photoelectric receiver — 257
    10.5 Problems — 258
    10.6 Projects — 258
    Appendix 10A: MATLAB listings — 258
    References — 260

11 Finite difference time domain (FDTD) formulation — 262
    11.1 General formulation — 262
    11.2 One-dimensional Yee implementation without dispersion — 266
    11.3 Boundary conditions in 1D — 272
    11.4 Two-dimensional Yee implementation without dispersion — 275
    11.5 Absorbing boundary conditions (ABC) in 2D — 277
    11.6 Dispersion — 280
    11.7 Problems — 280
    11.8 Projects — 281
    Appendix 11A: MATLAB listings — 281
    References — 286

12 Beam propagation method (BPM) — 288
- 12.1 Paraxial formulation — 288
- 12.2 General theory — 292
- 12.3 The 1 + 1 dimensional FD-BPM formulation — 299
- 12.4 Concluding remarks — 306
- 12.5 Problems — 307
- 12.6 Project — 308
- Appendix 12A: Details of derivation of the FD-BPM equation — 308
- Appendix 12B: MATLAB listings — 310
- References — 314

13 Some wavelength division multiplexing (WDM) devices — 316
- 13.1 Basics of WDM systems — 316
- 13.2 Basic WDM technologies — 317
- 13.3 Applications of BPM to photonic devices — 323
- 13.4 Projects — 325
- Appendix 13A: MATLAB listings — 325
- References — 329

14 Optical link — 331
- 14.1 Optical communication system — 331
- 14.2 Design of optical link — 333
- 14.3 Measures of link performance — 336
- 14.4 Optical fibre as a linear system — 338
- 14.5 Model of optical link based on filter functions — 340
- 14.6 Problems — 344
- 14.7 Projects — 344
- Appendix 14A: MATLAB listings — 345
- References — 348

15 Optical solitons — 351
- 15.1 Nonlinear optical susceptibility — 351
- 15.2 Main nonlinear effects — 352
- 15.3 Derivation of the nonlinear Schrödinger equation — 353
- 15.4 Split-step Fourier method — 357
- 15.5 Numerical results — 361
- 15.6 A few comments about soliton-based communications — 364
- 15.7 Problems — 364
- Appendix 15A: MATLAB listings — 365
- References — 366

16 Solar cells — 368
- 16.1 Introduction — 368
- 16.2 Principles of photovoltaics — 370

16.3	Equivalent circuit of solar cells	373
16.4	Multijunctions	376
	Appendix 16A: MATLAB listings	379
	References	381

17 Metamaterials — 384

17.1	Introduction	384
17.2	Veselago approach	388
17.3	How to create metamaterial?	389
17.4	Some applications of metamaterials	395
17.5	Metamaterials with an active element	400
17.6	Annotated bibliography	401
	Appendix 17A: MATLAB listings	401
	References	403

Appendix A Basic MATLAB — 406

A.1	Working session with m-files	407
A.2	Basic rules	409
A.3	Some rules about good programming in MATLAB	410
A.4	Basic graphics	412
A.5	Basic input-output	416
A.6	Numerical differentiation	417
A.7	Review questions	418
	References	418

Appendix B Summary of basic numerical methods — 420

B.1	One-variable Newton's method	420
B.2	Muller's method	422
B.3	Numerical differentiation	425
B.4	Runge-Kutta (RK) methods	432
B.5	Solving differential equations	433
B.6	Numerical integration	435
B.7	Symbolic integration in MATLAB	438
B.8	Fourier series	438
B.9	Fourier transform	441
B.10	FFT in MATLAB	443
B.11	Problems	446
	References	446

Index — 448

Preface

Photonics (also known as optoelectronics) is the technology of creation, transmission, detection, control and applications of light. It has many applications in various areas of science and engineering fields. Fibre optic communication is an important part of photonics. It uses light particles (photons) to carry information over optical fibre.

In the last 20 years we have witnessed the significant (and increasing) presence of photonics in our everyday life. The creation of the Internet and World Wide Web was possible due to tremendous technical progress created by photonics, development of photonic devices, improvement of optical fibre, wavelength division multiplexing (WDM) techniques, etc. The phenomenal growth of the Internet owes a lot to the field of photonics and photonic devices in particular.

This book serves as an attempt to introduce graduate students and senior undergraduates to the issues of computational photonics. The main motivation for developing an approach described in the present book was to establish the foundations needed to understand principles and devices behind photonics.

In this book we advocate a simulation-type approach to teach fundamentals of photonics. We provide a self-contained development which includes theoretical foundations and also the MATLAB code aimed at detailed simulations of real-life devices.

We emphasize the following characteristics of our very practical book:

- learning through computer simulations
- writing and analysing computer code always gives good sense of the values of all parameters
- our aim was to provide complete theoretical background with only basic knowledge assumed
- the book is self-contained in a sense that it starts from a very basic knowledge and ends with the discussion of several hot topics.

The author believes that one can learn a lot by studying not only a theory but also by performing numerical simulations (numerical experiments) using commercial software or developed in-house. Here, we adopt a view that the best way will be to develop a software ourselves (as opposed to commercial). Also there is one more component in the learning process, namely real-life experimentation. That part is outside the scope of the present book.

The goal was to equip students with solid foundations and underlying principles by providing a theoretical background and also by applying that theory to create simple programs in MATLAB which illustrate theoretical concepts and which allow students to conduct their own numerical experiments in order to get better understanding.

Alternatively, students can also use octave (www.octave.org) to run and to conduct numerical experiments using programs developed and distributed with this book. At the end, they can try to develop large computer programs by combining learned ideas. At this stage, I would like to remind the reader that according to Donald Knuth (cited in *Computer Algebra Handbook*, J. Grabmeier, E. Kaltofen and V. Weispfenning, eds., Springer, 2003) the tasks of increasing difficulty are:

i) publishing a paper
ii) publishing a book
iii) writing a large computer program.

The book has evolved from the lecture notes delivered in the Department of Electrical Engineering, University of Oulu, Finland, Institute of Physics, Wroclaw University of Technology and Department of Physics and Computer Science, Wilfrid Laurier University, Waterloo, Canada.

The book has been designed to serve a 12-week-long term with several extra chapters which allow the instructor more flexibility in choosing the material. An important part of the book is the MATLAB code provided to solve problems. Students can modify it and experiment with different parameters to see for themselves the role played by different factors. This is a very valuable element of the learning process.

The book also contains a small number of solved problems (labelled as Examples) which are intended to serve as a practical illustration of discussed topics. At the end of each chapter there is a number of problems intended to test the student's understanding of the material.

The book may serve as a reference for engineers, physicists, practising scientists and other professionals. It concentrates on operating principles of optical devices, as well as their models and numerical methods used for description. The covered material includes fibre, planar waveguides, laser diodes, detectors, optical amplifiers and semiconductor optical amplifiers, receivers, beam propagation method, some wavelength division devices, finite-difference time-domain method, linear and nonlinear pulses, solar cells and metamaterials.

The book can also be used as a textbook as it contains questions, problems (both solved and unsolved), working examples and projects. It can be used as a teaching aid at both undergraduate and graduate levels as a one-semester course in physics and electrical engineering departments.

The book contains 17 chapters plus two appendices. The first two chapters summarize basic topics in optics and electromagnetism. The book also contains a significant number of software exercises written in MATLAB, designed to enhance and help understanding the discussed topics. Many proposed projects have been designed with a view to implement them in MATLAB.

Some numerical methods as well as important practical devices were not considered. Those developments will be included in the second edition. The author would like to wish a potential reader the similar joy of reading the book and experimenting with programs as he had with writing it.

Finally, I welcome all type of comments from readers, especially concerning errors, inaccuracies and omissions. Please send your comments to *mwartak@wlu.ca*.

Requirements I do not assume any particular knowledge since I tried to cover all basic topics. Some previous exposure to optics and electromagnetic theory will be helpful.

Acknowledgements Thanks are due to Professors Harri Kopola and Risto Myllyla (University of Oulu) and Professor Jan Misiewicz (Wroclaw University of Technology) for arranging delivery of lectures. I wish also to thank my former students in Finland, Poland and Canada for comments on various parts of lecture notes. I am grateful to my wife for reading the whole manuscript and Dr Jacek Miloszewski for preparing the index and figure for the cover.

My special thank you goes to people at Cambridge, especially Dr Simon Capelin and Mrs Antoaneta Ouzounova for their patience and good advice, and Lynette James for her excellent work.

The book support grant from the Research Office of Wilfrid Laurier University is acknowledged. Support for my research from various Canadian organizations including federal NSERC as well as from province of Ontario is gratefully acknowledged.

I dedicate this book to my family.

1 Introduction

In this introductory chapter we will try to define computational photonics and to position it within a broad field of photonics. We will briefly summarize several subfields of photonics (with the main emphasis on optical fibre communication) to indicate potential possibilities where computational photonics can significantly contribute by reducing cost of designing new devices and speeding up their development.

1.1 What is photonics?

We start our discussion from a broader perspective by articulating what photonics is, what the current activities are and where one can get the most recent information.

Photonics is the field which involves electromagnetic energy, such as light, where the fundamental object is a photon. In some sense, photonics is parallel to electronics which involves electrons. Photonics is often referred to as optoelectronics, or as electro-optics to indicate that both fields have a lot in common. In fact, there is a lot of interplay of photonics and electronics. For example, a laser is driven by electricity to produce light or to modulate that light to transmit data.

Photonics applications use the photon in a similar way to that which electronic applications use the electron. However, there are several advantages of optical transmission of data over electrical. Furthermore, photons do not interact between themselves (which is both good and bad), so electromagnetic beams can pass through each other without interacting and/or causing interference.

Even with the telecommunication 'bubble' which happened some ten years ago, the subfield of photonics, namely optical fibre communication, is still a very important segment of photonics activities. As an example, a single optical fibre has the capacity to carry about 3 million telephone calls simultaneously. But, due to a crisis around 2000, many new applications of photonics have emerged or got noticed. Bio-photonics or medical photonics are amongst the most important ones.

Before we try to define computational photonics, let us concentrate for a moment on what photonics is at the time of writing (Winter, 2011). In the following paragraphs, we cite some information from the relevant conferences.

Some of the well-established conferences which summarize research and applications in a more traditional approach to photonics, and also dictate new directions are: Photonics West, held every January in California (here is the 2011 information [1]); Photonics East, taking place in the eastern part of the USA in the Fall; Optical Fibre Conference (OFC) taking place in March every year [2].

A new conference, the Photonics Global Conference (PGC), is a biennial event held since 2008 [3]. The aim of this conference is to foster interactions among broad disciplines in photonics with concentration on emerging directions. Symposia and Special Sessions for 2011 are summarized below:

- Symposia: 1. Optofluidics and Biophotonics. 2. Fibre-Based Devices and Applications. 3. Green Photonics. 4. High-Power Lasers and Their Industrial Applications. 5. Metamaterials and Plasmonics. 6. Nanophotonics. 7. Optical Communications and Networks.
- Special Sessions: 1. Quantum Communications. 2. Photonics Crystal Fibres and Their Applications. 3. Photonics Applications of Carbon Nanotube and Graphene. 4. Terahertz Technology. 5. Diffuse Optical Imaging.

By looking at the discussed topics, one can get some sense about current activities in photonics.

1.2 What is computational photonics?

Computational photonics is a branch of physics which uses numerical methods to study properties and propagation of light in waveguiding structures. (Here, light is used in a broad sense as a replacement for electromagnetic waves.) Within this field, an important part is played by studies of the behaviour of light and light-matter interactions using analytical and computational models. This emerging field of computational science is playing a critically important role in designing new generations of integrated optics modules, long haul transmission and telecommunication systems.

Generally, computational photonics is understood as the 'replacement' of the experimental method, where one is performing all the relevant 'experiments' on computer. Obviously, such an approach reduces development cost and speeds-up development of new products.

We will attempt to cover some of those developments. The field is, however, so broad that it is impossible to review all of its activities. Naturally, selection of the topics reflects author's expertise.

As a separate topic in photonics, one selects integrated photonics where the concentration is on waveguides, simulations of waveguide modes and photonic structures [4]. The central role is played there by a beam-propagation method which will be discussed later in the book.

1.2.1 Methods of computational photonics. Computational electromagnetics

According to Joannopoulos *et al.* [5], in a broad sense there are three categories of problems in computational photonics: frequency-domain eigensolvers, frequency-domain solvers and time-domain simulations. Those problems are extensively discussed in many sources including [5], also [6] and [7].

A recent general article by Gallagher [8] gives an introduction to the main algorithms used in photonic computer aided design (CAD) modelling along with discussion of their strengths and weaknesses. The main algorithms are:

- BPM – Beam Propagation Method
- EME – Eigenmode Expansion Methods
- FDTD – Finite Difference Time Domain.

The above methods are compared against speed, memory usage, numerical aperture, the refractive index contrasts in the device, polarization, losses, reflections and nonlinearity. The author's conclusion is that none of the discussed algorithms are universally perfect for all applications.

We finish this section by indicating that application of computational electromagnetics goes well beyond photonics, see the discussion by Jin [6]. In fact it has an extremely wide range of applications which goes from the analysis of electromagnetic wave scattering by various objects, antenna analysis and design, modelling and simulation of microwave devices, to numerical analysis of electromagnetic interference and electromagnetic compatibility.

1.2.2 Computational nano-photonics

This subfield has emerged only in recent years. As the dimensions of photonics devices shrink, it plays an increasingly important role. Nanostructures are generally considered as having sizes in the order of the wavelength of light. At those wavelengths, multiple scattering and near-field effects have a profound influence on the propagation of light and light-matter interaction. In turn, this leads to novel regimes for basic research as well as novel applications in various disciplines.

For instance, the modified dispersion relation of photonic crystals and photonic crystal fibres leads to novel nonlinear wave propagation effects such as giant soliton shifts and supercontinuum generation with applications in telecommunication, metrology and medical diagnostics. Owing to the complex nature of wave interference and interaction processes, experimental studies rely heavily on theoretical guidance both for the design of such systems as well as for the interpretation of measurements. In almost all cases, a quantitative theoretical description of such systems has to be based on advanced computational techniques that solve the corresponding numerically very large linear, nonlinear or coupled partial differential equations.

University research on photonic crystal structures applied to novel components in optical communication, boomed since it has become possible to create structures on the nanometre scale. Nanotechnology offers a potential for more efficient integrated optical circuitry at lower cost – from passive elements like filters and equalizers to active functions such as optical switching, interconnects and even novel lasers. In addition, the technology can provide new functionality and reduce overall optical communication cost thanks to smaller devices, higher bandwidth and low loss device characteristics that eliminate the need for amplification. The ultimate goal is to create three-dimensional photonic structures that may lead to progressively optical and one day all-optical computing.

A very recent overview on modelling in photonics, including discussion of important computational methods, was written by Obayya [9].

1.2.3 Overview of commercial software for photonics

An outstanding progress in photonics has been to a great extent possible due to the availability of reliable software. Some of the main commercial players are listed below:

1. At the time of writing, Optiwave (www.optiwave.com) provides comprehensive engineering design tools which benefit photonic, bio-photonic and system design engineers with a comprehensive design environment. Current Optiwave products include two groups: system and amplifier design, and component design.
   For system and amplifier design they offer two packages:
   - OptiSystem – suite for design of amplifier and optical communication systems.
   - OptiSPICE – the first optoelectronic circuit design software.
   For component design, they have the following products:
   - OptiBPM, based on the beam propagation method (BPM), offers design of complex optical waveguides which perform guiding, coupling, switching, splitting, multiplexing and demultiplexing of optical signals in photonic devices.
   - OptiFDTD, which is based on the finite-difference time-domain (FDTD) algorithm with second-order numerical accuracy and the most advanced boundary condition, namely uniaxial perfectly matched layer. Solutions for both electric and magnetic fields in temporal and spatial domain are obtained using the full-vector differential form of Maxwell's equations.
   - OptiFiber, which uses numerical mode solvers and other models specialized to fibres for calculating dispersion, losses, birefringence and polarization mode dispersion (PMD).
   - OptiGrating, which uses the coupled mode theory to model the light and enable analysis and synthesis of gratings.
2. RSoft (www.rsoftdesign.com) product family includes:
   - Component Design Suite to analyse complex photonic devices and components through industry-leading computer-aided design,
   - System Simulation to determine the performance of optical telecom and datacom links through comprehensive simulation techniques and component models, and
   - Network Modeling for cost-effective deployment of DWDM and SONET technologies while designing and optimizing an optical network.
3. Photon Design (www.photond.com)
   They offer several products for both passive and active component design, such as FIMMWAVE and CrystalWave.
   FIMMWAVE is a generic full-vectorial mode finder for waveguide structures. It combines methods based on semi-analytical techniques with other more numerical methods such as finite difference or finite element.
   FIMMWAVE comes with a range of user-friendly visual tools for designing waveguides, each optimized for a different geometry: rectangular geometries often

encountered in epitaxially grown integrated optics, circular geometries for the design of fibre waveguides, and more general geometries to cover, e.g. diffused waveguides or other unusual structures.

CrystalWave is a design environment for the layout and design of integrated optics components optimized for the design of photonic crystal structures. It is based on both FDTD and finite-element frequency-domain (FEFD) simulators and includes a mask file generator optimized for planar photonic crystal structures.

4. CST MICROWAVE STUDIO (www.cst.com/Content/Products/MWS/Overview.aspx) is a specialist tool for the 3D electromagnetic simulation of high frequency components. It enables the fast and accurate analysis of high frequency devices such as antennas, filters, couplers, planar and multi-layer structures.

There are several web resources devoted to photonics software and numerical modelling in photonics. We mention the following: Optical Waveguides: Numerical Modeling Website [10], Photonics resources page [11] and Photonics software [12].

1.3 Optical fibre communication

In this section we will outline the important role played by photonics in the communication over an optical fibre. Other sub-fields will be summarized in due course.

1.3.1 Short story of optical fibre communication

Light was used as a mean for communication probably since the origin of our civilization. Its modern applications in telecommunication, as developed in the twentieth century, can be traced to two main facts:

- availability of good optical fibre, and
- compact and reliable light sources.

Later in this section, we will discuss main developments in some detail. We start with a short history of communication with the emphasis on light communication. A recent popular history of fibre optics written for a broad audience was published by Hecht [13]. As an introductory work we recommend simple introduction to fundamentals of digital communications by Bateman [14].

1.3.2 Short history of communication

After discovery of electromagnetic waves due to work by Marconi, Tesla and many others, the radio was created operating in the 0.5–2 MHz frequency range with a bandwidth of 15 kHz. With the appearance of television which required bandwidth of about 6 MHz, the carrier frequencies moved to around 100 MHz. During the Second World War the

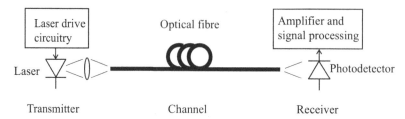

Fig. 1.1 Block diagram of a fibre system.

invention of radar pushed the frequencies to the microwave region (around GHz). Those frequencies (range of 2.4–5 GHz) are now actively used by cell phones as well as wireless links.

The big step toward higher frequencies was made possible in the 1960s with the invention of laser. The first one operated at the 694 nm wavelength. This corresponds to carrier frequency of approximately 5×10^{14} Hz (5×10^5 GHz = 500 THz). Utilization of only 1% of that frequency represents a signal channel of about 5 THz which can accommodate approximately 10^6 analogue video channels, each with 6 MHz bandwidth or 10^9 telephone calls at 5 kHz per call.

A typical communication channel which links two points, point-to-point link (as an example, see Fig. 1.1 for a generic fibre optics link), consists of an optical transmitter which is usually a semiconductor laser diode, an optical fibre intended to carry light beam and a receiver. A semiconductor laser diode can be directly modulated or an external modulator is used which produces modulated light beam which propagates in the optical fibre. Information is imposed into optical pulses. Then the light is input into the optical fibre where it propagates over some distance, and then it is converted into a signal which a human can understand.

All of those elements should operate efficiently at the corresponding carrier's frequencies (or wavelengths). In the 1960s only one element was in place: a transmitter (laser). The other two elements were nonexistent.

The next main step was proposed in 1966 by Charles Kao and George Hockham [15] who demonstrated the first silica-based optical fibre with sufficiently low enough propagation loss to enable its use as a communication medium. Soon, silica-based optical fibre became the preferred means of transmission in both long- and short-haul telecommunication networks. An all-optical network has the potential for a much higher data rate than combined electrical and optical networks and allows simultaneous transmission of multiple signals along one optical fibre link.

Below, we summarize selected developments of the early history of long-distance communication systems (after Hecht [13] and Einarsson [16]):

- TAT-1 (1956) First transatlantic telephone system contained two separate coaxial cables, one for each direction of transmission. The repeaters (based on vacuum tubes) were spliced into the cables at spacing of 70 km. Capacity of the transmission was 36 two-way voices.

- TAT-6 (1976) Capacity of 4200 two-way telephone channels. Repeaters (based on transistors) were separated at a distance of 9.4 km.
- TAT-7 (1983) System similar to TAT-6.
- TAT-8 (1988) First intercontinental optical fibre system between USA and Europe.
- HAW-4 (1988) System installed between USA and Hawaii; similar to TAT-8.
- TCP-3 Extension of HAW-4 to Japan and Guam.
- TAT-12 (1996) First transatlantic system in service with optical amplifiers.

A typical optical fibre used in TAT-8 system supported single-mode transmission [16]. It had outer diameter of 125 μm and core diameter of 8.3 μm. It operated at 1.3 μm wavelength. Digital information was transmitted at 295.6 Mbits. The repeaters which regenerated optical signals were spaced 46 km apart.

Some of the main developments that took place in relation to development of fibre communication are summarized below (compiled using information from Ref. [13]):

- October 1956: L. E. Curtiss and C. W. Peters described plastic-clad fibre at Optical Society of America meeting in Lake Placid, New York.
- Autumn 1965: C. K. Kao concludes that the fundamental limit on glass transparency is below 20 dB km^{-1} which would be practical for communication.
- Summer 1970: R. D. Maurer, D. Keck and P. Schultz from Corning Glass Works, USA make a single-mode fibre with loss of 16 dB km^{-1} at 633 nm by doping titanium into fibre core.
- April 22, 1977: General Telephone and Electronics. First live telephone traffic through fibre optics, at 6 Mbits^{-1} in Long Beach, CA.
- 1981: British Telecom transmits 140 Mbits s^{-1} through 49 km of single-mode fibre at 1.3 μm wavelength.
- Late 1981: Canada begins trial of fibre optics to homes in Elie, Manitoba.
- 1988: L. Mollenauer of Bell Labs demonstrates soliton transmission through 4000 km of single mode fibre.
- February 1991: M. Nakazawa, NTT Japan sends soliton signals through a million km of fibre.
- 1994: World Wide Web grows from 500 to 10 000 servers.
- 1996: Introduction of a commercial wavelength-division multiplexing (WDM) system.
- July 2000: Peak of telecom bubble. JDS Uniphase announces plans to merge with SDL Inc. in stock deal valued at $41 billion.
- 2001: NEC Corporation and Alcatel transmitted 10 terabits per second through a single fibre [17].
- December 2001: Failure of TAT-8 submarine cable.

The capacity of optical fibre communication systems grew very fast. Comparison with Moore's law for computers which states that computer power doubles every 18 months, indicates that fibre capacity grows faster [18]. In 1980, a typical fibre could carry about 45 Mb s^{-1}; in 2002 a single fibre was able to transmit more than 3.5 Tb s^{-1} of data. Over those 22 years the computational power of computers has increased 26 000 times whereas the fibre capacity increased 110 000 times over the same period of time.

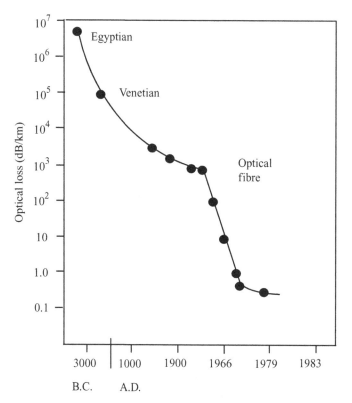

Fig. 1.2 Historical reduction of loss as a function of time from ancient times to the present. Copyright 1989 IEEE. Reprinted, with permission from S. R. Nagel, *IEEE Communications Magazine*, **25**, 33 (1987).

1.3.3 Development of optical fibre

Glass is the principal material used to fabricate optical fibre. To be useful as a transmission medium it has to be extremely clean and uniform. In Fig. 1.2 we illustrated the history of reduction losses in glass as a function of time (after [19]). To make practical use of glass as a medium used to send light, an important element, and in fact the very first challenge was to develop glass so pure that 1% of light entering at one end would be retained at the other end of a 1 km-thick block of glass. In terms of attenuation this 1% corresponds to about 20 decibels per kilometre (20 dB km^{-1}).

In 1966 Kao and Hockhman [15] suggested that the high loss in glass was due to impurities and was not the intrinsic property of glass. In fact, Kao suggested to use optical fibre as a transmission medium.

Kao's ideas were made possible in 1970 with the fabrication of low-loss glass fibre by the research group from Corning Glass Works [20]. That work established a potential of optical communications for providing a high bandwidth and long distance data transmission network. Corning Glass Works were the first to produce an optical glass with a transmission loss of 20 dB km^{-1}, which was thought to be the threshold value to permit efficient optical communication.

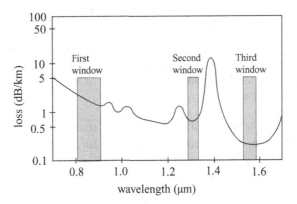

Fig. 1.3 Spectral loss of a typical low-loss optical fibre. Adapted from [22] with permission from the Institution of Engineering and Technology.

Before the Corning work, around 1966, the power loss for the best available glass was about 1000 dB km^{-1} which implies a 50% power loss in a distance of about 3 m [21]. Today, most optical fibres in use are so-called single-mode fibre able to maintain single mode of operation (see Chapter 5 for the relevant definitions). Commercially available fibres have losses of about 0.2 dB km^{-1} at the wavelength of 1550 nm and are capable of transmitting data at 2–10 Gbit s^{-1}.

The measured spectral loss of a typical low-loss optical fibre is shown in Fig. 1.3 [22]. The measured fibre had 9.4 μm core diameter and a 0.0028 refractive index difference between the core and cladding. Peaks were found at about 0.94, 1.24 and 1.39 μm. They were caused by OH vibrational absorption. The rise of the loss from 1.7 μm is mainly attributed to the intrinsic infrared absorption of the glass.

The locations of the first, second and third communications windows are also shown. The most common wavelengths used for optical communication span between 0.83–1.55 μm. Early technologies used the 0.8–0.9 μm wavelength band (referred to as the first window) mostly because optical sources and photodetectors were available at these wavelengths. Second telecommunication window, centered at around 1.3 μm, corresponds to zero dispersion of fibre. Dispersion is a consequence of the velocity propagation being different for different wavelengths of light. As a result, when pulse travels through a dispersive media it tends to spread out in time, the effect which ultimately limits speed of digital transmission. The third window, centered at around 1.55 μm, corresponds to the lowest loss. At that wavelength, the glass losses approach minimum of about 0.15 dB km^{-1}.

For the sake of comparison, the loss of about 0.2 dB km^{-1} corresponds to 50% power loss after propagation distance of about 15 km.

1.3.4 Comparison with electrical transmission

Historically, copper was a traditional medium used to transmit information by electrical means. Its data handling capacity is, however, limited. Only around 1970, after the fabrication of relatively low-loss optical fibre, it started to be used in an increasing scale. From practical perspective, the choice between copper or glass is based on several factors.

In general, optical fibre is chosen for systems which are used for longer distances and higher bandwidths, whereas electrical transmission is preferred in short distances and relatively low bandwidth applications. The main advantages of electrical transmission are associated with:

- relatively low cost of transmitters and receivers
- ability to carry electrical signals and also electrical power
- for not too large quantities, relatively low cost of materials
- simplicity of connecting electrical wires.

However, it has to be realized that optical fibre is much lighter than copper. For example, 700 km of telecommunication copper cable weighs 20 tonnes, whereas optical fibre cable of the same length weighs only around 7 kg [23].

The main advantages of optical fibre are:

- fibre cables experience effectively no crosstalk
- they are immune to electromagnetic interference
- lighter weight
- no electromagnetical radiation
- small cable size.

1.3.5 Governing standards

In order for various manufacturers to be able to develop components that function compatibly in fibre optic communication systems, a number of standards have been developed and published by the International Telecommunications Union (ITU) [24]. Other standards, produced by a variety of standards organizations, specify performance criteria for fibre, transmitters and receivers to be used together in conforming systems.

These and related standards are covered in a recent publication [25]. It is the work of over 25 world-renowned contributors and editors. It provides a reference to over a hundred standards and industry technical specifications for optical networks at all levels: from components to networking systems through global networks, as well as coverage of networks management and services.

The ITU has specified six transmission bands for fibre optic transmission [18]:

1. the O-band (original band) – 1260–1310 nm,
2. the E-band (extended band) – 1360–1460 nm,
3. the S-band (short band) – 1460–1530 nm,
4. the C-band (conventional band) – 1530–1565 nm,
5. the L-band (long band) – 1565–1625 nm,
6. the U-band (ultra band) – 1625–1675 nm.

A seventh band, not defined by the ITU, runs around 850 nm and is used in private networks.

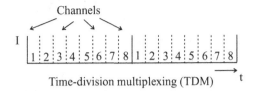
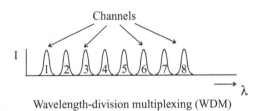

Fig. 1.4 Illustration of TDM and WDM formats.

Fig. 1.5 WDM transmission system.

1.3.6 Wavelength division multiplexing (WDM)

Wavelength division multiplexing (WDM) is a technique which allows optical signals with different wavelengths to propagate in a single optical fibre without affecting one another. The technique was invented to increase the information-carrying capacity of a fibre. WDM is not the only possible way to increase fibre capacity. In Fig. 1.4 we illustrate two basic formats, each allowing an increase of fibre capacity, time-division multiplexing (TDM) and wavelength-division multiplexing (WDM). In both formats messages from various channels are combined (multiplexed) onto a single information channel. This can be done in time or at various frequencies (wavelengths).

There is extensive literature on the principles, concepts and components for WDM, see [26], [27], [28], [29], [30] to just name a few. The WDM transmission system is shown in Fig. 1.5. Here, MUX denotes a multiplexer. Its role is to combine light from several sources, each operating at a particular wavelength carrying one channel, to the transmission fibre. At the receiving end, the demultiplexer (DMUX) separates different wavelengths.

1.3.7 Solitons

In the mid 1980s researchers started experimenting with solitons. The soliton phenomenon was first described by John Scott Russell, who in 1834 observed a solitary wave in a narrow water channel (the Union Canal in Scotland).

Solitons are pulses of special shape which travel without changing it. Pulse broadening is due to the fact that some wavelengths travel faster than other wavelengths, the effect known as dispersion. In fibre, there also exists nonlinearity. When taken together, dispersion and nonlinearity, they compensate for each other, thus producing conditions where pulse shape remains unchanged. To form a soliton, the initial pulse must have a particular peak energy and pulse shape.

In 1973, Akira Hasegawa of AT&T Bell Labs was the first to suggest that solitons could exist in optical fibre. In 1988, Linn Mollenauer and his team transmitted soliton pulses over 4000 km [31]. A recent overview of soliton-based optical communications was provided by Hasegawa [32] and by Mollenauer and Gordon [33].

1.4 Biological and medical photonics

A special issue of the *IEEE Journal of Selected Topics in Quantum Electronics* [34] summarizes hot topics in bio-photonics until circa 2005. Some of them are directly related to computational bio-photonics, such as elastic light scattering properties of tissues [35].

Two special issues of *IEEE Journal of Selected Topics in Quantum Electronics* on bio-photonics [36], [37] summarize recent progress and trends. Biological applications of stimulated parametric emission microscopy and stimulated Raman scattering microscopy are reviewed by Kajiyama *et al.* [38]. As an introduction to the subject we recommend the book by Prasad [39].

1.5 Photonic sensors

Photonic sensors have been the subject of intensive research over the last two decades. They are emerging as an important branch of photonics. Their development is based on analysing different components of the optical signal, like intensity, polarization, pulse shape or arrival time and also other phenomena like interference. Utilizing those various possibilities leads to different types of sensors. There are almost endless possibilities for their use. The best known applications are in civil and military environments for detection of a wide variety of biological, chemical and nuclear agents.

An important medium for sensing is the optical fibre and its recent variations which found their own ways of applications in sensing. These include: photonic band gap fibres (PBG), microstructure optical fibres (MOF), random hole optical fibres (RHOF) and hybrid ordered random hole optical fibres (HORHOF). They offer higher resolution, lower cost and/or expanded detection range capability for sources and detection schemes.

In this book we will not explicitly concentrate on analysis and/or simulations of photonic sensors, although some of the discussed properties of photonic devices can be directly applied in sensors.

As an example, the use of light emitting diodes (LED) in sensing has been summarized recently by O'Toole and Diamond [40]. The authors described development and advancement of LED-based chemical sensors and sensing devices.

Another example is device based on a high-finesse, whispering-gallery-mode disk resonator that can be used for the detection of biological pathogens [41]. The device operates by means of monitoring the change in transfer characteristics of the disk resonator when biological materials fall onto its active area.

The open access journal *Sensors* (ISSN 1424-8220; CODEN: SENSC9) [42] on science and technology of sensors and biosensors is published monthly online. A special issue: *Photonic Sensors for Chemical, Biological, and Nuclear Agent Detection* was published recently. It can be considered a good source of current developments and applications of new physical phenomena in sensing.

1.6 Silicon photonics

The application of silicon in photonics certainly needs a separate discussion. Silicon is a material of choice in electronics and it is also economical to build photonic devices on this material [43], see also the contribution by Soref [44] for an overview of optoelectronic integrated circuits. However, silicon is not a direct-gap material and therefore cannot efficiently generate light. Intensive work on those issues is taking place in many institutions, see the recent summary of the Intel-UC Santa Barbara collaboration on the hybrid silicon laser [45].

On another front, associated with computer technology, the continuation of Moore's Law and progress which it reflects is becoming increasingly dependent on ultra-fast data transfer between and within microprocessor. The existing electrical interconnects are expected to be replaced by high speed optical interconnects. They are seen as a promising way forward, and silicon photonics is seen as particularly useful, due to the ability to integrate electronic and optical components on the same silicon chip.

Particularly significant progress has been seen in the advances of optical modulators based on silicon, see the recent report by a leading collaboration [46].

In designing and optimizing those devices simulations play a very important role. Recent advances in modelling and simulation of silicon photonic devices are summarized by Passaro and De Leonardis [47]. They review recently developed simulation methods for submicrometre innovative Silicon-on-Insulator (SOI) guiding structures and photonic devices.

1.7 Photonic quantum information science

A new emerging direction within photonics is related to quantum communications and quantum information processing. Recent progress in this subfield is summarized by Politi *et al.* [48]. Here, we briefly outline recent main developments.

Photons potentially can play the crucial role in the quantum information processing due to their low-noise properties and ease of manipulation at the single qubit level. In recent years several quantum gates have been implemented using integrated photonics circuits.

1. A theoretical proposal for the implementation of the quantum Hadamard gate using photonic crystal structures [49]. An integrated optics Y-junction beam splitter has been proposed for realization of this gate. Reported numerical simulations support theoretical concepts.

2. A group from Bristol demonstrated the capability to build photonic quantum circuits in miniature waveguide devices [50] and optical fibres [51]. They reported recently on high-fidelity silica-on-silicon integrated optical realizations of key quantum photonic circuits, including two-photon quantum interference and a controlled-NOT logic gate [52]. They have also demonstrated controlled manipulation of up to four photons on-chip, including high-fidelity single qubit operations, using a lithographically patterned resistive phase shifter [53].

References

[1] Photonics West, http://spie.org/x2584.xml 3 July 2012.
[2] OFC Conference, www.ofcnfoec.org/home.aspx, 3 July 2012.
[3] Photonics Global Conference, www.photonicsglobal.org/index.php, 3 July 2012.
[4] C. R. Pollock and M. Lipson. *Integrated Photonics*. Kluwer Academic Publishers, Boston, 2003.
[5] J. D. Joannopoulos, S. G. Johnson, J. N. Winn, and R. D. Meade. *Photonic Crystals: Molding the Flow of Light*. Princeton University Press, Princeton and Oxford, 2008.
[6] J.-M. Jin. *Theory and Computation of Electromagnetic Fields*. Wiley-IEEE Press, Hoboken, NJ, 2010.
[7] A. Taflove and S. C. Hagness. *Computational Electrodynamics. The Finite-Difference Time-Domain Method*. Artech House, Boston, 2005.
[8] D. Gallagher. Photonic CAD matures. *IEEE LEOS Newsletter*, **2**:8–14, 2008.
[9] S. Obayya. *Computational Photonics*. Wiley, Chichester, 2011.
[10] Optical Waveguide Modeling, http://optical-waveguides-modeling.net/, 3 July 2012.
[11] Photonics resources page, www.ashokpm.org/photonics.htm, 3 July 2012.
[12] Photonics software, www.skywise711.com/lasers/software.html, 3 July 2012.
[13] J. Hecht. *City of Light: The Story of Fiber Optics*. Oxford University Press, Oxford, 1999.

[14] A. Bateman. *Digital Communications: Design for the Real World*. Addison-Wesley, Harlow, 1999.

[15] K. C. Kao and G. A. Hockman. Dielectric-fibre surface waveguides for optical frequencies. *Proc. IEE*, **133**:1151–8, 1966.

[16] G. Einarsson. *Principles of Lightwave Communications*. Wiley, Chichester, 1996.

[17] J. Hecht. Computers full of light: a short history of optical data communications. In C. DeCusatis, ed., *Handbook of Fiber Optic Data Communication*, pp. 3–17. Elsevier, Amsterdam, 2008.

[18] D. R. Goff, K. S. Hansen, and M. K. Stull. *Fiber Optic Video Transmission: The Complete Guide*. Focal Press, Oxford, 2002.

[19] S. R. Nagel. Optical fiber – the expanding medium. *IEEE Communications Magazine*, **25**:33–43, 1987.

[20] F. P. Kapron, D. B. Keck, and R. D. Maurer. Radiation losses in glass optical waveguides. *Appl. Phys. Lett.*, **17**:423–5, 1970.

[21] G. Ghatak and K. Thyagarajan. *Introduction to Fiber Optics*. Cambridge University Press, Cambridge, 1998.

[22] T. Miya, Y. Terunuma, T. Hosaka, and T. Miyashita. Ultimate low-loss single-mode fibre at 1.55 µm. *Electronics Letters*, **15**:106–8, 1979.

[23] Optical fiber, http://en.wikipedia.org/wiki/Optical_fiber, 3 July 2012.

[24] International Telecommunication Union, www.itu.int/en/pages/default.aspx, 3 July 2012.

[25] K. Kazi. *Optical Networking Standards: A Comprehensive Guide*. Springer, New York, 2006.

[26] R. T. Chen and L. S. Lome, eds., *Wavelength Division Multiplexing*. SPIE Optical Engineering Press, Bellington, WA, 1999.

[27] G. Keiser. *Optical Fiber Communications. Third Edition*. McGraw-Hill, Boston, 2000.

[28] S. V. Kartalopoulos. *DWDM. Networks, Devices and Technology*. IEEE Press, Wiley, Hoboken, NJ, 2003.

[29] J. Zheng and H. T. Mouftah. *Optical WDM Networks. Concepts and Design Principles*, IEEE Press, Wiley, Hoboken, NJ, 2004.

[30] J. C. Palais. *Fiber Optic Communications. Fifth Edition*. Prentice Hall, Upper Saddle River, NJ, 2005.

[31] L. F. Mollenauer and K. Smith. Demonstration of soliton transmission over more than 4000 km in fiber with loss periodically compensated by Raman gain. *Opt. Lett.*, **13**:675–7, 1988.

[32] A. Hasegawa. Soliton-based optical communications: An overview. *IEEE J. Select. Topics Quantum Electron.*, **6**:1161–72, 2000.

[33] L. F. Mollenauer and J. P. Gordon. *Solitons in Optical Fibers. Fundamentals and Applications*. Elsevier, Amsterdam, 2006.

[34] R. R. Jacobs. Introduction to the issue on biophotonics. *IEEE J. Select. Topics Quantum Electron.*, **11**:729–32, 2005.

[35] X. Li, A. Taflove, and V. Backman. Recent progress in exact and reduced-order modeling of light-scattering properties of complex structures. *IEEE J. Select. Topics Quantum Electron.*, **11**:759–65, 2005.

[36] I. K. Ilev, L. V. Wang, S. A. Boppart, S. Andersson-Engels, and B.-M. Kim. Introduction to the special issue on biophotonics. Part 1. *IEEE J. Select. Topics Quantum Electron.*, **16**:475–7, 2010.

[37] I. K. Ilev, L. V. Wang, S. A. Boppart, S. Andersson-Engels, and B.-M. Kim. Introduction to the special issue on biophotonics. Part 2. *IEEE J. Select. Topics Quantum Electron.*, **16**:703–5, 2010.

[38] S. Kajiyama, Y. Ozeki, K. Fukui, and K. Itoh. Biological applications of stimulated parametric emission microscopy and stimulated Raman scattering microscopy. In B. Javidi and T. Fournel, eds., *Information Optics and Photonics. Algorithms, Systems, and Applications*, pp. 89–100. Springer, New York, 2010.

[39] P. N. Prasad. *Introduction to Biophotonics*. Wiley, Hoboken, NJ, 2003.

[40] M. O'Toole and D. Diamond. Absorbance based light emitting diode optical sensors and sensing devices. *Sensors*, **8**:2453–79, 2008.

[41] R. W. Boyd and J. E. Heebner. Sensitive disk resonator photonic biosensor. *Applied Optics*, **40**:5742–7, 2001.

[42] Sensors – Open Access Journal, www.mdpi.com/journal/sensors, 3 July 2012.

[43] G. T. Reed and A. P. Knights. *Silicon Photonics: An Introduction*. Wiley, Chichester, 2004.

[44] R. Soref. Introduction: the opto-electronics integrated circuit. In G. T. Reed, ed., *Silicon Photonics: The State of the Art*, pp. 1–13. Wiley, Hoboken, NJ, 2008.

[45] J. E. Bowers, D. Liang, A. W. Fang, H. Park, R. Jones, and M. J. Paniccia. Hybrid silicon lasers: The final frontier to integrated computing. *Optics & Photonics News*, **21**:28–33, 2010.

[46] D. J. Thomson, F. Y. Gardes, G. T. Reed, F. Milesi, and J.-M. Fedeli. High speed silicon optical modulator with self aligned fabrication process. *Optics Express*, **18**:19064–9, 2011.

[47] V. M. N. Passaro and F. De Leonardis. Recent advances in modelling and simulation of silicon photonic devices. In G. Petrone and G. Cammarata, eds., *Modelling and Simulation*, pp. 367–90. I-Tech Education and Publishing, Vienna, 2008.

[48] A. Politi, J. Matthews, M. G. Thompson, and J. L. O'Brien. Integrated quantum photonics. *IEEE J. Select. Topics Quantum Electron.*, **15**:1673–84, 2009.

[49] S. Salemian and S. Mohammadnejad. Quantum Hadamard gate implementation using planar lightwave circuit and photonic crystal structures. *American Journal of Applied Sciences*, **5**:1144–8, 2008.

[50] A. Politi, M. J. Cryan, J. G. Rarity, S. Yu, and J. L. O'Brien. Silica-on-silicon waveguide quantum circuits. *Science*, **320**:646–9, 2008.

[51] A. S. Clark, J. Fulconis, J. G. Rarity, W. J. Wadsworth, and J. L. O'Brien. All-optical-fiber polarization-based quantum logic gate. *Phys. Rev. A*, **79**:030303, 2009.

[52] A. Laing, A. Peruzzo, A. Politi, M. Rodas Verde, M. Halder, T. C. Ralph, M. G. Thompson, and J. L. O'Brien. High-fidelity operation of quantum photonic circuits. *Appl. Phys. Lett.*, **97**:211109, 2010.

[53] J. C. F. Matthews, A. Politi, A. Stefanov, and J. L. O'Brien. Manipulation of multiphoton entanglement in waveguide quantum circuits. *Nature Photonics*, **3**:346–50, 2009.

2 Basic facts about optics

In this chapter, we review basic facts about geometrical optics. Geometrical optics is based on the concept of rays which represent optical effects geometrically. For an excellent introduction to this topic consult work by Pedrotti and Pedrotti [1].

2.1 Geometrical optics

2.1.1 Ray theory and applications

We visualize propagation of light by introducing the concept of *rays*. These rays obey several rules:

- In a medium they travel with velocity v given by

$$v = \frac{c}{\overline{n}} \quad (2.1)$$

where $c = 3 \times 10^8$ m/s is the velocity of light in a vacuum and the quantity $\overline{n}$ is known as refractive index. (We introduced here the notation $\overline{n}$ for refractive index instead of widely accepted n in the optics literature to avoid confusion with electron density, n which will be used in later chapters.) Some typical values of refractive indices are shown in Table 2.1.
- In a uniform medium rays travel in a straight path.
- At the plane of interface between two media, a ray is reflected at the angle equal to the angle of incidence, see Fig. 2.1. At the plane interface, the direction of transmitted ray is determined by Snell's law

$$\overline{n}_1 \sin \theta_1 = \overline{n}_2 \sin \theta_2 \quad (2.2)$$

where $\overline{n}_1$ and $\overline{n}_2$ are the values of refractive indices in both media and θ_1 and θ_2 are the angles of incidence and refraction, respectively.

The general relation between incoming ray A and the reflected ray B is

$$B = R \cdot A \quad (2.3)$$

with R reflection coefficient (complex). Reflection coefficients for two main types of waves which can propagate in planar waveguides are given by Fresnel formulas and will be determined in the next chapter.

Table 2.1 Refractive index for some materials.

Material	Refractive index
air	1.0
water	1.33
silica glass	1.5
GaAs	3.35
silicon	3.5
germanium	4.0

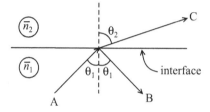

Fig. 2.1 Snell's law.

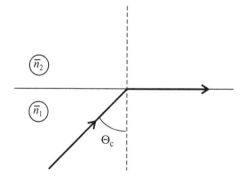

Fig. 2.2 Illustration of critical angle.

2.1.2 Critical angle

From Snell's law one can deduce the conditions under which light ray will stay in the same medium (e.g. dense) after reflection, see Fig. 2.2. The situation shown corresponds to $\theta_2 = \pi/2$. Angle θ_1 then becomes known as the critical angle, θ_c and its value is:

$$\sin \theta_c = \frac{\bar{n}_2}{\bar{n}_1} \qquad (2.4)$$

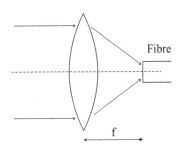

Fig. 2.3 Focusing light beam onto a fibre.

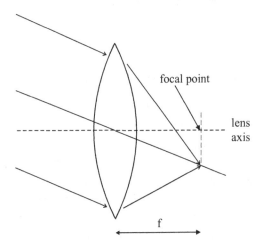

Fig. 2.4 Focusing an off-axis beam.

2.1.3 Lenses

We discuss here only *thin lenses*, defined when the vertical translation of a ray passing through lens is negligible. The ray enters and leaves the lens at approximately the same distance from the lens axis.

An important use of lenses in fibre optics is to focus light onto a fibre, see Fig. 2.3. Here, a collimated beam of light (parallel to the lens axis) is focused to a point. The point is known as the focal point and is located at a distance f, known as focal length from the lens. All the collimated rays converge to the focal point.

Focal distance f is related to the curvatures R_1 and R_2 of the spheres forming lens as

$$\frac{1}{f} = (\bar{n} - 1)\left(\frac{1}{R_1} + \frac{1}{R_2}\right) \tag{2.5}$$

If beam of light travels at some angle relative to the lens axis, Fig. 2.4, the rays will be focused in the focal plane.

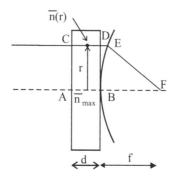

Fig. 2.5 A cylindrical disk with graded refractive index.

2.1.4 GRIN systems

GRIN systems whose name originates from GRadient in the INdex of refraction play the role similar to lenses. Here, the refractive index of the material is inhomogeneous. Spatial dependence of refractive index serves as yet another parameter in the design of an ideal lens.

The simple analysis of GRIN structure now follows [2], see Fig. 2.5. We have a flat cylindrical disk where the refractive index $\bar{n}(r)$ depends on radius r (distance from the axis), keeps its maximum at the optical axis and decreases outwards. In order to focus optical rays at the focal point F, the optical paths of all rays must be the same. (Optical path is defined as a product of geometrical distance in a given region and the value of refractive index in that region.) Optical path difference of the central ray and the one going through region with $\bar{n}(r)$ between planar wave front on the left and spherical wavefront on the right should be zero

$$\bar{n}_{\max} \cdot d \approx \bar{n}(r) \cdot d + DE \qquad (2.6)$$

Distance DE is determined as follows: $EF = f$, where f is the focal length, $DF = DE + f$ and $DF \approx \sqrt{r^2 + f^2}$. Combining those relations, we have

$$DE = \sqrt{r^2 + f^2} - f \qquad (2.7)$$

Use the expansion

$$\sqrt{r^2 + f^2} = f\sqrt{1 + \frac{r^2}{f^2}} \approx f\left(1 + \frac{r^2}{2f^2}\right) \qquad (2.8)$$

Apply the above expansion in (2.7) and obtain

$$DE \approx f + \frac{r^2}{2f^2} - f = \frac{r^2}{2f^2} \qquad (2.9)$$

Finally, the dependence of refractive index is

$$\bar{n}(r) = \bar{n}_{\max} - \frac{r^2}{2df^2} \qquad (2.10)$$

The above equation tells us that in order to focus parallel light by the GRIN plate, the index of refraction should depend parabolically on radius.

2.2 Wave optics

Light is an electromagnetic wave and consists of an electric field vector and a perpendicular magnetic field vector oscillating at a very high rate (of the order of 10^{14} Hz) and travelling in space along direction which is perpendicular to both electric and magnetic field vectors. (These topics are discussed in more detail in the next chapter.) Here we state that the electric field vector $\mathbf{E}(\mathbf{r},t)$ obeys the following wave equation:

$$\nabla^2 \mathbf{E}(\mathbf{r},t) - \frac{1}{c^2}\frac{\partial^2 \mathbf{E}(\mathbf{r},t)}{\partial t^2} = 0 \qquad (2.11)$$

The above wave equation is time invariant; that is it does not change under the replacement $t \to -t$. We will soon utilize that fact in establishing the so-called Stokes relations.

The one-dimensional wave equation in a medium is

$$\frac{1}{v^2}\frac{\partial^2}{\partial t^2}E(z,t) = \frac{\partial^2}{\partial z^2}E(z,t)$$

where v is the velocity of light in the medium. An exact solution can be expressed in terms of two waves E^+ and E^- travelling in the positive and negative z-directions

$$E(z,t) = E^+(z - vt) + E^-(z + vt)$$

Dispersion relation for the above wave equation is obtained by substituting $E(z,t) = E_0 \sin(\omega t - kz)$ into the wave equation, performing differentiation and dividing both sides by $E(z,t)$. The obtained dispersion relation which relates k and ω is

$$\omega = \pm v \cdot k$$

It shows a linear dependence between angular frequency ω of the propagating wave and its wavenumber k. Angular frequency ω is related to frequency as $\omega = 2\pi f$ whereas wavenumber k is determined as

$$k = \frac{\omega}{v} = \frac{2\pi}{\lambda}$$

Here λ is the wavelength of the light wave and v is the phase velocity of the wave. The factor $\omega t - kz$ is known as the *phase* of the wave.

Example When electromagnetic wave propagates in a medium where losses are significant, the attenuation must be incorporated into an expression for electric field of the wave. The relevant expression is

$$E(z,t) = E_0 e^{-\alpha z} \sin(\omega t - kz)$$

where α is the attenuation coefficient. The intensity of light is proportional to the square of its electric field. Therefore, the power of the light beam decreases as exp($-2\alpha L$) during the propagation distance L. If we measure the distance in km, the unit of α is km^{-1}. Power reduction in decibels (dB) is determined as

$$dB = 10 \log_{10} e^{-2\alpha L} \tag{2.12}$$

Find the relation between the attenuation coefficient and the power change in dB/km.

Solution
One has

$$\begin{aligned} Loss(dB) &= 10 \log_{10} e^{-2\alpha L} \\ &= -20 \cdot \alpha \cdot L \cdot \log e \\ &= -8.685 \cdot \alpha \cdot L \end{aligned}$$

Therefore

$$Loss(dB)/L(km) = -8.685 \cdot \alpha$$

or

$$dB/km = -8.685 \cdot \alpha \tag{2.13}$$

where α is in km^{-1}.

2.2.1 Phase velocity

Consider a plane monochromatic wave propagating along the z-axis in an infinite medium, see Fig. 2.6

$$E(z, t) = E_0 \cos(kz - \omega t) \tag{2.14}$$

Select one point, say at the amplitude crest, and analyse its movement. If we assume that this point maintains constant phase as

$$kz - \omega t = const$$

we can determine velocity of that point. From the above relation

$$k \frac{dz}{dt} - \omega = 0$$

or

$$v_p = \frac{dz}{dt} = \frac{\omega}{k} \tag{2.15}$$

The quantity v_p is known as *phase velocity* and it represents the velocity at which a surface of constant phase moves in the medium. Velocity in Eq. (2.1) should be interpreted as phase velocity and Eq. (2.1) be replaced by

$$v_p = \frac{c}{n} \tag{2.16}$$

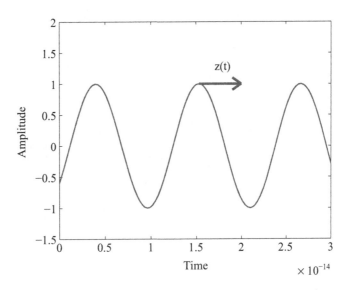

Fig. 2.6 Illustration of phase velocity.

2.2.2 Group velocity

The group velocity describes the speed of propagation of a pulse in a medium. To see this fact, consider propagation of two plane waves with different parameters but equal amplitudes

$$E_1(z,t) = E_0 \cos(k_1 z - \omega_1 t), \quad E_2(z,t) = E_0 \cos(k_2 z - \omega_2 t) \quad (2.17)$$

Assume that their frequencies and wavenumbers differ by $2\Delta\omega$ and $2\Delta k$ and write

$$\omega_1 = \omega + \Delta\omega, \quad \omega_2 = \omega - \Delta\omega$$
$$k_1 = k + \Delta k, \quad k_2 = k - \Delta k$$

Superposition of those waves results in a wave $E(z,t)$ which is

$$\begin{aligned} E(z,t) &= E_1(z,t) + E_2(z,t) \\ &= E_0 \{\cos[(k+\Delta k)z - (\omega+\Delta\omega)t] + \cos[(k-\Delta k)z - (\omega-\Delta\omega)t]\} \end{aligned}$$

Using trigonometric identity

$$2\cos\alpha\cos\beta = \cos(\alpha+\beta) + \cos(\alpha-\beta)$$

allows us to write expression for $E(z,t)$ as

$$E(z,t) = 2E_0 \cos(kz - \omega t)\cos(\Delta k z - \Delta\omega t) \quad (2.18)$$

Eq. (2.18) represents a wave at carrier frequency ω that is modulated by a sinusoidal envelope at the beat frequency $\Delta\omega$, see Fig. 2.7. Phase velocity of the carrier wave is

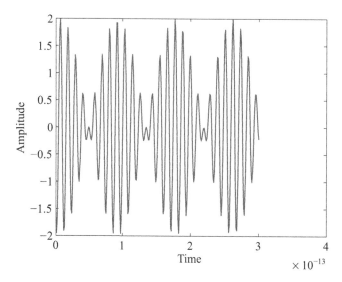

Fig. 2.7 Illustration of group velocity.

$v_p = \omega/k$. The envelope moves with velocity v_g which is

$$v_g = \frac{\Delta\omega}{\Delta k} \rightarrow \frac{d\omega}{dk} \qquad (2.19)$$

In a free space $\omega = kc$ which is known as dispersion relation, the group velocity is $v_g = \frac{d\omega}{dk} = c$. In a free space (vacuum), the phase and group velocities are identical. In media, the dispersion relation, i.e. $\omega = \omega(k)$ dependence is usually a complicated function.

Example Write a MATLAB program to visualize phase and group velocities based on Eqs. 2.15 and 2.19.

Solution

The MATLAB code for phase velocity is provided in the Listing 2A.1 and in the Listing 2A.2 for group velocity. The results are shown in Fig. 2.6 and Fig. 2.7.

2.2.3 Stokes relations

Stokes relations relate reflectivities and transmittivities of light at the plane of interface between two media. Assume that electric field E which is represented by the corresponding ray interacts with the interface and is partially reflected and partially transmitted, see Fig. 2.8. Here r, t are reflection and transmission coefficients for fields (not intensities) propagating from medium 2 to medium 1 (here, from top to bottom). Similarly, r', t' denote the reflection and transmission coefficients for fields propagating from medium 1 to medium 2.

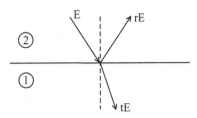

Fig. 2.8 Transmission and reflection at the interface used in deriving Stokes relations.

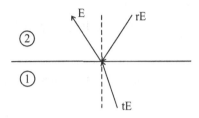

Fig. 2.9 Transmission and reflection at the interface used in deriving Stokes relations. Time-reversed process.

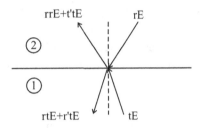

Fig. 2.10 Transmission and reflection at the interface used in deriving Stokes relations. Full processes.

At the beginning of this section we have established the property of time invariance of the wave equation. Time reversal invariance tells us that the above picture (reflection and transmission) should also work in reverse time, see Fig. 2.9 where we combined rays tE and rE to form ray E.

However, when one inputs two rays (tE and rE) at the interface between two media 1 and 2, one from above and one from below, there should exist partial reflection and transmission for both rays, as is illustrated in Fig. 2.10.

A comparison of Fig. 2.9 and Fig. 2.10 reveals physically identical situations, and therefore the following relations hold:

$$r^2 E + t't E = E$$
$$rt E + r't E = 0$$

From the above we obtain Stokes relations

$$t't = 1 - r^2$$
$$r' = -r$$

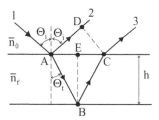

Fig. 2.11 Interference in a single film with light incident at arbitrary angle θ_i.

2.2.4 Interference in dielectric film

Interference in dielectric film is an important phenomenon behind operation of optical devices, like filters. It is also a fundamental effect for the operation of the Fabry-Perot interferometer.

To analyse this interference, consider a single layer dielectric film having refractive index $\overline{n}_f$ which is in contact with the medium having refractive index $\overline{n}_0$ from above, see Fig. 2.11. For such a configuration, we will derive conditions for constructive and destructive interference. We will do this by calculating the optical path difference.

The incident beam 1 creates on reflection and transmission two emerging beams 2 and 3. The phase difference between beams 2 and 3 as measured at points C and D is due to optical path difference Δ (optical path is defined as a product of geometrical path times refractive index) and is

$$\Delta = \overline{n}_f(AB + BC) - \overline{n}_0 AD \tag{2.20}$$

From Fig. 2.11 one can determine the following trigonometric relations:

$$\frac{AD}{AC} = \cos(90^o - \theta_i) = \sin\theta_i \tag{2.21}$$

$$\frac{h}{AB} = \cos\theta_t \tag{2.22}$$

$$\frac{AE}{h} = \tan\theta_t \tag{2.23}$$

and also

$$AC = 2 \cdot AE \tag{2.24}$$

Combining Eqs. (2.23) and (2.24), one obtains

$$AC = 2 \cdot t \cdot \tan\theta_t \tag{2.25}$$

Substituting (2.25) into (2.21) we obtain geometrical path for AD part

$$AD = 2 \cdot h \cdot \tan\theta_t \cdot \sin\theta_i \tag{2.26}$$

Geometrical path $AB + AC$ is obtained using Eq. (2.22) as

$$AB + AC = 2 \cdot AB = 2 \cdot \frac{t}{\cos\theta_t} \tag{2.27}$$

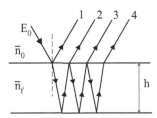

Fig. 2.12 Multiple interference in a parallel plate.

Combining Eqs. (2.26) and (2.27) with (2.20) one finds the optical path difference as

$$\Delta = \bar{n}_f \cdot 2 \cdot \frac{t}{\cos\theta_t} - \bar{n}_0 \cdot 2 \cdot t \cdot \frac{\sin\theta_t}{\cos\theta_t} \cdot \sin\theta_i \tag{2.28}$$

$$= \frac{2h\bar{n}_f}{\cos\theta_t}\left\{1 - \frac{\bar{n}_0}{\bar{n}_f}\sin\theta_t\sin\theta_i\right\} \tag{2.29}$$

Using Snell's law

$$\bar{n}_0 \sin\theta_i = \bar{n}_f \sin\theta_t$$

and the following trigonometric identity

$$\cos^2\theta_t + \sin^2\theta_t = 1$$

the curly bracket in Eq. (2.29) gives

$$1 - \frac{\bar{n}_0}{\bar{n}_f}\sin\theta_t\sin\theta_i = 1 - \frac{1}{\bar{n}_f}\sin\theta_t \bar{n}_f \sin\theta_t$$
$$= 1 - \sin^2\theta_t = \cos^2\theta_t$$

Employing the last result in (2.29), we finally obtain for *the optical path difference*

$$\Delta = 2 \cdot h \cdot \bar{n}_f \cdot \cos\theta_t \tag{2.30}$$

Thus, the optical path difference is expressed in terms of the angle of refraction. For normal incidence $\theta_i = \theta_t = 0$ and $\Delta = 2\bar{n}_f h$. The *phase difference* δ acquired by light ray due to travel in the film corresponding to optical path difference Δ is

$$\delta = k \cdot \Delta = \frac{2\pi}{\lambda_0} \cdot \Delta \tag{2.31}$$

2.2.5 Multiple interference in a parallel plate

Based on discussion from the previous section, we will now consider interference of multiple beams in a parallel plate, see Fig. 2.12. Here, E_0 is the amplitude of the incoming light, r, t are the coefficients of external reflections and transmissions, and r', t' are coefficients of internal reflections and transmissions.

Phase difference between two successive reflected beams which exists due to travel in the film having refractive index $\bar{n}_f$ is, Eq. (2.31)

$$\delta = k \cdot \Delta$$

where Δ is given by Eq. (2.30). Analysing Fig. 2.12, one can write the following relations for outgoing beams

$$
\begin{aligned}
E_0 e^{i\omega t} & \quad \text{incident beam} \\
E_1 &= r E_0 e^{i\omega t} & \quad \text{first reflected beam} \\
E_2 &= tt'r' E_0 e^{i\omega t - i\delta} & \quad \text{second reflected beam} \\
E_3 &= tt'r'^3 E_0 e^{i\omega t - i2\delta} & \quad \text{third reflected beam} \\
E_4 &= tt'r'^5 E_0 e^{i\omega t - i3\delta} & \quad \text{fourth reflected beam}
\end{aligned}
$$

kth reflected beam is

$$E_k = tt' r'^{(2k-1)} E_0 e^{i\omega t - i(k-1)\delta}$$

Sum of all reflected beams, E_R is

$$
\begin{aligned}
E_R &= \sum_{k=1}^{\infty} E_k = r E_0 e^{i\omega t} + \sum_{k=2}^{\infty} tt' r'^{(2k-3)} E_0 e^{i\omega t} e^{-i(k-2)\delta} \\
&= E_0 e^{i\omega t} \left\{ r + tt' r' e^{-i\delta} \sum_{k=2}^{\infty} r'^{(2k-4)} e^{-i(k-2)\delta} \right\}
\end{aligned}
$$

To perform the above infinite sum, use an elementary result for the sum of the geometric series

$$S = \sum_{k=2}^{\infty} x^{k-2} = 1 + x + x^2 + \cdots$$

For $|x| < 1$, the sum is $S = 1/(1-x)$. In our case here, $x = r'^2 e^{-i\delta}$, and total reflection is thus

$$E_R = E_0 e^{i\omega t} \left\{ r + \frac{tt' r' e^{-i\delta}}{1 - r'^2 e^{-i\delta}} \right\}$$

Using Stokes relations, we obtain

$$
\begin{aligned}
E_R &= E_0 e^{i\omega t} \left\{ r - \frac{(1 - r^2) r e^{-i\delta}}{1 - r^2 e^{-i\delta}} \right\} \\
&= E_0 e^{i\omega t} \frac{r(1 - e^{-i\delta})}{1 - r^2 e^{-i\delta}}
\end{aligned}
$$

Light intensity of the reflected beam is proportional to the irradiance I_R which is

$$
\begin{aligned}
I_R \sim |E_R|^2 &= E_0^2 r^2 \left\{ \frac{e^{i\omega t}(1 - e^{-i\delta})}{1 - r^2 e^{-i\delta}} \right\} \left\{ \frac{e^{-i\omega t}(1 - e^{i\delta})}{1 - r^2 e^{i\delta}} \right\} \\
&= E_0^2 r^2 \frac{2(1 - \cos \delta)}{1 + r^4 - 2r^2 \cos \delta}
\end{aligned}
$$

Introduce I_i as the irradiance of the incident beam and I_T as the irradiance of the transmitted beam. We have

$$\frac{I_R}{I_i} = \frac{|E_R|^2}{|E_i|^2}$$

and

$$I_R + I_T = I_i$$

which originates from energy conservation in nonabsorbing films. Using the above relations, one obtains

$$I_R = \frac{2r^2(1 - \cos\delta)}{1 + r^4 - 2r^2 \cos\delta} I_i \tag{2.32}$$

and

$$I_T = \frac{(1 - r^2)^2}{1 + r^4 - 2r^2 \cos\delta} I_i \tag{2.33}$$

2.2.6 Fabry-Perot (FP) interferometer

The FP interferometer is formed by two parallel plates with some medium (often air) between them. Expressions for transmission and reflection in such a system have been derived in the previous section.

The expression for transmission can be used to determine the *transmittance T* (or *Airy function*) as follows:

$$T \equiv \frac{I_T}{I_i} = \frac{(1 - r^2)^2}{1 + r^4 - 2r^2 \cos\delta}$$
$$= \frac{1 - 2r^2 + r^4}{1 + r^4 - 2r^2 + 4r^2 \sin^2 \frac{\delta}{2}}$$
$$= \frac{1}{1 + \frac{4r^2}{(1-r^2)^2} \sin^2 \frac{\delta}{2}}$$

using trigonometric relation $\cos\delta = 1 - 2\sin^2 \frac{\delta}{2}$.

Define *coefficient of finesse*

$$F \equiv \frac{4r^2}{(1 - r^2)^2} \tag{2.34}$$

Transmittance is thus expressed as

$$T = \frac{1}{1 + F \sin^2 \frac{\delta}{2}} \tag{2.35}$$

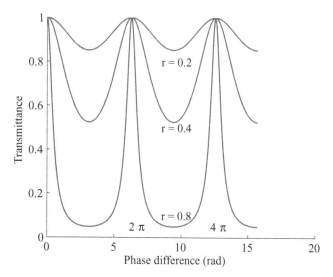

Fig. 2.13 Fabry-Perot fringe profile.

Example Write a MATLAB program to analyse the transmittance of an FP etalon given by Eq. (2.35).

Solution

The MATLAB code is shown in Appendix, Listing A 2.3 and the FP fringe profile is shown in Fig. 2.13. The figure shows plots of transmission as a function of phase difference δ for various values of external reflection r.

2.3 Problems

1. Use Snell's law to derive an expression for critical angle θ_c. Compute the value of θ_c for a water-air interface ($\bar{n}_{water} = 1.33$).
2. Isotropic light source located at distance d under water illuminates circular area of a radius 5 m. Determine d.
3. A tank of water is covered with a 1 cm thick layer of linseed oil ($\bar{n}_{oil} = 1.48$) above which is air. What angle must a beam of light, originating in the tank, make at the water-oil interface if no light is to escape?
4. Analyse the travel of a light ray through a parallel glass plate assuming refractive index of glass $\bar{n}$. Determine displacement of the outgoing ray.
5. A point source S is located on the axis of, and 30 cm from a plane-convex thin lens which has a radius of 5 cm. The glass lens is immersed in air. Determine the location of the image (a) when the flat surface is toward S and (b) when the curved surface is toward S.

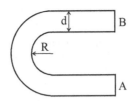

Fig. 2.14 Propagation of light in U-tube.

6. Analyse multiple interference in a parallel plate FP etalon formed by two parallel plates with an active medium inside with gain g [3]. Derive an expression for a signal gain.
7. A glass rod of rectangular cross section is bent into U-shape as shown in Fig. 2.14 [4]. A parallel beam of light falls perpendicularly on the flat surface A. Determine the minimum value of the ratio R/d for which all light entering the glass through surface A will emerge from the glass through surface B. Assume that the refractive index of glass is 1.5.
8. The Sellmeier dispersion equation is an empirical expression for the refractive index of a dielectric (like glass) in terms of the wavelength λ. One of the most popular models is [5], [6]

$$\bar{n}^2 = 1 + G_1 \frac{\lambda^2}{\lambda^2 - \lambda_1^2} + G_2 \frac{\lambda^2}{\lambda^2 - \lambda_2^2} + G_3 \frac{\lambda^2}{\lambda^2 - \lambda_3^2} \quad (2.36)$$

Here G_1, G_2, G_3 and $\lambda_1, \lambda_2, \lambda_3$ are constants (called Sellmeier coefficients) which are determined by fitting the above expression to the experimental data. Some of the Sellmeier parameters are given in Table 2.2 (data combined from [5], [7] and [6]).

Table 2.2 Sellmeier parameters for SiO_2.

Parameter	G_1	G_2	G_3	λ_1 (μm)	λ_2 (μm)	λ_3 (μm)
SiO_2	0.696749	0.408218	0.890815	0.0690660	0.115662	9.900559

Write a MATLAB program to plot refractive index as a function of λ from 0.5 μm to 1.8 μm for pure silica.

9. Using the Sellmeier relation, plot reciprocal group velocity and group velocity for SiO_2. Assume $1.36 < \lambda[\mu m] < 1.65$.

Appendix 2A: MATLAB listings

In Table 2.3 we provided a list of MATLAB files created for Chapter 2 and a short description of each function.

Table 2.3 List of MATLAB functions for Chapter 2.

Listing	Function name	Description
2A.1	*phase_velocity.m*	Plots phase velocity
2A.2	*group_velocity.m*	Plots group velocity
2A.3	*FP_transmit.m*	Plots transmittance of Fabry-Perot etalon

Listing 2A.1 Program phase_velocity.m. MATLAB program which plots phase velocity.

```
% File name: phase_velocity.m
% Illustrative plot of phase velocity
clear all
N_max = 101;                       % number of points for plot
t = linspace(0,30d-15,N_max);      % creation of theta arguments
%
c = 3d14;                          % velocity of light in microns/s
n = 3.4;                           % refractive index
v_p = c/n;                         % phase velocity
lambda = 1.0;                      % microns
k = 2*pi/lambda;
frequency = v_p/lambda;
z = 0.6;                           % distance in microns
omega = 2*pi*frequency;
A = sin(k*z - omega*t);
plot(t,A,'LineWidth',1.5)
xlabel('Time','FontSize',14);
ylabel('Amplitude','FontSize',14);
set(gca,'FontSize',14);            % size of tick marks on both axes
axis([0 30d-15 -1.5 2])
text(17d-15, 1.3, 'z(t)','Fontsize',16)
line([1.53d-14,2d-14],[1,1],'LineWidth',3.0)    % drawing arrow
line([1.9d-14,2d-14],[1.1,1],'LineWidth',3.0)
line([1.9d-14,2d-14],[0.9,1],'LineWidth',3.0)
pause
close all
```

Listing 2A.2 Function group_velocity.m. MATLAB program which plots group velocity.

```
% File name: group_velocity.m
% Illustrative plot of group velocity by superposition
% of two plane waves
clear all
N_max = 300;                       % number of points for plot
```

```
t = linspace(0,300d-15,N_max);      % creation of theta arguments
%
c = 3d14;                           % velocity of light in microns/s
n = 3.4;                            % refractive index
v_p = c/n;                          % phase velocity
lambda = 1.0;                       % microns
frequency = v_p/lambda;
z = 0.6;                            % distance in microns
omega = 2*pi*frequency;
k = 2*pi/lambda;
Delta_omega = omega/15.0;
Delta_k = k/15.0;
%
omega_1 = omega + Delta_omega;
omega_2 = omega - Delta_omega;
k_1 = k + Delta_k;
k_2 = k - Delta_k;
A = 2*cos(k*z - omega*t).*cos(Delta_k*z - Delta_omega*t);
plot(t,A,'LineWidth',1.3)
xlabel('Time','FontSize',14)
ylabel('Amplitude','FontSize',14)
set(gca,'FontSize',14);             % size of tick marks on both axes
pause
close all
```

Listing 2A.3 Function FP_transmit.m. Program plots transmittance of Fabry-Perot etalon.

```
% FP_transmit.m
% Plot of transmittance of FP etalon
clear all
N_max = 401;                        % number of points for plot
t = linspace(0,1,N_max);            % creation of theta arguments
delta = (5*pi)*t;                   % angles in radians
%
hold on
for r = [0.2 0.4 0.8]               % reflection coefficient
    F = 4*r^2/(1-r^2).^2;           % coefficient of finesse
    T = 1./(1 + F*(sin(delta/2)).^2);
    plot(delta,T,'LineWidth',1.5)
end
%
% Redefine figure properties
ylabel('Transmittance','FontSize',14)
xlabel('Phase difference (rad)','FontSize',14)
```

```
set(gca,'FontSize',14);              % size of tick marks on both axes
text(6.1, 0.05, '2 \pi','Fontsize',14)
text(12.2, 0.05, '4 \pi','Fontsize',14)
text(8.5, 0.1, 'r = 0.8','Fontsize',14)
text(8.5, 0.5, 'r = 0.4','Fontsize',14)
text(8.5, 0.8, 'r = 0.2','Fontsize',14)
%
pause
close all
```

References

[1] F. L. Pedrotti and L. S. Pedrotti. *Introduction to Optics. Third Edition*. Prentice Hall, Upper Saddle River, NJ, 2007.

[2] E. Hecht. *Optics. Fourth Edition*. Addison-Wesley, San Francisco, 2002.

[3] Y. Yamamoto. Characteristics of AlGaAs Fabry-Perot cavity type laser amplifiers. *IEEE J. Quantum Electron.*, **16**:1047–52, 1980.

[4] L. Yung-Kuo, ed. *Problems and Solutions on Optics*. World Scientific, Singapore, 1991.

[5] J. W. Fleming. Dispersion in $GeO_2 - SiO_2$ glasses. *Applied Optics*, **23**:4486–93, 1984.

[6] S. O. Kasap. *Optoelectronics and Photonics: Principles and Practices*. Prentice Hall, Upper Saddle River, NJ, 2001.

[7] G. Ghosh, M. Endo, and T. Iwasaki. Temperature-dependent Sellmeier coefficients and chromatic dispersions for some optical fiber glasses. *J. Lightwave Technol.*, **12**:1338–42, 1994.

3 Basic facts from electromagnetism

In this chapter we review foundations of electromagnetic theory and basic properties of light.

3.1 Maxwell's equations

Our discussion is based on Maxwell's equations which in differential form are [1], [2]

$$\nabla \times \mathbf{E} = -\frac{\partial \mathbf{B}}{\partial t} \tag{3.1}$$

$$\nabla \times \mathbf{H} = \mathbf{J} + \frac{\partial \mathbf{D}}{\partial t} \tag{3.2}$$

$$\nabla \cdot \mathbf{D} = \rho_v \tag{3.3}$$

$$\nabla \cdot \mathbf{B} = 0 \tag{3.4}$$

where

$\mathbf{E}$ is electric field intensity $[V/m]$
$\mathbf{B}$ is magnetic flux density $[T]$
$\mathbf{H}$ is magnetic field intensity $[A/m]$
$\mathbf{D}$ is electric flux density $[C/m^2]$
$\mathbf{J}$ is electric current density $[A/m^2]$
ρ_v is volume charge density $[C/m^3]$.

The operator ∇ in Cartesian coordinates is

$$\nabla = \left[\frac{\partial}{\partial x}, \frac{\partial}{\partial y}, \frac{\partial}{\partial z}\right] \tag{3.5}$$

The above relations are supplemented with constitutive relations

$$\mathbf{D} = \varepsilon \mathbf{E} \tag{3.6}$$

$$\mathbf{B} = \mu \mathbf{H} \tag{3.7}$$

$$\mathbf{J} = \sigma \mathbf{E} \tag{3.8}$$

where $\varepsilon = \varepsilon_0 \varepsilon_r$ is the dielectric permittivity $[F/m]$, $\mu = \mu_0 \mu_r$ is permeability $[H/m]$, σ is electric conductivity, ε_r is the relative dielectric constant. For optical problems considered here, $\mu_r = 1$.

First, we recall two mathematical theorems, Gauss's theorem

$$\oint_S \mathbf{F} \cdot d\mathbf{s} = \int_V \nabla \cdot \mathbf{F} dv \qquad (3.9)$$

where S is the closed surface area defining volume V, and Stokes's theorem

$$\oint_L \mathbf{F} \cdot dl = \int_A \nabla \times \mathbf{F} \cdot d\mathbf{s} \qquad (3.10)$$

where contour L defines surface area A. With the help of the above theorems, Maxwell's equations in differential form can be transformed into an integral form. Integral forms of Maxwell's equations are

$$\oint_S \mathbf{D} \cdot d\mathbf{s} = \int_V \rho_v dv \qquad (3.11)$$

$$\oint_S \mathbf{B} \cdot d\mathbf{s} = 0 \qquad (3.12)$$

$$\oint_L \mathbf{E} \cdot d\mathbf{l} = -\int_A \frac{\partial \mathbf{B}}{\partial t} \qquad (3.13)$$

$$\oint_L \mathbf{H} \cdot d\mathbf{l} = \int_A \frac{\partial \mathbf{D}}{\partial t} + I \qquad (3.14)$$

Properties of the medium are mostly determined by ε, μ and σ. Further, for the dielectric medium the main role is played by ε. The medium is known as linear if ε is independent of E; otherwise it is nonlinear. If it does not depend on position in space, the medium is said to be homogeneous; otherwise it is inhomogeneous. If properties are independent of direction, the medium is isotropic; otherwise it is anisotropic.

3.2 Boundary conditions

Boundary conditions are derived from an integral form of Maxwell's equations. For that purpose we separate all vectors into two components, one parallel to the interface and one normal to the interface. The derivation of boundary conditions is facilitated by using the contour and cylindrical shapes as shown in Fig. 3.1. It will be done independently for the electric and magnetic fields.

3.2.1 Electric boundary conditions

First, analyse transversal components. Integrate Eq. (3.13) over closed loop C, see Fig. 3.1 and then set $\Delta h \to 0$

$$\int_{ABCDA} \mathbf{E} \cdot d\mathbf{l} = -\mathbf{E}_1 \cdot d\mathbf{l} + \mathbf{E}_2 \cdot d\mathbf{l} = -E_{1t}\Delta w + E_{2t}\Delta w = 0$$

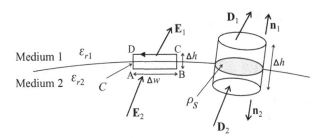

Fig. 3.1 An interface between two media. Contour and volume used to derive boundary conditions for fields between two different dielectrics are shown.

Therefore

$$E_{1t} = E_{2t} \tag{3.15}$$

The tangential component of the electric field is therefore continuous across the boundary between any two dielectric media. Using general relation (3.6) one obtains the boundary condition for tangential component of vector **D**

$$\frac{D_{1t}}{\varepsilon_{r1}} = \frac{D_{2t}}{\varepsilon_{r2}} \tag{3.16}$$

In order to obtain conditions for normal components, consider the cylinder shown in Fig. 3.1. Here, $\mathbf{n}_1$ and $\mathbf{n}_2$ are normal vectors pointing outwards of the top and bottom surfaces into corresponding dielectrics. Apply Gauss's law with integration over surface S of the cylinder

$$\int_S \mathbf{D} \cdot d\mathbf{s} = \int_{top} \mathbf{D}_1 \cdot \mathbf{n}_1 ds + \int_{bottom} \mathbf{D}_2 \cdot \mathbf{n}_2 ds = \rho_s \Delta s$$

The contribution from the outer surface of the cylinder vanishes in the limit $\Delta h \to 0$. Use the fact that $\mathbf{n}_2 = -\mathbf{n}_1$ and have

$$\mathbf{n}_1 \cdot (\mathbf{D}_1 - \mathbf{D}_2) = \rho_s \tag{3.17}$$

or

$$\varepsilon_{r1} E_{2n} - \varepsilon_{r2} E_{1n} = \rho_s \tag{3.18}$$

when using general relation (3.6). Normal component of vector **D** is not continuous across boundary (unless $\rho_s = 0$).

3.2.2 Magnetic boundary conditions

When deriving boundary conditions for magnetic fields, we use approach similar as for electric fields. We use Eq. (3.12) where integration is over the cylinder

$$\int_S \mathbf{B} \cdot d\mathbf{s} = 0$$

Table 3.1 Summary of boundary conditions for electric and magnetic fields.

Field components	General form	Specific form
Tangential E	$\mathbf{n}_2 \times (\mathbf{E}_1 - \mathbf{E}_2) = 0$	$E_{1t} = E_{2t}$
Normal D	$\mathbf{n}_2 \cdot (\mathbf{D}_1 - \mathbf{D}_2) = \rho_s$	$D_{1t} - D_{2t} = \rho_s$
Tangential H	$\mathbf{n}_2 \times (\mathbf{H}_1 - \mathbf{H}_2) = \mathbf{J}_s$	$H_{2t} = H_{1t}$
Normal B	$\mathbf{n}_2 \cdot (\mathbf{B}_1 - \mathbf{B}_2) = 0$	$B_{1n} = B_{2n}$

or

$$B_{1n} = B_{2n} \tag{3.19}$$

Using general relation (3.7), one obtains condition for magnetic field **H**

$$\mu_{r1} H_{1n} = \mu_{r2} H_{2n} \tag{3.20}$$

The above relations tell us that normal component of **B** is continuous across the boundary.

To obtain how transversal magnetic components behave across interface, apply Ampere's law for contour C and then let $\Delta h \to 0$. One obtains

$$\int_C \mathbf{H} \cdot d\mathbf{l} = \int_A^B \mathbf{H}_2 \cdot d\mathbf{l} - \int_C^D \mathbf{H}_1 \cdot d\mathbf{l} = I$$

Here I is the net current crossing the surface of the loop. As we let Δh approach zero, the surface of the loop approaches a thin line of length Δw. Hence, the total current flowing through this thin line is $I = J_s \cdot \Delta w$, where J_s is the magnitude of the normal component of the surface current density traversing the loop. We can therefore express the above equation as

$$H_{2t} - H_{1t} = J_s \tag{3.21}$$

Utilizing unit vector $\mathbf{n}_2$, the above relation can be written as

$$\mathbf{n}_2 \times (\mathbf{H}_1 - \mathbf{H}_2) = \mathbf{J}_s \tag{3.22}$$

where $\mathbf{n}_2$ is the normal vector pointing away from medium 2, see Fig. 3.1. $\mathbf{J}_s$ is the surface current.

In Table 3.1 we summarize boundary conditions between two dielectrics for the electric and magnetic fields. The behaviour of various field components is shown schematically in Fig. 3.2.

3.3 Wave equation

Here, we will derive the wave equation for a source-free medium where $\varrho_v = 0$ and $\mathbf{J} = 0$. The wave equation is obtained by applying $\nabla \times$ operation to both sides of Eq. (3.1). One

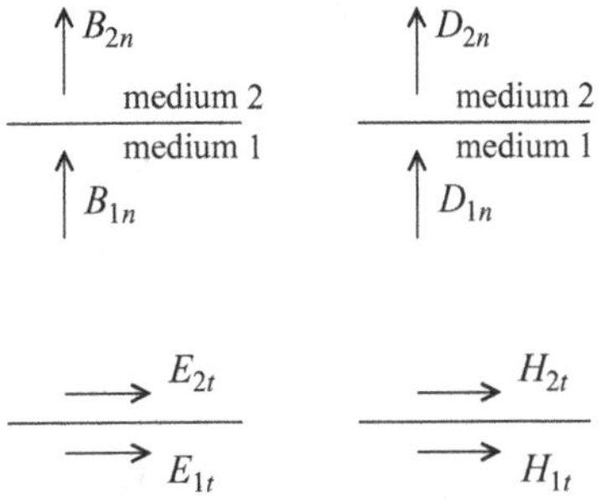

Fig. 3.2 Boundary conditions for fields.

obtains

$$\nabla \times \nabla \times \mathbf{E} = -\frac{\partial}{\partial t}(\nabla \times \mathbf{B}) = -\mu\frac{\partial}{\partial t}(\nabla \times \mathbf{H}) = -\mu\frac{\partial}{\partial t}\frac{\partial \mathbf{D}}{\partial t} = -\mu\varepsilon\frac{\partial^2 \mathbf{D}}{\partial t^2}$$

where we have used Maxwell Eq. (3.2) and relation (3.6). Next apply the following mathematical formula

$$\nabla \times \nabla \times \mathbf{E} = \nabla(\nabla \cdot \mathbf{E}) - \nabla^2 \mathbf{E} \tag{3.23}$$

With the help of Maxwell Eq. (3.3), one finds finally

$$\nabla^2 \mathbf{E} - \mu\varepsilon\frac{\partial^2}{\partial t^2}\mathbf{E} \tag{3.24}$$

which is the desired wave equation. The quantity $\mu\varepsilon$ is related to velocity of light in a vacuum as (assuming $\mu_r = 1$)

$$\mu\varepsilon = \mu_0\varepsilon_0\varepsilon_r = \frac{\bar{n}^2}{c^2} \tag{3.25}$$

where $\bar{n}$ is the refractive index of the medium.

3.4 Time-harmonic fields

In many practical situations fields have sinusoidal time dependence and are known as time-harmonic. This fact is expressed as

$$\mathbf{E}(\mathbf{r}, t) = Re\{\mathbf{E}(\mathbf{r})\, e^{j\omega t}\} \tag{3.26}$$

where $\mathbf{E}(\mathbf{r})$ is the phasor form of $\mathbf{E}(\mathbf{r}, t)$ and is in general complex. $Re\{...\}$ indicates 'taking the real part in' quantity in brackets. Finally, ω is the angular frequency in [rad/s]. In what follows, all fields will be represented in phasor notation.

Applying the time-harmonic assumption (3.26) to source-free Maxwell's equations results in

$$\nabla \times \mathbf{E} = -j\omega\mu\mathbf{H} \tag{3.27}$$

$$\nabla \times \mathbf{H} = j\omega\varepsilon\mathbf{E} \tag{3.28}$$

Applying the time-harmonic assumption again to wave equation (3.24) gives

$$\nabla^2 \mathbf{E} + k^2 \mathbf{E} = 0 \tag{3.29}$$

where $k = \omega\sqrt{\mu\varepsilon}$. Explicitly, the above wave equation is

$$\left(\frac{\partial^2}{\partial x^2} + \frac{\partial^2}{\partial y^2} + \frac{\partial^2}{\partial z^2} + k^2\right) E_i = 0 \tag{3.30}$$

with $i = x, y, z$.

Example As an example, consider propagation of a uniform plane wave characterized by a uniform electric field with nonzero component E_x. Assume also

$$\frac{\partial^2 E_x}{\partial x^2} = 0, \quad \frac{\partial^2 E_x}{\partial y^2} = 0 \tag{3.31}$$

The wave equation reduces to

$$\frac{\partial^2 E_x}{\partial z^2} + k^2 E_x = 0 \tag{3.32}$$

and has the following forward propagating solution:

$$E_x(z) = E_0 e^{jkz} \tag{3.33}$$

Magnetic field is determined from the Maxwell's Equation (3.27)

$$\nabla \times \mathbf{E} = \begin{vmatrix} \mathbf{a}_x & \mathbf{a}_y & \mathbf{a}_z \\ 0 & 0 & \frac{\partial}{\partial z} \\ E_x(z) & 0 & 0 \end{vmatrix} = -j\omega\mu \left(\mathbf{a}_x H_x + \mathbf{a}_y H_y + \mathbf{a}_z H_z\right) \tag{3.34}$$

Here $\mathbf{a}_x, \mathbf{a}_y, \mathbf{a}_z$ are unit vectors along x, y, z axes, respectively. From the above equation, one finds

$$H_x = 0 \tag{3.35}$$

$$H_y = \frac{1}{j\omega\mu} \frac{\partial E_x(z)}{\partial z}$$

$$H_z = 0$$

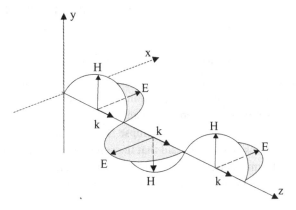

Fig. 3.3 Visualization of the electric and magnetic fields.

Using the solution for $E_x(z)$, one finally obtains the expression for magnetic field

$$\mathbf{H} = \mathbf{a}_y H_y(z) \qquad (3.36)$$

$$= \frac{1}{-j\omega\mu}(-jkE_x)\mathbf{a}_y \qquad (3.37)$$

$$= \mathbf{a}_y \frac{k}{\omega\mu} E_x(z)$$

$$= \mathbf{a}_y \frac{1}{Z} E_x(z)$$

where the impedance Z of the medium is defined as $Z = \sqrt{\frac{\mu}{\varepsilon}}$. The propagating wave is shown in Fig. 3.3.

We finish this section by providing a useful relation between electric and magnetic fields. Assuming time-harmonic and plane-wave dependence of both fields as

$$\mathbf{E} \sim \exp(i\omega t - i\mathbf{k} \cdot \mathbf{r})$$

and using constitutive relation (3.7), Maxwell Eq. (3.1) takes the form

$$\mathbf{k} \times \mathbf{E} = \omega\mu_0 \mathbf{H}$$

Introducing unit vector $\widehat{k}$ along wave vector $\mathbf{k}$ and the expression for wave number $k = \omega\bar{n}\sqrt{\mu_0\varepsilon_0}$, one finds

$$\bar{n}\widehat{k} \times \mathbf{E} = Z_0 \mathbf{H}$$

where Z_0 is the impedance of the free-space.

3.5 Polarized waves

Here we will discuss the concept of polarization of electromagnetic waves. Polarization characterizes the curve which **E** vector makes (in the plane orthogonal to the direction of propagation) at a given point in space as a function of time. In the most general case, the curve produced is an ellipse and, accordingly, the wave is called elliptically polarized. Under certain conditions, the ellipse may be reduced to a circle or a segment of a straight line. In those cases it is said that the wave's polarization is circular or linear, respectively. Since the magnetic field vector is related to the electric field vector, it does not need separate discussion. First, consider a single electromagnetic wave.

3.5.1 Linearly polarized waves

Consider an electromagnetic wave characterized by electric field vector **E** directed along the x-axis

$$\mathbf{E} = \mathbf{a}_x E_0 \cos(\omega t - kz + \phi) \tag{3.38}$$

It is known as a linearly polarized plane wave with the electric field vector oscillating in the x direction. Wave propagates in the $+z$ direction. In Eq. (3.38) $\omega = 2\pi \nu$ is the angular frequency and k is the propagation constant defined as

$$k = \frac{\omega}{v}$$

where $v = c/\bar{n}$ is the velocity of the electromagnetic wave in the medium having refractive index $\bar{n}$. ϕ is known as a phase of electromagnetic wave.

The electromagnetic wave can also be written in the complex representation as

$$\mathbf{E} = \mathbf{a}_x E_0 e^{i(\omega t - kz + \phi)} \tag{3.39}$$

The actual field as described by Eq. (3.38) is obtained from (3.39) by taking *real* part. A more general expression for the electromagnetic wave is

$$\mathbf{E} = \hat{\mathbf{e}} E_0 e^{i(\omega t - \mathbf{k} \cdot \mathbf{r} + \phi)}$$

which is known as the plane polarized wave. Here unit vector $\hat{\mathbf{e}}$ lies in the plane known as plane of polarization. It is perpendicular to vector **k** which describes direction of propagation

$$\mathbf{k} \cdot \hat{\mathbf{e}} = 0$$

3.5.2 Circularly and elliptically polarized waves

In general, when we have an arbitrary number of plane waves propagating in the same direction they add up to a complicated wave. In the simplest case, one has only two such plane waves. To be more specific, consider two plane waves oscillating along orthogonal directions. They are linearly polarized having the same frequencies and propagating in the same direction

$$\mathbf{E}_1 = E_x \mathbf{a}_x = \mathbf{a}_x E_{0x} \cos(\omega t - kz) \tag{3.40}$$

$$\mathbf{E}_2 = E_y \mathbf{a}_y = \mathbf{a}_y E_{0y} \cos(\omega t - kz + \phi) \tag{3.41}$$

We want to know the type of the resulting wave and the curve traced by the tip of the total electric vector $\mathbf{E} = \mathbf{E}_1 + \mathbf{E}_2$

$$\mathbf{E} = \mathbf{E}_1 + \mathbf{E}_2 = E_0 \{\cos(\omega t - kz) + \cos(\omega t - kz + \phi)\}$$

First, eliminate $\cos(\omega t - kz)$ term. From Eq. (3.40)

$$\cos(\omega t - kz) = \frac{E_x}{E_{0x}} \tag{3.42}$$

Use trigonometric identity

$$\cos(\alpha - \beta) = \cos\alpha \cos\beta + \sin\alpha \sin\beta$$

to express Eq. (3.41) as follows:

$$E_y = E_{0y} \left\{ \cos(\omega t - kz) \cos\delta + [1 - \cos^2(\omega t - kz)]^{1/2} \sin\delta \right\}$$

Substitute Eq. (3.42) in the above and have

$$\frac{E_y}{E_{0y}} = \frac{E_x}{E_{0x}} \cos\phi + \left(1 - \frac{E_x^2}{E_{0x}^2}\right)^{1/2} \sin\phi$$

Squaring both sides gives

$$\left(\frac{E_y}{E_{0y}} - \frac{E_x}{E_{0x}} \cos\phi\right)^2 = \left(1 - \frac{E_x^2}{E_{0x}^2}\right) \sin^2\phi$$

or

$$\frac{E_y^2}{E_{0y}^2} - 2\cos\phi \frac{E_y}{E_{0y}} \frac{E_x}{E_{0x}} + \frac{E_x^2}{E_{0x}^2} \cos^2\phi + \frac{E_x^2}{E_{0x}^2} \sin^2\phi = \sin^2\phi$$

Finally, the above equation gives

$$\left(\frac{E_y}{E_{0y}}\right)^2 + \left(\frac{E_x}{E_{0x}}\right)^2 - 2\left(\frac{E_y}{E_{0y}}\right)\left(\frac{E_x}{E_{0x}}\right) \cos\phi = \sin^2\phi \tag{3.43}$$

This is general equation of an ellipse. Thus the endpoint of $\mathbf{E}(z, t)$ will trace an ellipse at a given point in space. It is said that the wave is elliptically polarized.

When phase $\phi = \frac{\pi}{2}$, the resultant total electric field is

$$\left(\frac{E_y}{E_{0y}}\right)^2 + \left(\frac{E_x}{E_{0x}}\right)^2 = 1$$

which describes right elliptically polarized wave since as time increases the end of electric vector $\mathbf{E}$ rotates clockwise on the circumference of an ellipse. Typical situations are illustrated in Fig. 3.4 where we show elliptic, circular and linear polarizations.

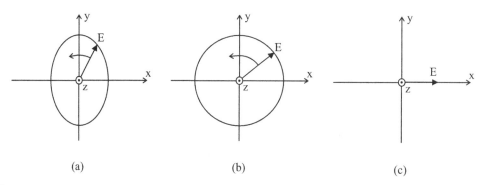

Fig. 3.4 Typical states of polarization: (a) elliptic, (b) circular and (c) linear.

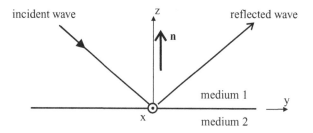

Fig. 3.5 Picture used in defining plane of incidence.

3.6 Fresnel coefficients and phases

In this section we discuss electromagnetic wave undergoing reflection at the boundary between two dielectrics, see Fig. 3.5. The plane of incidence is defined as the plane formed by unit vector **n** normal to the interface between two media and the directions of propagation of the incident and reflected waves.

We will derive so-called Fresnel coefficients [3] which are reflection and transmission coefficients expressed in terms of the angle of incidence and material properties (ε dielectric constants) of the two dielectrics. Further, Fresnel phases are determined from Fresnel coefficients using the following definition [1]:

$$r = e^{-2j\phi} \tag{3.44}$$

Both reflection coefficients r and phases ϕ will be calculated for both types of modes, TE and TM.

3.6.1 TE polarization

Referring to Fig. 3.6 the E_{1i}, E_{1r}, E_{2t} are complex values of incident, reflected and transmitted electric fields in medium 1 and 2. The incident electric field in medium 1 (E_{1i}) is parallel to the interface between both media. Such orientation is known as TE polarization. It is

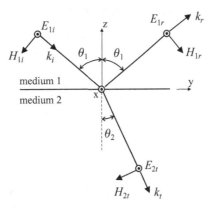

Fig. 3.6 Fresnel reflection for TE polarization. Directions of vectors for TE polarized wave.

often said that electric field vector **E** is normal to the plane of incidence. Such configuration is also known as s-polarization.

Boundary conditions require that the tangential components of the total field **E** and the total field **H** on both sides of the interface be equal. Those conditions result in the following equations:

$$E_{1i} + E_{1r} = E_{2t} \tag{3.45}$$

$$-H_{1i}\cos\theta_1 + H_{1r}\cos\theta_1 = -H_{2t}\cos\theta_2$$

We also have the following relations involving impedances in both media:

$$\frac{E_{1i}}{H_{1i}} = Z_1, \quad \frac{E_{1r}}{H_{1r}} = Z_1, \quad \frac{E_{2t}}{H_{2t}} = Z_2 \tag{3.46}$$

For a non-magnetic medium ($\mu_r = 1$) the impedance can be written as $Z = \frac{Z_0}{\bar{n}}$, with $\bar{n}$ being the refractive index of the medium and

$$Z_0 = \sqrt{\frac{\mu_0}{\varepsilon_0}}$$

is the impedance of a free space. Replacing magnetic fields in the second equation of Eq. (3.45) by the electric fields using Eq. (3.46) gives

$$E_{1i} + E_{1r} = E_{2t} \tag{3.47}$$

$$-\frac{E_{1i}}{Z_1}\cos\theta_1 + \frac{E_{1r}}{Z_1}\cos\theta_1 = -\frac{E_{2t}}{Z_2}\cos\theta_2$$

Reflection coefficient is defined as

$$r_{TE} = \frac{E_{1r}}{E_{1i}} \tag{3.48}$$

From Eqs. (3.47) by eliminating E_{2t} and using definition (3.48) one obtains 5

$$r_{TE} = \frac{Z_2\cos\theta_1 - Z_1\cos\theta_2}{Z_2\cos\theta_1 + Z_1\cos\theta_2}$$

Replacing impedances Z by refractive indices $\bar{n}$, we finally have

$$r_{TE} = \frac{\bar{n}_1 \cos\theta_1 - \bar{n}_2 \cos\theta_2}{\bar{n}_1 \cos\theta_1 + \bar{n}_2 \cos\theta_2}$$

$$= \frac{\bar{n}_1 \cos\theta_1 - \sqrt{\bar{n}_2^2 - \bar{n}_1^2 \cos\theta_1}}{\bar{n}_1 \cos\theta_1 + \sqrt{\bar{n}_2^2 - \bar{n}_1^2 \cos\theta_1}} \quad (3.49)$$

using Snell's law. For angles θ_1 such that $\bar{n}_2^2 - \bar{n}_1^2 \cos\theta_1 < 0$ the reflection coefficient becomes complex. For such cases, we write it as

$$r_{TE} = \frac{\bar{n}_1 \cos\theta_1 - j\sqrt{\bar{n}_1^2 \cos\theta_1 - \bar{n}_2^2}}{\bar{n}_1 \cos\theta_1 + j\sqrt{\bar{n}_1^2 \cos\theta_1 - \bar{n}_2^2}} \equiv \frac{a - jb}{a + jb}$$

Such complex number can be expressed as

$$r_{TE} = \frac{e^{-j\phi_{TE}}}{e^{j\phi_{TE}}} = e^{-2j\phi_{TE}}$$

where we have defined $a + jb = e^{j\phi_{TE}}$. Finally, using the definition of Fresnel phase, Eq. (3.44) one obtains

$$\tan\phi_{TE} = \frac{\sqrt{\bar{n}_1^2 \sin^2\theta_1 - \bar{n}_2^2}}{\bar{n}_1 \cos\theta_1} \quad (3.50)$$

The above equation represents phase shift during reflection for TE polarized electromagnetic wave. For a more detailed discussion about phase change on reflection consult books by Pedrotti [4] and Liu [5].

Example (a) Consider normal incidence of light on an air-silica interface. Compute the fraction of reflected and transmitted power. Also, express the transmitted loss in decibels. Assume refractive index of silica to be 1.45. (b) Repeat the calculations for Si which has $\bar{n} = 3.50$. (c) Consider coupling of *GaAs* optical source with a refractive index of 3.6 to a silica fibre which has refractive index of 1.48. Assume close physical contact of the fibre end and the source.

Solution
(a) The corresponding coefficient known as reflectance [4] is

$$R = |r|^2 = \left(\frac{\bar{n}_1 - \bar{n}_2}{\bar{n}_1 + \bar{n}_2}\right)^2 \quad (3.51)$$

Substituting values for air and silica, one has

$$R = \left(\frac{1.45 - 1.00}{1.45 + 1.00}\right)^2 = 0.03$$

so about 3% of the light is reflected. The remainder, 97%, is transmitted.

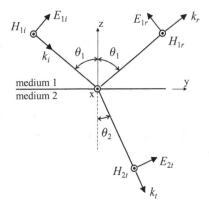

Fig. 3.7 Fresnel reflection for TM polarization. Directions of vectors for a TM polarized wave.

The transmission loss in dB is

$$-10\log_{10} 0.97 = 0.13 [dB]$$

This result shows that there is about $0.2 dB$ loss when light enters glass from air.

(b) For Si, using the same procedure we obtain

$$R = \left(\frac{1.00 - 3.50}{1.00 + 3.50}\right)^2 = 0.309$$

This means that about 31% of light is reflected.

(c) The Fresnel reflection at the interface is

$$R = \left(\frac{3.60 - 1.48}{3.60 + 1.48}\right)^2 = 0.174$$

Therefore about 17.4% of the created optical power is emitted back into the source. The power coupled into optical fibre is

$$P_{\text{coupled}} = (1 - R) P_{\text{emitted}}$$

The power loss, α in decibels is

$$\alpha = -10\log_{10}\frac{P_{\text{coupled}}}{P_{\text{emitted}}} = -10\log_{10}(1 - R)$$
$$= -10\log_{10}((1 - 0.174)) - 10\log_{10}(0.826) = 0.83 [\text{dB}]$$

3.6.2 TM polarization

Field configuration used to analyse reflection for TM polarization (magnetic field parallel to the interface) is shown in Fig. 3.7. Here, electric field vector E is parallel to the plane of incidence. Such configuration is also known as p-polarization.

We will derive coefficient of reflection r_{TM} for TM mode. As before the E_{1i}, E_{1r}, E_{2t} are (complex) values of incident, reflected and transmitted electric fields in medium 1 and 2. Similar notation holds for magnetic vectors. Boundary conditions require that tangential components are continuous across interface. The relevant conditions are

$$E_{1i}\cos\theta_1 - E_{1r}\cos\theta_1 = E_{2t}\cos\theta_2 \quad (3.52)$$

$$H_{1i} + H_{1r} = H_{2t}$$

Using Eqs. (3.46) for impedances to eliminate magnetic fields in the previous equations results in the following:

$$E_{1i}\cos\theta_1 - E_{1r}\cos\theta_1 = E_{2t}\cos\theta_2 \quad (3.53)$$

$$\frac{E_{1i}}{Z_1} + \frac{E_{1r}}{Z_1} = \frac{E_{2t}}{Z_2}$$

Reflection coefficient is defined as

$$r_{TM} = \frac{E_{1r}}{E_{1i}} \quad (3.54)$$

From Eqs. (3.53)

$$r_{TM} = \frac{Z_1\cos\theta_1 - Z_2\cos\theta_2}{Z_1\cos\theta_1 + Z_2\cos\theta_2}$$

It can be also expressed in terms of refractive index

$$r_{TM} = \frac{\bar{n}_2\cos\theta_1 - \bar{n}_1\cos\theta_2}{\bar{n}_2\cos\theta_1 + \bar{n}_1\cos\theta_2}$$

$$= \frac{\bar{n}_2\cos\theta_1 - \sqrt{\bar{n}_1^2 - \bar{n}_2^2}\cos\theta_1}{\bar{n}_2\cos\theta_1 + \sqrt{\bar{n}_1^2 - \bar{n}_2^2}\cos\theta_1} \quad (3.55)$$

TM phase is obtained in the same way as for TE polarization. Final result is

$$\tan\phi_{TM} = \frac{\bar{n}_1^2}{\bar{n}_2^2} \times \frac{\sqrt{\bar{n}_1^2\sin^2\theta_1 - \bar{n}_2^2}}{\bar{n}_1\cos\theta_1} \quad (3.56)$$

Again, for more information consult [4] and [5].

3.7 Polarization by reflection from dielectric surfaces

In interpreting the formulas for r_{TE} and r_{TM} we distinguish between two situations:

(1) for $\bar{n}_1 < \bar{n}_2$ or $\bar{n} = \frac{\bar{n}_2}{\bar{n}_1} > 1$, one defines so-called external reflection,
(2) for $\bar{n}_1 > \bar{n}_2$ or $\bar{n} = \frac{\bar{n}_2}{\bar{n}_1} < 1$, one defines so-called internal reflection.

Example of (1) is air-to-glass reflection and example of (2) is the glass-to-air reflection. A plot of coefficients of reflection for $\bar{n} = 1.50$ is shown in Fig. 3.8. MATLAB code is

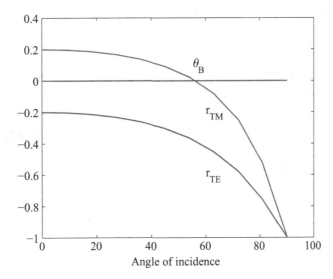

Fig. 3.8 Reflections of TE and TM modes for external reflection with $n = 1.50$. Brewster's angle is also shown.

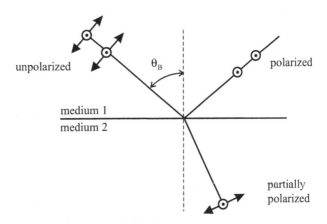

Fig. 3.9 Polarization at Brewster's angle.

shown in Appendix, Listing 3A.1. The so-called Brewster's angle for which $r_{TM} = 0$ is also shown. For more information, see the book by Pedrotti and Pedrotti [4].

One can obtain polarization by reflection at Brewster's angle, see Fig. 3.9. In the figure we illustrate properties of the reflected and transmitted light incident at the surface at Brewster's angle.

Unpolarized light is incident at the interface at Brewster's angle θ_B. Upon reflection one obtains polarized TM light. At the same time, the refracted light is only partially polarized since r_{TE} is nonzero.

At Brewster's angle θ_B of incidence, coefficient of reflection r_{TM} (also known as $r_\parallel$ since **E** is parallel to the plane of incidence) is zero; i.e. $r_\parallel = 0$. The above happens when the sum of the angles of incidence and refraction is equal to $\pi/2$

$$\theta_1 + \theta_2 = \frac{\pi}{2}$$

No such angle exists for the *s* polarization (TE). Thus if an unpolarized light is incident at Brewster's angle, the reflected light will be linearly polarized, in fact s-polarized.

3.7.1 Expression for Brewster's angle

Expression for r_{TM} is

$$r_{TM} = \frac{\bar{n}_2 \cos\theta_1 - \frac{\bar{n}_1}{\bar{n}_2}\sqrt{\bar{n}_2^2 - \bar{n}_1^2 \sin^2\theta_1}}{\bar{n}_2 \cos\theta_1 + \frac{\bar{n}_1}{\bar{n}_2}\sqrt{\bar{n}_2^2 - \bar{n}_1^2 \sin^2\theta_1}}$$

Vanishing of r_{TM} corresponds to $\theta_1 = \theta_B$. We have

$$\bar{n}_2 \cos\theta_B - \frac{\bar{n}_1}{\bar{n}_2}\sqrt{\bar{n}_2^2 - \bar{n}_1^2 \sin^2\theta_B} = 0$$

from which we find an expression for Brewster's angle

$$\tan\theta_B = \frac{\bar{n}_2}{\bar{n}_1}$$

3.8 Antireflection coating

In Chapter 9 on semiconductor optical amplifiers (SOA) we will discuss why there is a need to reduce (or completely eliminate) the effect of reflections. Therefore, the practical question is how to eliminate reflections at the interface between two dielectrics.

As seen before, light passing through a boundary between two dielectrics is lost due to reflection, see Eq. (3.51). The effect depends on the difference of the values of refractive indices between neighbouring layers. One can observe that for equal refractive indices there will be no reflection but also no refraction, which is not a very interesting possibility.

A practical method of reducing reflections is to use several layers with properly selected values of refractive indices. For a single layer, interference of two reflected waves can lead to elimination of reflection, but only at a particular wavelength.

Let us analyse this situation in more detail. Consider reflections of two waves R_1 and R_2, see Fig. 3.10. One of the conditions for destructive interference is that the amplitudes of both reflected waves should be equal. From Eq. (3.51), reflection of ray R_1 at interface 'a' is described by reflection coefficient

$$R = \left(\frac{\bar{n}_0 - \bar{n}_f}{\bar{n}_0 + \bar{n}_f}\right)^2$$

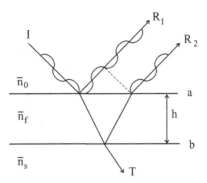

Fig. 3.10 Illustration of reflection from single dielectric film layer. Case of destructive interference is shown.

whereas reflection of ray R_2 at interface 'b' is described by coefficient

$$R = \left(\frac{\overline{n}_s - \overline{n}_f}{\overline{n}_s + \overline{n}_f}\right)^2$$

Both coefficients should be equal which gives

$$\left(\frac{\overline{n}_0 - \overline{n}_f}{\overline{n}_0 + \overline{n}_f}\right)^2 = \left(\frac{\overline{n}_s - \overline{n}_f}{\overline{n}_s + \overline{n}_f}\right)^2$$

Assuming that the upper medium is air, i.e. $\overline{n}_0 = 1$, from the above one finds expression for the value of refractive index for a layer as

$$\overline{n}_f = \sqrt{\overline{n}_s} \tag{3.57}$$

Therefore, at a particular wavelength there will be no reflection once the refractive index of coating layer is given by (3.57). If there is a need to design structures producing no reflection over some frequency band, one must design multilayer structure consisting of layers with different refractive indices. In order to design such structures a more realistic description is necessary. Typical calculations are based on transfer matrix approach.

3.8.1 Transfer matrix approach

We now describe the transfer matrix approach, also known as the characteristic matrix approach [6].

Consider a single dielectric layer with refractive index $\overline{n}$ deposited on a substrate with refractive index $\overline{n}_s$. Above it is a medium with refractive index $\overline{n}_0$, see Fig. 3.11.

We consider an s-polarized wave where electric field vector **E** is perpendicular to the plane of incidence. Boundary conditions dictate that tangential components of **E** and **H** fields are continuous across the interfaces, i.e. at positions $z = z_1$ and $z = z_2$.

We introduce the convention that $\mathbf{E}^+(z)$ denotes light propagating in the positive z direction whereas $\mathbf{E}^-(z)$ denotes light propagating in the negative z direction. Close to interface described by $z = z_1$, due to continuity of tangential component one writes

$$\mathbf{E}(z_1) = \mathbf{E}^+(z_1^+) + \mathbf{E}^-(z_1^+) = \mathbf{E}^+(z_1^-) + \mathbf{E}^-(z_1^-) \tag{3.58}$$

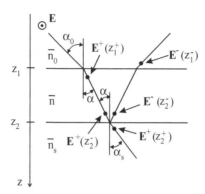

Fig. 3.11 Illustration of electromagnetic fields at the boundaries used in the derivation of transfer matrix.

where z_1^+ signifies value of coordinate z slightly larger than z_1, etc. Similarly, close to $z = z_2$

$$\mathbf{E}(z_2) = \mathbf{E}^+(z_2^+) + \mathbf{E}^-(z_2^+) = \mathbf{E}^+(z_2^-) + \mathbf{E}^-(z_2^-) \quad (3.59)$$

For future use we introduce $d = z_2 - z_1$. When field $\mathbf{E}(z)$ propagates in the film over distance d, it acquires phase δ derived previously (see Eq. 2.31)

$$\delta = \frac{2\pi}{\lambda_0} \bar{n} d \cos\alpha \quad (3.60)$$

Electric fields that are close to interfaces are therefore related as

$$\mathbf{E}^+(z_2^-) = \mathbf{E}^+(z_1^+) e^{-i\delta} \quad (3.61)$$

and

$$\mathbf{E}^-(z_2^-) = \mathbf{E}^-(z_1^+) e^{i\delta} \quad (3.62)$$

To determine the behaviour of a magnetic field, we use the fact that fields $\mathbf{E}$ and $\mathbf{H}$ are related as

$$\mathbf{H} = \frac{\bar{n}}{Z_0} \widehat{k} \times \mathbf{E}$$

where $\widehat{k}$ is the unit vector along the wave vector $\mathbf{k}$ and $Z_0 = \sqrt{\frac{\mu_0}{\varepsilon_0}}$ is the impedance of the free space. The condition of continuity of tangential component of $\mathbf{H}$ vector gives:

at $z = z_1$

$$Z_0 \mathbf{H}(z_1) = \left[\mathbf{E}^+(z_1^-) - \mathbf{E}^-(z_1^-)\right] \bar{n}_0 \cos\alpha_0 = \left[\mathbf{E}^+(z_1^+) - \mathbf{E}^-(z_1^+)\right] \bar{n} \cos\alpha \quad (3.63)$$

at $z = z_2$

$$Z_0 \mathbf{H}(z_2) = \left[\mathbf{E}^+(z_2^-) - \mathbf{E}^-(z_2^-)\right] \bar{n} \cos\alpha = \mathbf{E}^+(z_2^+) \bar{n}_s \cos\alpha \quad (3.64)$$

Substitute Eqs. (3.61) and (3.62) involving phase δ into (3.59) and (3.64) to obtain

$$\mathbf{E}(z_2) = \mathbf{E}^+(z_1^+) e^{-i\delta} + \mathbf{E}^-(z_1^+) e^{i\delta}$$

$$\mathbf{H}(z_2) = \frac{\bar{n} \cos\alpha}{Z_0} \left[\mathbf{E}^+(z_1^+) e^{-i\delta} - \mathbf{E}^-(z_1^+) e^{i\delta}\right]$$

Solutions of the above equations are

$$2\mathbf{E}^-(z_1^+)e^{i\delta} = \mathbf{E}(z_2) - \frac{Z_0}{\bar{n}\cos\alpha}\mathbf{H}(z_2)$$

$$2\mathbf{E}^+(z_1^+)e^{-i\delta} = \mathbf{E}(z_2) + \frac{Z_0}{\bar{n}\cos\alpha}\mathbf{H}(z_2)$$

With the help of the above solutions and using Eqs. (3.58) and (3.63) we can evaluate fields at z_1 and z_2

$$\mathbf{E}(z_1) = \mathbf{E}(z_2)\cos\delta + \frac{Z_0}{\bar{n}\cos\alpha}\mathbf{H}(z_2)i\sin\delta$$

and

$$Z_0\mathbf{H}(z_1) = \mathbf{E}(z_2)i\sin\delta\,\bar{n}\cos\alpha + Z_0\mathbf{H}(z_2)\cos\delta$$

The above equations can be written in matrix form

$$\begin{bmatrix}\mathbf{E}(z_1)\\Z_0\mathbf{H}(z_1)\end{bmatrix} = \begin{bmatrix}\cos\delta & \frac{i\sin\delta}{\bar{n}\cos\alpha}\\i\bar{n}\sin\delta\cos\alpha & \cos\delta\end{bmatrix}\begin{bmatrix}\mathbf{E}(z_2)\\Z_0\mathbf{H}(z_2)\end{bmatrix} = \overleftrightarrow{M}\begin{bmatrix}\mathbf{E}(z_2)\\Z_0\mathbf{H}(z_2)\end{bmatrix} \quad (3.65)$$

For normal incidence, i.e. $\alpha = 0$, the transfer matrix is

$$\overleftrightarrow{M} = \begin{bmatrix}\cos\delta & \frac{i\sin\delta}{\bar{n}}\\i\bar{n}\sin\delta & \cos\delta\end{bmatrix} \quad (3.66)$$

Special cases of importance are:
1. *The quarter-wave layer* when $\bar{n}d = \lambda_0/4$; $\delta = \pi/2$ and matrix $\overleftrightarrow{M}$ is

$$\overleftrightarrow{M}_{\lambda/4} = \begin{bmatrix}0 & \frac{i}{\bar{n}}\\i\bar{n} & 0\end{bmatrix} \quad (3.67)$$

2. *The half-wave layer* when $\bar{n}d = \lambda_0/2$; $\delta = \pi$ and matrix $\overleftrightarrow{M}$ is

$$\overleftrightarrow{M}_{\lambda/2} = \begin{bmatrix}-1 & 0\\0 & -1\end{bmatrix} \quad (3.68)$$

3.9 Bragg mirrors

The Bragg mirror, also known as the Bragg reflector, consists of identical layers of dielectrics with high and low values of refractive indices as shown in Fig. 3.12. The main interest in fabricating such structures is that they have extremely high reflectivities at optical and infrared frequencies. They are important elements of VCSELs where high reflectivity and bandwidth are required. A typical structure forming Bragg mirror consists of N layers of dielectrics with refractive indices $\bar{n}_L$ (low refractive index) and $\bar{n}_H$ (high refractive index). The ratio of those values, the so-called contrast ratio, plays an important role.

The structure is known as a quarter-wave dielectric stack, which means that the optical thicknesses are quarter-wavelength long; that is $\bar{n}_H \cdot a_H = \bar{n}_L \cdot a_L = \lambda_0/4$ at some wavelength

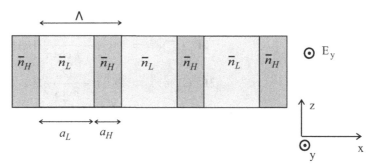

Fig. 3.12 Schematic of a seven-layer dielectric (Bragg) mirror; shown are $N = 3$ periods.

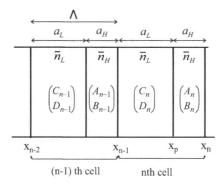

Fig. 3.13 Notation used in the analysis of Bragg mirrors.

λ_0. The structure consists of an odd number of layers with the high index layer being the first and the last layers [7].

Here, we outline a formalism which is used in the design of Bragg mirrors. We also illustrate its applications with an example. We restrict our discussion to s-polarization (TE modes). Electric field vector is directed along y-axis $\mathbf{E} = E(x) \cdot \hat{a}_y$, see Fig. 3.12 where we showed the orientation of electric fields and analysed structure.

We develop transfer matrix notation which is based on the transfer of fields from cell n to cell $n-1$. The cell is formed by two neighbouring layers which create a basic periodic structure which is then repeated some number of times. The electric field will now be expressed in the three layers of interest, see Fig. 3.13 for notation.

In cell n one expresses the electric field as

$$E_y(x) = \begin{cases} A_n\, e^{-jk_H(x-x_n)} + B_n\, e^{jk_H(x-x_n)}, & x_p < x < x_n \\ C_n\, e^{-jk_L(x-x_p)} + D_n\, e^{jk_L(x-x_p)}, & x_n < x < x_p \end{cases} \quad (3.69)$$

where $k_H^2 = \left(\frac{\bar{n}_H \omega}{c}\right)^2$ and $k_L^2 = \left(\frac{\bar{n}_L \omega}{c}\right)^2$. One has also the following relations for coordinates of layers: $x_n = n \cdot \Lambda$, $x_{n-1} = (n-1) \cdot \Lambda$, $x_p = n \cdot \Lambda - a_H$. Here $\Lambda = a_H + a_L$ is the period of the structure.

The magnetic field vector is determined from the Maxwell equation, Eq. (3.1). For our geometry, one has the relation between electric and magnetic fields

$$\frac{\partial E_y(x)}{\partial x} = -j\omega\mu_0 H_z \tag{3.70}$$

From continuity of electric field at x_p and x_{n-1} one obtains

$$\begin{aligned} x = x_p & \quad C_n + D_n = A_n e^{jk_H \cdot a_H} + B_n e^{-jk_H \cdot a_H} \\ x = x_{n-1} & \quad A_{n-1} + B_{n-1} = C_n e^{jk_L \cdot a_L} + D_n e^{-jk_L \cdot a_L} \end{aligned} \tag{3.71}$$

Similarly, continuity of magnetic field produces equations

$$\begin{aligned} x = x_p & \quad -k_L C_n + k_L D_n = -k_H A_n e^{jk_H \cdot a_H} + k_H B_n e^{-jk_H \cdot a_H} \\ x = x_{n-1} & \quad -k_H A_{n-1} + k_H B_{n-1} = -k_L C_n e^{jk_L \cdot a_L} + k_L D_n e^{-jk_L \cdot a_L} \end{aligned} \tag{3.72}$$

The previous four equations can be written in matrix forms as follows:

$$\begin{bmatrix} 1 & 1 \\ k_L & -k_L \end{bmatrix} \begin{bmatrix} C_n \\ D_n \end{bmatrix} = \begin{bmatrix} e^{-jk_H a_H} & e^{jk_H a_H} \\ k_H e^{-jk_H a_H} & -k_H e^{jk_H a_H} \end{bmatrix} \begin{bmatrix} A_n \\ B_n \end{bmatrix}$$

and

$$\begin{bmatrix} 1 & 1 \\ k_H & -k_H \end{bmatrix} \begin{bmatrix} A_{n-1} \\ B_{n-1} \end{bmatrix} = \begin{bmatrix} e^{jk_L a_L} & e^{-jk_L a_L} \\ k_L e^{jk_L a_L} & -k_L e^{-jk_L a_L} \end{bmatrix} \begin{bmatrix} A_n \\ B_n \end{bmatrix}$$

The above equations transfer fields from cell n to cell $n-1$. In the next step amplitudes $\begin{bmatrix} C_n \\ D_n \end{bmatrix}$ will be eliminated and the transfer process will be described in terms of amplitudes $\begin{bmatrix} A_n \\ B_n \end{bmatrix}$. By doing so, one obtains the practical equation

$$\begin{bmatrix} A_{n-1} \\ B_{n-1} \end{bmatrix} = \overleftrightarrow{T} \begin{bmatrix} A_n \\ B_n \end{bmatrix} \tag{3.73}$$

Matrix $\overleftrightarrow{T}$ can be expressed as

$$\overleftrightarrow{T} = \overleftrightarrow{X}_H \cdot \overleftrightarrow{T}_L \cdot \overleftrightarrow{X}_L \cdot \overleftrightarrow{T}_H \tag{3.74}$$

Matrix $\overleftrightarrow{T}_L$ describes propagation in a uniform medium having refractive index $\bar{n}_L$ and thickness a_L and has the form

$$\overleftrightarrow{T}_L = \begin{bmatrix} e^{jk_L a_L} & 0 \\ 0 & e^{-jk_L a_L} \end{bmatrix} \tag{3.75}$$

Similarly, matrix $\overleftrightarrow{T}_H$ describes propagation in a uniform medium having refractive index $\bar{n}_H$ and thickness a_H. Matrices $\overleftrightarrow{X}_H$ and $\overleftrightarrow{X}_L$ describe behaviour at interfaces and are

$$\overleftrightarrow{X}_H = \begin{bmatrix} \dfrac{k_H + k_L}{2k_H} & \dfrac{k_H - k_L}{2k_H} \\ \dfrac{k_H - k_L}{2k_H} & \dfrac{k_H + k_L}{2k_H} \end{bmatrix} \tag{3.76}$$

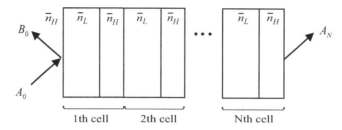

Fig. 3.14 Notation used in the definition of reflection from Bragg mirror.

and

$$\overleftrightarrow{X}_L = \begin{bmatrix} \dfrac{k_L + k_H}{2k_L} & \dfrac{k_L - k_H}{2k_L} \\ \dfrac{k_L - k_H}{2k_L} & \dfrac{k_L + k_H}{2k_L} \end{bmatrix} \tag{3.77}$$

By referring to the Fig. 3.13 propagation (to the left) from an arbitrary point in the structure, say point x_n involves matrix $\overleftrightarrow{T}_H$ (propagation over a uniform region with $\bar{n}_H$, then propagation over an interface from region having $\bar{n}_H$ to $\bar{n}_L$ (described by matrix $\overleftrightarrow{X}_L$), propagation over uniform region with $\bar{n}_L$ (described by matrix $\overleftrightarrow{T}_L$) and finally, propagation over an interface from $\bar{n}_L$ to $\bar{n}_H$ (described by matrix $\overleftrightarrow{X}_H$).

Matrix $\overleftrightarrow{T}$ can be expressed as

$$\overleftrightarrow{T} = \begin{bmatrix} a & b \\ c & d \end{bmatrix} \tag{3.78}$$

where the elements are

$$a = e^{ia_H k_H} \left[\cos a_L k_L + \frac{i}{2}\left(\frac{k_H}{k_L} + \frac{k_L}{k_H}\right) \sin a_L k_L \right] \tag{3.79}$$

$$b = \frac{i}{2} e^{-ia_H k_H} \left(\frac{k_L}{k_H} - \frac{k_H}{k_L}\right) \sin a_L k_L \tag{3.80}$$

$$c = \frac{i}{2} e^{ia_H k_H} \left(\frac{k_H}{k_L} - \frac{k_L}{k_H}\right) \sin a_L k_L \tag{3.81}$$

$$d = e^{-ia_H k_H} \left[\cos a_L k_L - \frac{i}{2}\left(\frac{k_H}{k_L} + \frac{k_L}{k_H}\right) \sin a_L k_L \right] \tag{3.82}$$

Matrix $\overleftrightarrow{T}$ is unimodular matrix, i.e. real square matrix with determinant $ad - bc = \pm 1$ [8].

Referring to Fig. 3.14, the amplitude reflection coefficient is defined as

$$r = \frac{B_0}{A_0} \tag{3.83}$$

Choosing a column vector of layer one as the zeroth unit cell and remembering that all cells are identical, one obtains for N cells

$$\begin{bmatrix} A_0 \\ B_0 \end{bmatrix} = \begin{bmatrix} a & b \\ c & d \end{bmatrix}^N \begin{bmatrix} A_n \\ B_n \end{bmatrix} = \overleftrightarrow{T}^N \begin{bmatrix} A_n \\ B_n \end{bmatrix}$$

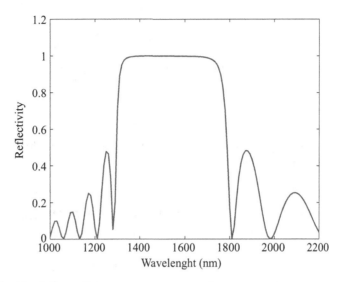

Fig. 3.15 Spectrum of reflectivity of a Bragg mirror.

Since matrix $\overleftrightarrow{T}$ is a unimodular matrix, one has the following relation [8]

$$\begin{bmatrix} a & b \\ c & d \end{bmatrix}^N = \begin{bmatrix} a \cdot U_{N-1}(x) - U_{N-2}(x) & b \cdot U_{N-1}(x) \\ c \cdot U_{N-1}(x) & d \cdot U_{N-1}(x) - U_{N-2}(x) \end{bmatrix}$$

where $x = \frac{1}{2}(a+d)$ and $U_N(x)$ are Chebyshev polynomials of the second kind

$$U_m(x) = \frac{\sin\left[(m+1)\cos^{-1} x\right]}{\sqrt{1-x^2}} = \frac{\sin\left[(m+1)\theta\right]}{\sin \theta}$$

where $x = \cos \theta$. Using the above relations, the amplitude reflection coefficient is evaluated as

$$\begin{bmatrix} A_0 \\ B_0 \end{bmatrix} = \begin{bmatrix} a \cdot U_{N-1}(x) - U_{N-2}(x) & b \cdot U_{N-1}(x) \\ c \cdot U_{N-1}(x) & d \cdot U_{N-1}(x) - U_{N-2}(x) \end{bmatrix} \begin{bmatrix} A_N \\ 0 \end{bmatrix}$$

for cell N. From above relation, one finds

$$r = \frac{B_0}{A_0} = \frac{c \cdot U_{N-1}(x)}{a \cdot U_{N-1}(x) - U_{N-2}(x)}$$

The definition of reflectance is [9] $R = |r|^2$. After a little algebra one finds

$$R = \frac{|c|^2}{|c|^2 + \left(\frac{\sin K\Lambda}{\sin NK\Lambda}\right)^2} \tag{3.84}$$

Spectrum of reflectivity as given by Eq. (3.84) for $N = 10$ periods of Bragg reflector is shown in Fig. 3.15. The MATLAB code is shown in Appendix, Listing 3A.2. The following parameters were adopted: $\bar{n}_H = 2.25$, $\bar{n}_L = 1.45$, $a_H = 167$ nm and $a_L = 259$ nm.

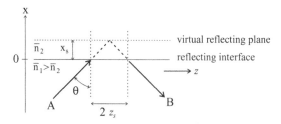

Fig. 3.16 The lateral displacement of a light beam on total reflection at the interface between two dielectric media (Goos-Hänchen shift).

3.10 Goos-Hänchen shift

Detailed examination of the internal reflection of a light beam on a planar dielectric interface shows that the reflected beam is shifted laterally from the trajectory predicted by the simple ray analysis, as shown in Fig. 3.16. The lateral displacement arises because wave energy actually penetrates beyond the interface into the lower index media before turning around. Total lateral shift is $d = 2z_s$ whereas x_s is penetration depth.

To determine lateral shift d, we follow Kogelnik [10]. Consider incident beam A as consisting of two plane waves incident at two slightly different angles with wave vectors $\beta \pm \Delta\beta$. The complex amplitude $A(z)$ of the incident wave packet at the interface $x = 0$ is

$$\begin{aligned} A(z) &= e^{-j(\beta+\Delta\beta)z} + e^{-j(\beta-\Delta\beta)z} \\ &= e^{-j\beta z}\left(e^{-j\Delta\beta z} + e^{j\Delta\beta z}\right) \\ &= 2\cos(\Delta\beta z)e^{-j\beta z} \end{aligned} \quad (3.85)$$

Reflected beam B is

$$B = R \cdot A \quad (3.86)$$

where $R = \exp(2j\phi)$. Phase shift ϕ that occurs upon reflection is different for both types of polarization and has been determined before, see Eqs. (3.50) and (3.56). Phase shift ϕ is a function of angle θ and propagation constant β. Assume a small value of $\Delta\beta$ and expand

$$\phi(\beta + \Delta\beta) = \phi(\beta) + \frac{d\phi}{d\beta}\Delta\beta \equiv \phi(\beta) + \Delta\phi$$

Using the above expansion and (3.85) allows us to write reflected beam B as

$$\begin{aligned} B &= R \cdot A = e^{2j\phi(\beta\pm\Delta\beta)}\left\{e^{-j(\beta+\Delta\beta)z} + e^{-j(\beta-\Delta\beta)z}\right\} \\ &= e^{-j(\beta z - 2\phi)}\left\{e^{j(\Delta\beta z - 2\Delta\phi)} + e^{-j(\Delta\beta z - 2\Delta\phi)}\right\} \\ &= 2e^{-j(\beta z - 2\phi)}\cos(\Delta\beta - 2z_s) \end{aligned} \quad (3.87)$$

where $z_s = \frac{d\phi}{dz}$. The net effect of the phase shift is to displace the beam along the z axis by a distance $2z_s$. Distance z_s is different for both types of modes. For TE modes [11]

$$k_0 z_s = \frac{\tan\theta}{\left(\beta^2 - n_c^2\right)^{1/2}}$$

whereas for TM modes

$$k_0 z_s = \frac{\tan\theta}{\left(\beta^2 - n_c^2\right)^{1/2}} \frac{1}{\beta^2/n_f^2 + \beta^2/n_c^2 - 1}$$

3.11 Poynting theorem

Electromagnetic waves carry with them electromagnetic power. We will now deliver a relation between the rate of such energy transfer and the electric and magnetic field intensities. We start with Maxwell's equations (3.1) and (3.2)

$$\nabla \times \mathbf{E} = -\frac{\partial \mathbf{B}}{\partial t} \tag{3.88}$$

$$\nabla \times \mathbf{H} = \frac{\partial \mathbf{D}}{\partial t} + \mathbf{J} \tag{3.89}$$

Use the following mathematical identity:

$$\nabla \cdot (\mathbf{E} \times \mathbf{H}) = \mathbf{H} \cdot (\nabla \times \mathbf{E}) - \mathbf{E}(\nabla \times \mathbf{H}) \tag{3.90}$$

Substituting Maxwell's equations into (3.90) and using constitutive relations, one obtains

$$\begin{aligned}\nabla \cdot (\mathbf{E} \times \mathbf{H}) &= -\mathbf{H} \cdot \mu \frac{\partial \mathbf{H}}{\partial t} - \mathbf{E} \cdot \mathbf{J} - \mathbf{E} \cdot \mu \frac{\partial \mathbf{E}}{\partial t} \\ &= -\mu \frac{1}{2} \frac{\partial \mathbf{H} \cdot \mathbf{H}}{\partial t} - \sigma E^2 - \varepsilon \frac{1}{2} \frac{\partial \mathbf{E} \cdot \mathbf{E}}{\partial t} \\ &= -\frac{\partial}{\partial t}\left(\frac{1}{2}\varepsilon E^2 + \frac{1}{2}\mu H^2\right) - \sigma E^2\end{aligned} \tag{3.91}$$

Integrate (3.91) over volume V and apply Gauss's theorem, and then one finds

$$\int_V \nabla \cdot (\mathbf{E} \times \mathbf{H}) dv = -\frac{\partial}{\partial t}\int_V \left(\frac{1}{2}\varepsilon E^2 + \frac{1}{2}\mu H^2\right) dv - \int_V \sigma E^2 dv \tag{3.92}$$

The Poynting vector $\mathbf{P}$ is defined as

$$\mathbf{P} = \mathbf{E} \times \mathbf{H} \tag{3.93}$$

Eq. (3.92) can be written as

$$-\oint_S \mathbf{P} \cdot d\mathbf{s} = \frac{\partial}{\partial t}\int_V (w_e + w_m)\, dv + \int_V p_\sigma dv \tag{3.94}$$

where

$w_e = \frac{1}{2}\varepsilon E^2$ is electric energy density
$w_m = \frac{1}{2}\mu H^2$ is magnetic energy density
$p_\sigma = \sigma E^2$ is Ohmic power density

From the above equation, one can interpret vector **P** as representing the power flow per unit area.

3.12 Problems

1. Using the expression derived for a transfer matrix, consider an antireflection structure consisting of two-layer quarter-wave-thickness films made of CeF_3 (low index layer with refractive index $\bar{n}_1 = 1.65$) and high-index layer of ZrO_2 (refractive index $\bar{n} = 2.1$) deposited on glass substrate with refractive index $\bar{n}_s = 1.52$. Assume that on top there is air with $\bar{n}_0 = 1$. Plot the spectral reflectance of such structure.
2. Show that a linearly polarized plane wave can be decomposed into a right-hand and a left-hand circularly polarized waves of equal amplitudes.
3. Derive an expression for a transfer matrix assuming p-polarization. In this case the vector of electric field **E** is parallel to the plane of incidence.
4. Determine the reflectance for normal incidence and plot it versus path difference/wavelength for various values of refractive index.
5. Write a MATLAB program to illustrate propagation of EM wave elliptically polarized. The wave will consist of two perpendicular waves of unequal amplitudes. Visualize propagation in 3D. Analyse the state of polarization (SOP) by changing phase difference ϕ between individual waves.
6. Analyse a Bragg mirror for TM polarization. Derive matrix elements of matrix $\overleftrightarrow{T}$ for this polarization [9].
7. Assume that for some materials refractive index at a particular wavelength is negative. Discuss the consequences of such an assumption. Consider modification of Snell's law.
8. What percentage of the incoming irradiance is reflected at an air-glass ($\bar{n}_{glass} = 1.5$) interface for a beam of natural light incident at $70°$?
9. Write an expression for a right circularly polarized wave propagation in the positive z-direction such that its **E**-field points in the negative x-direction at $z = 0$ and $t = 0$.
10. Verify that linear light is a special case of elliptical light.

3.13 Project

1. Write a MATLAB program to determine reflectance for a multilayered structure consisting of different dielectric materials of various thicknesses. Analyse the special case of $d = \lambda/2$.

Table 3.2 List of MATLAB functions for Chapter 3.

Listing	Function name	Description
3A.1	*reflections_TE_TM.m*	Plots reflections based on Fresnel equations for TE and TM modes
3A.2	*Bragg_an.m*	Plots reflectivity spectrum of Bragg mirror for TE mode using an analytical method

2. Develop MATLAB functions to analyse propagation of TM modes for arbitrary multilayered waveguiding structure. Use a similar approach as described above for TE modes.

Appendix 3A: MATLAB listings

In Table 3.2 we provide a list of MATLAB files created for Chapter 3 and a short description of each function.

Listing 3A.1 Program reflections_TE_TM.m. Program plots reflection coefficients for TE and TM modes.

```
% reflections_TE_TM.m
% Plot of reflections based on Fresnel equations
% TE and TM reflections
% cases of external reflection
clear all
n = 1.50;                          % relative refractive index
N_max = 11;                        % number of points for plot
t = linspace(0,1,N_max);           % creation of theta arguments
theta = (pi/2)*t;                  % angles in radians
% Plot reflection coefficients
num_TE = cos(theta) - sqrt(n^2-(sin(theta)).^2 );
den_TE = cos(theta) + sqrt(n^2-(sin(theta)).^2 );
r_TE = num_TE./den_TE;             % Plot for TE mode
num_TM = n^2*cos(theta) - sqrt(n^2-(sin(theta)).^2 );
den_TM = n^2*cos(theta) + sqrt(n^2-(sin(theta)).^2 );
r_TM = num_TM./den_TM;             % Plot for TM mode
%
angle_degrees = (theta./pi)*180;
x_line = [0 max(angle_degrees)];   % needed to draw horizontal line
y_line = [0 0];                    % passing through zero
plot(angle_degrees,r_TE,angle_degrees,r_TM, x_line, y_line, '-',...
    'LineWidth',1.5)
```

```
xlabel('Angle of incidence','FontSize',14)
set(gca,'FontSize',14);                    % size of tick marks on both axes
text(60, -0.6, 'r_{TE}','Fontsize',14)
text(60, -0.2, 'r_{TM}','Fontsize',14)
text(55, 0.1, '\theta_{Brewster}','Fontsize',14)
pause
close all
```

Listing 3A.2 Program Bragg_an.m. Determines spectrum of reflectivity of a Bragg mirror for TE mode using an analytical method.

```
% File name: bragg_an.m
% Determines reflectivity spectrum of Bragg mirror for TE mode
% using analytical method
clear all
N = 10;                         % number of periods
n_L = 1.45;                     % refractive index
n_H = 2.25;                     % refractive index
a_L = 259;                      % thickness (nm)
a_H = 167;                      % thickness (nm)
Lambda = a_L + a_H;             % period of the structure
lambda = 1000:10:2200;
k_L = 2*pi*n_L./lambda;
k_H = 2*pi*n_H./lambda;
%
a=exp(1i*a_H*k_H).*(cos(k_L*a_L)+(1i/2)*(k_H./k_L+k_L./k_H).*sin(k_L*a_L));
d=exp(-1i*a_H*k_H).*(cos(k_L*a_L)-(1i/2)*(k_H./k_L+k_L./k_H).*sin(k_L*a_L));
b = exp(-1i*a_H*k_H).*((1i/2)*(k_L./k_H - k_H./k_L).*sin(k_L*a_L));
c = exp(1i*a_H*k_H).*((1i/2)*(k_H./k_L - k_L./k_H).*sin(k_L*a_L));
%
K = (1/Lambda)*acos((a+d)/2);
tt = (sin(K*Lambda)./sin(N*K*Lambda)).^2;
denom = abs(c).^2 + tt;
R = abs(c).^2./denom;
plot(lambda,R,'LineWidth',1.5)
axis([1000 2200 0 1.2])
xlabel('Wavelenght (nm)','FontSize',14)
ylabel('Reflectivity','FontSize',14)
set(gca,'FontSize',14);                    % size of tick marks on both axes
pause
close all
```

References

[1] J. D. Kraus. *Electromagnetics, Fourth Edition*. McGraw-Hill, New York, 1992.

[2] D. K. Cheng. *Field and Wave Electromagnetics. Second Edition*. Addison-Wesley Publishing Company, Reading, MA, 1989.

[3] Z. Popovic and B. D. Popovic. *Introductory Electromagnetics*. Prentice-Hall, Upper Saddle River, NJ, 2000.

[4] F. L. Pedrotti and L. S. Pedrotti. *Introduction to Optics. Third Edition*. Prentice Hall, Upper Saddle River, NJ, 2007.

[5] J.-M. Liu. *Photonic Devices*. Cambridge University Press, Cambridge, 2005.

[6] A. Thelen. *Design of Optical Interference Coatings*. McGraw-Hill, New York, 1989.

[7] R. S. Geels. *Vertical-Cavity Surface-Emitting Lasers: Design, Fabrication and Characterization*. PhD thesis, University of California, Santa Barbara, 1991.

[8] E. W. Weisstein, Unimodular Matrix. From *MathWorld–A Wolfram Web Resource*, http://mathworld.wolfram.com/UnimodularMatrix.html, 3 July 2012.

[9] P. Yeh, A. Yariv, and C.-S. Hong. Electromagnetic propagation in periodic stratified media. I. general theory. *J. Opt. Soc. Am.*, **67**:423–38, 1977.

[10] H. Kogelnik. Theory of optical waveguides. In T. Tamir, ed., *Springer Series in Electronics and Photonics. Guided-Wave Optoelectronics*, volume 26, pp. 7–88. Springer-Verlag, Berlin, 1988.

[11] C. R. Pollock and M. Lipson. *Integrated Photonics*. Kluwer Academic Publishers, Boston, 2003.

4 Slab waveguides

Slab waveguides are important ingredients of both passive and active devices. Therefore, discussion and understanding of their properties are critical. The following topics will be discussed in this chapter:

- ray optics of the slab waveguide
- electromagnetic description
- three-layer problem for symmetric guiding structure
- TE modes in multilayer waveguide.

4.1 Ray optics of the slab waveguide

In this section we discuss applications of ray optics concepts to analyse the propagation of light in slab waveguides. We start with summarizing the concept of numerical aperture (NA).

4.1.1 Numerical aperture

Consider light entering a waveguide from air, having the refractive index $\bar{n}_0$ to waveguiding structure with refractive indices $\bar{n}_1$ and $\bar{n}_2$ (see Fig. 4.1). We represent light within ray optics. We want to understand the conditions under which light will propagate in the middle layer with the value of refractive index $\bar{n}_1$.

Ray 1 (dashed line) which enters waveguide at large incident angle θ_1 propagates through the middle layer and penetrates the upper layer with refractive index $\bar{n}_2$. Since this ray does not propagate in the middle layer, it is effectively lost.

As we gradually decrease incident angle θ, we reach the situation where a ray slides across the interface (case for ray 3). In such a case incident angle θ is known as the acceptance angle θ_a. The internal angle of incidence at point D is then ϕ_c and it is determined from the relation

$$\sin \phi_c = \frac{\bar{n}_2}{\bar{n}_1} \qquad (4.1)$$

For the incident ray (like ray 2) which enters the interface at an angle smaller than the acceptance angle $\theta < \theta_a$, it will propagate in the middle layer. Let us analyse the existing situation in more detail.

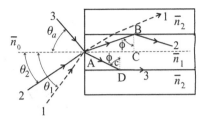

Fig. 4.1 Light entering slab waveguide.

From triangle ABC, one finds $\bar{\theta}_2 = \pi/2 - \phi$ (angle $\bar{\theta}_2$ not shown in Fig. 4.1). From Snell's law for ray 3, we obtain

$$\bar{n}_0 \sin \theta_1 = \bar{n}_1 \sin \theta_2 = \bar{n}_1 \sin(\pi/2 - \phi) = \bar{n}_1 \cos \phi = \bar{n}_1 \left(1 - \sin^2 \phi\right)^{1/2}$$

From the above, we have (using Eq. 4.1) for ray 2 (propagating at the acceptance angle, i.e. for $\theta_1 = \theta_a$)

$$\bar{n}_0 \sin \theta_a = \bar{n}_1 \left(1 - \sin^2 \phi_c\right)^{1/2} = \left(\bar{n}_1^2 - \bar{n}_2^2\right)^{1/2}$$

Numerical aperture (NA) is defined as

$$NA \equiv \bar{n}_0 \sin \theta_a = \left(\bar{n}_1^2 - \bar{n}_2^2\right)^{1/2} \tag{4.2}$$

We introduce relative refractive index difference Δ as

$$\Delta \equiv \frac{\bar{n}_1^2 - \bar{n}_2^2}{2\bar{n}_1^2}$$

The approximate expression for Δ is obtained by observing that $\bar{n}_1 \approx \bar{n}_2$, which allows us to write $\bar{n}_1 + \bar{n}_2 \approx 2\bar{n}_1$. The approximate formula for Δ is therefore

$$\Delta = \frac{(\bar{n}_1 + \bar{n}_2)(\bar{n}_1 - \bar{n}_2)}{2\bar{n}_1^2} \approx \frac{2\bar{n}_1(\bar{n}_1 - \bar{n}_2)}{2\bar{n}_1^2} = \frac{\bar{n}_1 - \bar{n}_2}{\bar{n}_1}$$

In terms of Δ, NA can be approximately expressed as

$$NA = \bar{n}_1 (2\Delta)^{1/2}$$

4.1.2 Guided modes

We will first discuss the propagation of a light ray in a waveguide formed by the film of dielectric surrounded from below by substrate and by cover layer from above, see Fig. 4.2. In such a structure light travels in a zig-zag fashion through the film. Light is monochromatic and coherent and is characterized by ω angular frequency and λ free-space wavelength. The following relation holds $k = \frac{2\pi}{\lambda}$, where k is known as the wavenumber. Electric field of the propagating wave is

$$E \sim e^{-jk\bar{n}_f(\pm x \cos \phi + z \sin \phi)}$$

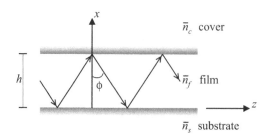

Fig. 4.2 Light propagation in a slab waveguide in a ray approximation.

Let β be a propagation constant of the guided mode of the slab. One has

$$\beta = \frac{\omega}{v_p} = k\bar{n}_f \sin\phi \quad (4.3)$$

where v_p is the phase velocity. Please note that not all angles are allowed (sometimes none). Referring to Fig. 4.2, rays at $\phi = 90°$ travel horizontally along the waveguide. Generally, rays which undergo total internal reflection can travel within the waveguide. That condition can be fulfilled for rays travelling at angles fulfilling the condition (all angles are determined with respect to normal)

$$\phi_c \leq \phi \leq 90°$$

or

$$\bar{n}_f \sin\phi_c \leq \bar{n}_f \sin\phi \leq \bar{n}_f \sin 90° = \bar{n}_f$$

where we have also multiplied the inequality by $\bar{n}_f$. Using Eqs. (4.1) and (4.3), one obtains

$$\bar{n}_s \leq N \leq \bar{n}_f \quad (4.4)$$

Here we introduced 'effective guide index' N defined as

$$N = \frac{\beta}{k} = \bar{n}_f \sin\phi \quad (4.5)$$

4.1.3 Transverse resonance condition

The condition is expressed as 'the sum of all phase shifts during propagation must be a multiple of 2π'. The following are the contributions to the phase change:

$k\bar{n}_f h \cos\phi$ phase change during first transverse passage through the film
$-2\phi_c$ phase shift on total reflection from cover
$-2\phi_s$ phase shift on total reflection from substrate

Combining the above contributions, one obtains transverse resonance condition as

$$2k\bar{n}_f h \cos\phi - 2\phi_s - 2\phi_c = 2\pi\nu, \quad \nu = 0, \pm 1, \pm 2, \ldots \quad (4.6)$$

which is the dispersion equation of the guide. Variable ν identifies mode number.

Expression for the Fresnel phase for TE polarization was derived in Chapter 3, Eq. (3.50). Modifying notation slightly (angle θ_1 is replaced by ϕ), the relation reads

$$\tan \phi = \frac{\sqrt{\bar{n}_1^2 \sin^2 \phi - \bar{n}_2^2}}{\bar{n}_1 \cos \phi}$$

Adopting the above expression for our present geometry, one writes for phases on reflections:

1. for reflection on the film-cover interface

$$\tan \phi_c = \frac{\sqrt{\bar{n}_f^2 \sin^2 \phi - \bar{n}_c^2}}{\bar{n}_f \cos \phi} = \frac{\sqrt{\bar{n}_f^2 \sin^2 \phi - \bar{n}_c^2}}{\sqrt{\bar{n}_f^2 - \bar{n}_f^2 \sin^2 \phi}} \tag{4.7}$$

2. for reflection on the film-substrate interface

$$\tan \phi_s = \frac{\sqrt{\bar{n}_f^2 \sin^2 \phi - \bar{n}_s^2}}{\bar{n}_f \cos \phi} = \frac{\sqrt{\bar{n}_f^2 \sin^2 \phi - \bar{n}_s^2}}{\sqrt{\bar{n}_f^2 - \bar{n}_f^2 \sin^2 \phi}} \tag{4.8}$$

4.1.4 Transverse condition: normalized form

The above transverse condition will now be cast in a normalized form which is more suitable for numerical work. For that, let us introduce the following variables: variable V

$$V \equiv k \cdot h \sqrt{\bar{n}_f^2 - \bar{n}_s^2} \tag{4.9}$$

normalized guide index, b

$$b \equiv \frac{N^2 - \bar{n}_s^2}{\bar{n}_f^2 - \bar{n}_s^2} \tag{4.10}$$

and asymmetry parameter for TE modes, a

$$a \equiv \frac{\bar{n}_s^2 - \bar{n}_c^2}{\bar{n}_f^2 - \bar{n}_s^2} \tag{4.11}$$

For TE modes, use transverse resonance condition and Fresnel phases, and obtain

$$V\sqrt{1-b} = v \cdot \pi + \tan^{-1} \sqrt{\frac{b}{1-b}} + \tan^{-1} \sqrt{\frac{b+a}{1-b}} \tag{4.12}$$

Example Derive transverse resonance condition in a normalized form, Eq. (4.12).

Solution
Using definitions (4.10) and (4.11), one can prove the following relations:

$$\frac{b}{1-b} = \frac{N^2 - \bar{n}_s^2}{\bar{n}_f^2 - N^2}$$

and
$$\frac{b+a}{1-b} = \frac{N^2 - \bar{n}_c^2}{\bar{n}_f^2 - N^2}$$

and also
$$1 - b = \frac{\bar{n}_f^2 - N^2}{\bar{n}_f^2 - \bar{n}_s^2}$$

Using the above relations, the expressions for phases, Eqs. (4.7) and (4.8) can be written as
$$\phi_c = \tan^{-1}\sqrt{\frac{N^2 - \bar{n}_c^2}{\bar{n}_f^2 - N^2}} = \tan^{-1}\sqrt{\frac{b+a}{1-b}}$$

and
$$\phi_s = \tan^{-1}\sqrt{\frac{N^2 - \bar{n}_s^2}{\bar{n}_f^2 - N^2}} = \tan^{-1}\sqrt{\frac{b}{1-b}}$$

One also has
$$\bar{n}_f \cos\phi = \sqrt{\bar{n}_f^2 - \bar{n}_f^2 \sin^2\phi} = \sqrt{\bar{n}_f^2 - N^2}$$

Substituting the above formulas into (4.6), we obtain
$$V\sqrt{1-b} = \nu \cdot \pi + \tan^{-1}\sqrt{\frac{b}{1-b}} + \tan^{-1}\sqrt{\frac{b+a}{1-b}}$$

A MATLAB program to analyse the transverse resonance condition (4.12) which plots normalized guide index b versus variable V, is provided in the Appendix, Listing 4A.1. It allows us to input different values of the asymmetry parameter a.

A plot of normalized guide index b versus variable V for three values of parameter a ($a = 0, 8, 50$) is shown in Fig. 4.3.

We finish this section with a typical simplified expression for effective index N. From a definition of b one obtains
$$b(\bar{n}_f^2 - \bar{n}_s^2) + \bar{n}_s^2 \equiv N^2$$

Perform the algebraic steps
$$N^2 = \left(1 + b\frac{\bar{n}_f^2 - \bar{n}_s^2}{\bar{n}_s^2}\right)\bar{n}_s^2$$

$$N = \sqrt{1 + b\frac{\bar{n}_f^2 - \bar{n}_s^2}{\bar{n}_s^2}}\,\bar{n}_s$$

$$N \simeq \left(1 + b\frac{1}{2}\frac{\bar{n}_f^2 - \bar{n}_s^2}{\bar{n}_s^2}\right)\bar{n}_s$$

$$N = \left(1 + b\frac{1}{2}\frac{(\bar{n}_f - \bar{n}_s)(\bar{n}_f + \bar{n}_s)}{\bar{n}_s^2}\right)\bar{n}_s$$

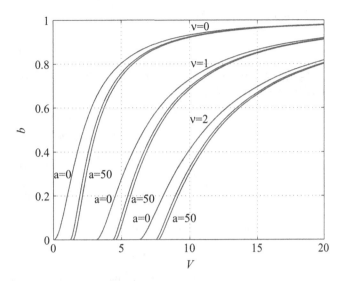

Fig. 4.3 Normalized $b - V$ diagram of a planar waveguide for various degrees of asymmetry.

Finally, one finds

$$N \approx \overline{n}_s + \frac{1}{2} b \left(\overline{n}_f - \overline{n}_s \right) \tag{4.13}$$

4.2 Fundamentals of EM theory of dielectric waveguides

4.2.1 General discussion

Assuming time-harmonic fields and using constitutive relations, the source-free Maxwell equations are

$$\nabla \times \mathbf{E} = -j\omega\mu\mathbf{H} \tag{4.14}$$
$$\nabla \times \mathbf{H} = j\omega\varepsilon\mathbf{E} \tag{4.15}$$

Boundary conditions were derived in Chapter 3. Separate fields into transversal and longitudinal components as

$$\mathbf{E} = \mathbf{E}_t + \mathbf{E}_z, \quad \mathbf{H} = \mathbf{H}_t + \mathbf{H}_z \tag{4.16}$$

where

$$\mathbf{E}_t = \left[E_x, E_y, 0 \right]$$

is the transverse part and

$$\mathbf{E}_z = [0, 0, E_z]$$

is the longitudinal part. Also

$$\nabla = \nabla_t + \mathbf{a}_z \frac{\partial}{\partial z}, \quad \mathbf{a}_z = [0, 0, 1] \qquad (4.17)$$

where $\mathbf{a}_z$ is a unit vector along z-direction. Substituting (4.16) and (4.17) into Eqs. (4.14) and (4.15), one obtains

$$\nabla_t \times \mathbf{E}_t = -j\omega\mu\mathbf{H}_z \qquad (4.18)$$

$$\nabla_t \times \mathbf{H}_t = j\omega\varepsilon\mathbf{E}_z \qquad (4.19)$$

$$\nabla_t \times \mathbf{E}_z + \mathbf{a}_z \times \frac{\partial \mathbf{E}_t}{\partial z} = -j\omega\mu\mathbf{H}_t \qquad (4.20)$$

$$\nabla_t \times \mathbf{H}_z + \mathbf{a}_z \times \frac{\partial \mathbf{H}_t}{\partial z} = j\omega\varepsilon\mathbf{E}_t \qquad (4.21)$$

Modes in a waveguide are characterized by a dielectric constant

$$\varepsilon(x, y) = \varepsilon_0 \bar{n}^2(x, y) \qquad (4.22)$$

where $n(x, y)$ is the refractive index profile in the transversal plane. Write the fields as

$$\begin{aligned} \mathbf{E}(x, y, z) &= \mathbf{E}_\nu(x, y) e^{-j\beta_\nu z} \\ \mathbf{H}(x, y, z) &= \mathbf{H}_\nu(x, y) e^{-j\beta_\nu z} \end{aligned} \qquad (4.23)$$

where we have introduced index ν which label modes and β_ν is the propagation constant of the mode ν. After substitution of Eqs. (4.23) into (4.18)–(4.21) one obtains

$$\nabla_t \times \mathbf{E}_{t\nu}(x, y) = -j\omega\mu\mathbf{H}_{z\nu}(x, y) \qquad (4.24)$$

$$\nabla_t \times \mathbf{H}_{t\nu}(x, y) = j\omega\varepsilon\mathbf{E}_{z\nu}(x, y) \qquad (4.25)$$

$$\nabla_t \times \mathbf{E}_{z\nu}(x, y) - j\beta_\nu \mathbf{a}_z \times \mathbf{E}_{t\nu}(x, y) = -j\omega\mu\mathbf{H}_{t\nu}(x, y) \qquad (4.26)$$

$$\nabla_t \times \mathbf{H}_{z\nu}(x, y) - j\beta_\nu \mathbf{a}_z \times \mathbf{H}_{t\nu}(x, y) = j\omega\varepsilon\mathbf{E}_{t\nu}(x, y) \qquad (4.27)$$

From the analysis of the above equations, the existence of several types of modes can be recognized. Those will be described in more details later on. Generally, the main modes are:

- guided modes (bound states) – discrete spectrum of β_ν
- radiation modes – belong to continuum
- evanescent modes - $\beta_\nu = -j\alpha_\nu$; they decay as $\exp(-\alpha_\nu z)$.

4.2.2 Explicit form of general equations

Using the following general formulas

$$\nabla_t \times E_t = \begin{vmatrix} \mathbf{a}_x & \mathbf{a}_y & \mathbf{a}_z \\ \frac{\partial}{\partial x} & \frac{\partial}{\partial y} & 0 \\ E_x & E_y & 0 \end{vmatrix}$$

$$\nabla_t \times E_z = \begin{vmatrix} \mathbf{a}_x & \mathbf{a}_y & \mathbf{a}_z \\ \frac{\partial}{\partial x} & \frac{\partial}{\partial y} & 0 \\ 0 & 0 & E_z \end{vmatrix}$$

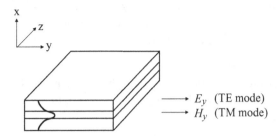

Fig. 4.4 Planar wide waveguide.

$$\mathbf{a}_z \times E_t = \begin{vmatrix} \mathbf{a}_x & \mathbf{a}_y & \mathbf{a}_z \\ 0 & 0 & 1 \\ E_x & E_y & 0 \end{vmatrix}$$

where $\mathbf{a}_x$, $\mathbf{a}_y$, $\mathbf{a}_z$ are unit vectors along the corresponding directions, we write general Eqs. (4.24)–(4.27) in expanded form (dropping mode index)

$$\left[0, 0, \frac{\partial E_y}{\partial x} - \frac{\partial E_x}{\partial y}\right] = -j\omega\mu\,[0, 0, H_z] \tag{4.28}$$

$$\left[0, 0, \frac{\partial H_y}{\partial x} - \frac{\partial H_x}{\partial y}\right] = j\omega\varepsilon\,[0, 0, E_z] \tag{4.29}$$

$$\left[\frac{\partial E_z}{\partial y}, -\frac{\partial E_z}{\partial x}, 0\right] - j\beta\,[-E_y, E_x, 0] = -j\omega\mu\,[H_x, H_y, 0] \tag{4.30}$$

$$\left[\frac{\partial H_z}{\partial y}, -\frac{\partial H_z}{\partial x}, 0\right] - j\beta\,[-H_y, H_x, 0] = j\omega\varepsilon\,[E_x, E_y, 0] \tag{4.31}$$

In the following sections the previous general equations will be used to analyse specific situations.

4.3 Wave equation for a planar wide waveguide

For the wide waveguide where its dimension in the y-direction is much larger than thickness, the field configuration along that direction remains approximately constant, see Fig. 4.4. Therefore, we can consider only confinement in the x-direction and set

$$\frac{\partial}{\partial y} = 0$$

Also, the refractive index assumes only x-dependence

$$\bar{n} = \bar{n}(x)$$

Such a waveguide supports two modes:

- transverse electric TE, where $E_y \neq 0$ and $E_x = E_z = 0$
- transverse magnetic TM, where $H_y \neq 0$ and $H_x = H_z = 0$.

From general formulas given by Eqs. (4.28)–(4.31), one obtains a set of three equations describing TE modes and also three equations describing TM modes. Those are written below for both types of modes separately.

TE modes

To describe TE modes we use only equations which involve E_y and its derivatives. One observes that when $E_x = E_z = 0$, from Eq. (4.30) $H_y = 0$. Equations describing those modes are

$$\beta E_y = -\omega \mu H_x \tag{4.32}$$

$$\frac{\partial H_z}{\partial x} + j\beta H_x = -j\omega \varepsilon E_y \tag{4.33}$$

$$\frac{\partial E_y}{\partial x} = -j\omega \mu H_z \tag{4.34}$$

The last step will be to eliminate H_x and H_z. We will differentiate the third equation, then substitute for $\frac{\partial H_z}{\partial x}$ using the second equation and finally eliminate H_x using first equation. We obtain a wave equation for TE modes

$$\frac{\partial^2 E_y}{\partial x^2} = \left(\beta^2 - \bar{n}^2 k^2\right) E_y \tag{4.35}$$

where $k = \frac{\omega}{c} = \omega\sqrt{\varepsilon_0 \mu_0}$.

TM modes

We proceed in a similar way as for the TE modes. From Eq. (4.31) when $H_x = H_z = 0$, one obtains that $E_y = 0$. We keep only equations which involve H_y and its derivatives. Equations which describe TM modes are

$$\beta H_y = \omega \varepsilon E_x$$

$$\frac{\partial H_y}{\partial x} = j\omega \varepsilon E_z$$

$$\frac{\partial E_z}{\partial x} + j\beta E_x = j\omega \mu H_y$$

We will use the fact that $\varepsilon = \varepsilon_0 \bar{n}^2$ and eliminate E_x and E_z. Here one must be more careful because $\bar{n}^2$ can be x dependent. From the first equation we determine E_x and from the second equation we determine E_z. Substituting the results into the third equation gives the wave equation for TM mode

$$\bar{n}^2 \frac{\partial}{\partial x}\left(\frac{1}{\bar{n}^2} \partial x \frac{\partial H_y}{\partial x}\right) = \left(\beta^2 - \bar{n}^2 k^2\right) H_y \tag{4.36}$$

4.4 Three-layer symmetrical guiding structure (TE modes)

We will analyse the three-layer symmetrical structure for TE modes. The structure is as shown in Fig. 4.4. The details are illustrated in Fig. 4.5. A film layer of thickness $2a$ and

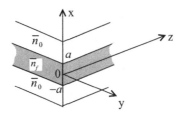

Fig. 4.5 Planar three-layer symmetric waveguide.

refractive index $\bar{n}_f$ is surrounded by a substrate and cladding layers have refractive indices equal to $\bar{n}_0$.

We introduce the following notation:

$$\kappa_f^2 = \bar{n}_f^2 k^2 - \beta^2 \tag{4.37}$$

and

$$\gamma^2 = \beta^2 - \bar{n}_0^2 k^2 \tag{4.38}$$

Here, γ is determined by the value of refractive index around the film. In general, γ assumes values which are dictated by the values of refractive indices in the cladding and substrate. Therefore, one deals with more than one value of γ.

The wave equation is

$$\frac{d^2 E_y(x)}{dx^2} = \left(\beta^2 - \bar{n}^2 k^2\right) E_y(x)$$

We discuss two separate solutions: odd and even modes.

Odd modes:

We assume the following guiding solution:

$$\begin{array}{lll} E_y(x) = A_c e^{-\gamma(x-a)}, & a < x & \text{(cladding)} \\ E_y(x) = B \sin \kappa_f x, & -a < x < a & \text{(film)} \\ E_y(x) = A_s e^{\gamma(x+a)}, & x < -a & \text{(substrate)} \end{array} \tag{4.39}$$

At the interfaces, i.e. for $x = \pm a$, the field E_y and its derivative $\frac{dE_y}{dx}$ must be continuous. The continuity at $x = a$ gives the following equations:

$$A_c = B \sin \kappa_f a$$
$$-\gamma A_c = \kappa_f B \cos \kappa_f a$$

From above equations, one obtains

$$-\gamma = \kappa_f \cot \kappa_f a \tag{4.40}$$

The continuity at $x = -a$ gives an identical equation. By introducing variable

$$y = \kappa_f a$$

Eq. (4.40) can be written as

$$-y \cot y = \gamma a$$

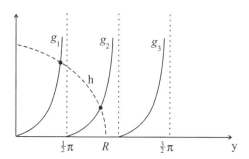

Fig. 4.6 Graphical solution of the transcendental equation for even modes.

From the definition for γ we obtain

$$-\gamma a = a\sqrt{n^2 k^2 - \beta^2} = \sqrt{a^2 \bar{n}^2 k^2 - a^2 \beta^2} \qquad (4.41)$$

Similarly, from the definition for κ_f we obtain

$$a^2 \kappa_f^2 = a^2 n \bar{n}_f^2 k^2 - a^2 \beta^2$$

Determining $a^2 \beta^2$ from the above and substituting into Eq. (4.41), one finds

$$-y \cot y = \sqrt{R^2 - y^2} \qquad (4.42)$$

where we have defined

$$R^2 = a^2 k^2 \left(\bar{n}_f^2 - \bar{n}^2 \right) \qquad (4.43)$$

Transcendental Eq. (4.42) must be solved numerically.

Even modes:

For even modes we assume the following guiding solution:

$$\begin{array}{lll} E_y(x) = A_c e^{-\gamma(x-a)}, & a < x & \text{(cladding)} \\ E_y(x) = B \cos \kappa_f x, & -a < x < a & \text{(film)} \\ E_y(x) = A_s e^{\gamma(x+a)}, & x < -a & \text{(substrate)} \end{array} \qquad (4.44)$$

The remaining steps are done in the exactly same way as for odd modes. The resulting transcendental equation for propagation constant is

$$y \tan y = \sqrt{R^2 - y^2} \qquad (4.45)$$

The function R is defined by Eq. (4.43). Eq. (4.45) is represented in Fig. 4.6. Intersections between line h which represents the right hand side of Eq. (4.45) and lines $g's$ which represent the left hand side of Eq. (4.45), determine propagation constants. Those must be found numerically. For numerical searches, the following functions are introduced:

$$f_{\text{even}}(y) = y \tan y - \sqrt{R^2 - y^2} \qquad (4.46)$$

and

$$f_{\text{odd}}(y) = -y \cot y - \sqrt{R^2 - y^2} \qquad (4.47)$$

The above functions are used in numerical search, since e.g. $f_{even}(y_1) = 0$ corresponds to the solution y_1 from which propagation constant β_1 is determined. The algorithm is described in the next section.

4.4.1 The algorithm

The following main steps constitute the numerical algorithm.
1. Search intervals for even and odd functions are formed as follows:

even functions

$$y_i = n\pi, y_f = \min\left(n\pi + \frac{\pi}{2} - 10^{-3}, R\right)$$

odd functions

$$y_i = \frac{\pi}{2} + n\pi, y_f = \min\left((n+1)\pi - 10^{-3}, R\right)$$

where y_i and y_f are initial and final values of the appropriate search interval. The small constant (10^{-3}) has been introduced to avoid working close to singularity.

2. Search is done for zeros in each interval.
3. Even and odd solutions are obtained independently in separate searches.
4. The solutions are found using MATLAB function *fzero()* as

$$y_{temp} = fzero\left(func, [y_i, y_f]\right)$$

where the first argument contains search function and the second one contains interval for search.

4.5 Modes of the arbitrary three-layer asymmetric planar waveguide in 1D

Consider the modified version of the structure shown in Fig. 4.5. The structure has different values of refractive indices of substrate and cladding and it is known as asymmetric planar waveguide. Here, $\bar{n}_c$ signifies refractive index of cladding, $\bar{n}_f$ of film, and $\bar{n}_s$ that of substrate. For an asymmetric slab, $\bar{n}_c \neq \bar{n}_s$. Define the following quantities:

$$\begin{aligned}\kappa_c^2 &= \bar{n}_c^2 k^2 - \beta^2 \equiv -\gamma_c^2 \\ \kappa_f^2 &= \bar{n}_f^2 k^2 - \beta^2 \\ \kappa_s^2 &= \bar{n}_s^2 k^2 - \beta^2 \equiv -\gamma_s^2\end{aligned} \quad (4.48)$$

where γ_i describes transverse decay and κ_i contains propagation constants. i takes the values c, f, s as appropriate. In the next subsection we discuss TE modes for this three-layer structure.

4.5.1 TE modes

For guided TE modes the following solutions exist:

$$E_y(x) = A_c e^{-\gamma_c(x-a)} \qquad a < x \qquad \text{(cover)}$$
$$E_y(x) = A \cos \kappa_f x + B \sin \kappa_f x \qquad -a < x < a \quad \text{(film)} \qquad (4.49)$$
$$E_y(x) = A_s e^{\gamma_s(x+a)} \qquad x < -a \qquad \text{(substrate)}$$

Derivatives are determined as follows:

$$\frac{dE_y(x)}{dx} = -\gamma_c A_c e^{-\gamma_c(x-a)} \qquad a < x \qquad \text{(cover)}$$
$$\frac{dE_y(x)}{dx} = -\kappa_f A \cos \kappa_f x + \kappa_f B \sin \kappa_f x \quad -a < x < a \quad \text{(film)} \qquad (4.50)$$
$$\frac{dE_y(x)}{dx} = \gamma_s A_s e^{\gamma_s(x+a)} \qquad x < -a \qquad \text{(substrate)}$$

Boundary conditions dictate that

$$E_y \text{ and } \frac{dE_y(x)}{dx} \text{ are continuous for } x = a \text{ and for } x = -a \qquad (4.51)$$

When applying boundary conditions for E_y and $\frac{dE_y(x)}{dx}$, we get the following equations:

$$\text{for } x = -a \quad A \cos \kappa_f a - B \sin \kappa_f a = A_s$$
$$\kappa_f A \sin \kappa_f a + \kappa_f B \cos \kappa_f a = \gamma_s A_s$$
$$\text{for } x = a \quad A_c = A \cos \kappa_f a + B \sin \kappa_f a$$
$$-\gamma_c A_c = -\kappa_f A \sin \kappa_f a + \kappa_f B \cos \kappa_f a$$

The above equations can be written in a matrix form

$$\begin{bmatrix} \cos \kappa_f a & -\sin \kappa_f a & -1 & 0 \\ \kappa_f \sin \kappa_f a & \kappa_f \cos \kappa_f a & -\gamma_s & 0 \\ \cos \kappa_f a & \sin \kappa_f a & 0 & -1 \\ -\kappa_f \sin \kappa_f a & \kappa_f \cos \kappa_f a & 0 & \gamma_c \end{bmatrix} \begin{bmatrix} A \\ B \\ A_s \\ A_c \end{bmatrix} = 0 \qquad (4.52)$$

For the above homogeneous system to have nontrivial solution, the main determinant should vanish.

$$\begin{vmatrix} \cos \kappa_f a & -\sin \kappa_f a & -1 & 0 \\ \kappa_f \sin \kappa_f a & \kappa_f \cos \kappa_f a & -\gamma_s & 0 \\ \cos \kappa_f a & \sin \kappa_f a & 0 & -1 \\ -\kappa_f \sin \kappa_f a & \kappa_f \cos \kappa_f a & 0 & \gamma_c \end{vmatrix} = 0 \qquad (4.53)$$

The above determinant is evaluated as follows. Let us expand it over last column and we get

$$\gamma_c \begin{vmatrix} \cos \kappa_f a & -\sin \kappa_f a & -1 \\ \kappa_f \sin \kappa_f a & \kappa_f \cos \kappa_f a & -\gamma_s \\ \cos \kappa_f a & \sin \kappa_f a & 0 \end{vmatrix} + \begin{vmatrix} \cos \kappa_f a & -\sin \kappa_f a & -1 \\ \kappa_f \sin \kappa_f a & \kappa_f \cos \kappa_f a & -\gamma_c \\ -\kappa_f \sin \kappa_f a & \kappa_f \cos \kappa_f a & 0 \end{vmatrix} = 0$$

Evaluating both determinants, we obtain

$$\sin^2 \kappa_f a - \cos^2 \kappa_f a + \frac{\kappa_f}{\gamma_c} \sin \kappa_f a \cos \kappa_f a - \frac{\gamma_s}{\kappa_f} \sin \kappa_f a \cos \kappa_f a = 0$$

which can be expressed as

$$\tan^2 \kappa_f a - 1 + \frac{\kappa_f}{\gamma_c} \tan \kappa_f a - \frac{\gamma_s}{\kappa_f} \tan \kappa_f a = 0 \tag{4.54}$$

It is the general equation for a three-layer asymmetric waveguide. For a symmetric waveguide

$$\gamma_s = \gamma_c = \gamma$$

and one can write Eq. (4.54) as

$$\left(\tan \kappa_f a - \frac{\gamma}{\kappa_f}\right)\left(\tan \kappa_f a + \frac{\kappa_f}{\gamma}\right) = 0 \tag{4.55}$$

which describes even and odd modes discussed in the previous section (symmetric modes).

Note: The above modes are not normalized for power. Power P carried by a mode per unit guide width is determined as follows:

$$\begin{aligned} P &= -2 \int_{-\infty}^{+\infty} dx E_y H_x \\ &= \frac{2\beta}{\omega\mu} \int_{-\infty}^{+\infty} dx E_y^2 \\ &= N\sqrt{\frac{\varepsilon_0}{\mu_0}} E_f^2 \cdot h_{eff} \\ &= E_f \cdot H_f \cdot h_{eff} \end{aligned} \tag{4.56}$$

where $h_{eff} \equiv 2a + \frac{1}{\gamma_s} + \frac{1}{\gamma_c}$ is the effective thickness of the waveguide.

4.5.2 Field profiles for TE modes

Analysis of the general determinant given by Eq. (4.54) for the asymmetric waveguide is complicated. If one wants to obtain formulas suitable for numerical evaluations and also suitable for obtaining field profiles, a better approach is to eliminate all constants appearing in Eq. (4.49). Thus, one will have a field profile expressed in terms of one constant only, say A_s. One then needs to determine derivatives which should be continuous across interfaces. From continuity of derivatives, the relevant transcendental equation used to obtain propagation constants is obtained.

We consider asymmetric structure where the substrate-film discontinuity is at $x = 0$ and film-cover interface is located at $x = h$. From continuity of the TE field at $x = 0$ and $x = h$ one finds

$$A_c = A \cos \kappa_f h + B \sin \kappa_f h \tag{4.57}$$

and

$$A_s = A \tag{4.58}$$

Table 4.1 Asymmetric three-layer waveguide.

Refractive index	Thickness
$n_c = 1.40$	—
$n_f = 1.50$	5 µm
$n_s = 1.45$	—

From continuity of derivatives at the above points

$$-\gamma_c A_c = -\kappa_f A \sin \kappa_f h + \kappa_f B \cos \kappa_f h \tag{4.59}$$

and

$$\gamma_s A_s = \kappa_f B \tag{4.60}$$

From Eqs. (4.58) and (4.60) the constants A and B can be expressed in terms of A_s. Substitution of those expressions into Eq. (4.57) gives constant A_c in terms of A_s. Using those results, we can replace constants A, B and A_c in Eq. (4.49). The resulting equations are

$$\begin{aligned} E_y(x) &= A_s \left(\cos \kappa_f h + \tfrac{\gamma_s}{\kappa_f} \sin \kappa_f h \right) \exp\left[-\gamma_c (x-h)\right] & h < x \\ E_y(x) &= A_s \left(\cos \kappa_f x + \tfrac{\gamma_s}{\kappa_f} \sin \kappa_f x \right) & 0 < x < h \\ E_y(x) &= A_s \exp(\gamma_s x) & x < 0 \end{aligned} \tag{4.61}$$

From the above, the derivatives with respect to x are

$$\begin{aligned} E_y'(x) &= -A_s \gamma_c \left(\cos \kappa_f h + \tfrac{\gamma_s}{\kappa_f} \sin \kappa_f h \right) \exp\left[-\gamma_c (x-h)\right] & h < x \\ E_y'(x) &= A_s \left(-\kappa_f \sin \kappa_f x + \gamma_s \cos \kappa_f x \right) & 0 < x < h \\ E_y'(x) &= A_s \gamma_s \exp(\gamma_s x) & x < 0 \end{aligned} \tag{4.62}$$

Applying continuity of derivatives at $x = h$, one finds

$$-\gamma_c \left(\cos \kappa_f h + \frac{\gamma_s}{\kappa_f} \sin \kappa_f h \right) = -\kappa_f \sin \kappa_f h + \gamma_s \cos \kappa_f h$$

From the above, it follows the following transcendental equation:

$$\tan \kappa_f h = \frac{\gamma_s + \gamma_c}{\kappa_f - \gamma_c \gamma_s / \kappa_f} \tag{4.63}$$

The above equation is used in numerical search for propagation constants. A typical plot of Eq. (4.63) is shown in Fig. 4.7. Using those propagation constants, from Eq. (4.61) one finds field profiles.

Example We consider the three-layer asymmetric planar structure described in Table 4.1 (from [1]) operating with light of wavelength $\lambda = 1$ µm.

Table 4.2 Propagation constants for a three-layer asymmetric structure defined in Table 4.1.

Mode	Propagation constant $\beta\,(\mu m^{-1})$
TE_0	9.40873
TE_1	9.36079
TE_2	9.28184
TE_3	9.17521

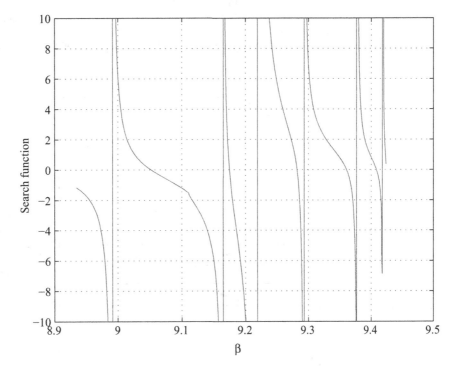

Fig. 4.7 Graphical plot of Eq. (4.63) for the three-layer asymmetric waveguide defined in Table 4.1.

The structure is analysed with MATLAB code from Appendix 4A.2.1. The resulting propagation constants are summarized in Table 4.2. There is a good agreement with published results [1]. Field profiles for two modes are plotted in Fig. 4.8.

4.6 Multilayer slab waveguides: 1D approach

In this section we discuss slab waveguides consisting of more than three layers. Multilayers require the repeated applications of boundary conditions at the layer

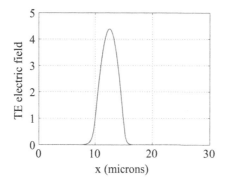
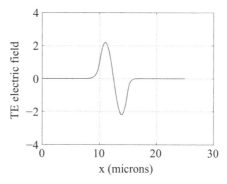

Fig. 4.8 Field profiles for an asymmetric three-layer structure. Fundamental mode TE_0 (left). Mode TE_1 (right).

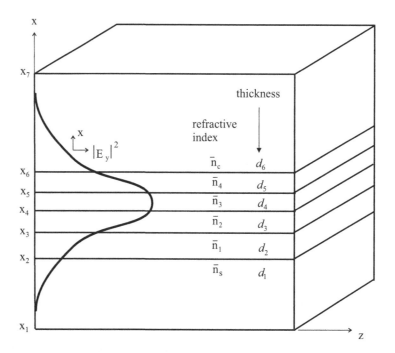

Fig. 4.9 Multilayer guiding structure. Electric field profile is also shown.

interfaces. Schematically, the structure is as shown in Fig. 4.9. We make the following assumptions:

- 1D problem
- $\bar{n} = \bar{n}(x)$ is the refractive index
- $\frac{\partial}{\partial y} = 0$

Discussion of TE and TM modes is done separately. We start with TE modes.

4.6.1 TE mode

The initial equations were derived previously, Eqs. (4.32)–(4.34). We write them here for easy reference:

$$\begin{aligned}\frac{\partial E_y}{\partial x} &= -j\omega\mu_0 H_z \\ -\beta E_y &= \omega\mu_0 H_x \\ \frac{\partial H_z}{\partial x} + j\beta H_x &= -j\omega\mu_0\varepsilon_0\varepsilon_r E_y\end{aligned} \quad (4.64)$$

The above system can be combined together to obtain a wave equation, as

$$\begin{aligned}\frac{\partial^2 E_y}{\partial x^2} &= -j\omega\mu_0 \frac{\partial H_z}{\partial x} && \text{from first Eq.} \\ &= -j\omega\mu_0\left(-j\beta H_x - j\omega\varepsilon_0\varepsilon_r E_y\right) && \text{from third Eq.} \\ &= -j\omega\mu_0\left\{-j\beta\left(\frac{-\beta E_y}{\omega\mu_0}\right) - j\omega\varepsilon_0\varepsilon_r E_y\right\} && \text{from second Eq.} \\ &= \beta^2 E_y - \omega^2\mu_0\varepsilon_0\varepsilon_r E_y\end{aligned}$$

Finally, one writes wave equation for 'i' layer as

$$\frac{\partial^2}{\partial x^2} E_{yi}(x) = \left(\beta^2 - k_0^2 \bar{n}_i^2\right) E_{yi}(x) \quad (4.65)$$

where $k_0^2 = \mu_0\varepsilon_0\omega^2$ and $\varepsilon_{ri} = \bar{n}_i^2$. Introduce

$$\kappa_i^2 = k_0^2 \bar{n}_i^2 - \beta^2 \quad (4.66)$$

The solution to the wave equation is

$$E_{yi}(x) = A_i e^{-j\kappa_i(x-x_{i-1})} + B_i e^{j\kappa_i(x-x_{i-1})} \quad (4.67)$$

At the interfaces of layers, the boundary condition of TE mode are

$$E_i(x)\big|_{x=x_{i-1}} = E_{i-1}(x)\big|_{x=x_{i-1}} \quad (4.68)$$

$$\frac{\partial E_i(x)}{\partial x}\bigg|_{x=x_{i-1}} = \frac{\partial E_{i-1}}{\partial x}\bigg|_{x=x_{i-1}} \quad (4.69)$$

It is more convenient to work with another set of variables $U_i(x)$ and $V_i(x)$ defined as

$$U_i(x) = E_{yi}(x) \quad (4.70)$$
$$V_i(x) = \omega\mu_0 H_{zi}(x) \quad (4.71)$$

From Eqs. (4.64) one obtains equations for new variables. Directly from the first of Eqs. (4.64)

$$\frac{dU_i(x)}{dx} = -jV_i(x) \quad (4.72)$$

Using the second equation to eliminate H_x in the third of Eqs. (4.64), one has

$$\frac{\partial H_z}{\partial x} + j\beta \frac{(-\beta)}{\omega\mu_0} E_y = -j\omega\mu_0\varepsilon_0\varepsilon_r E_y$$

or

$$\omega\mu_0 \frac{\partial H_z}{\partial x} - j\beta^2 E_y = -j\omega^2\mu_0\varepsilon_0\varepsilon_r E_y$$
$$= -jk_0^2 \bar{n}_i^2 E_y$$

For layer $i-th$

$$\frac{dV_i(x)}{dx} - j\beta^2 U_i(x) = -jk_0^2 \bar{n}_i^2 U_i(x)$$

or

$$\frac{dV_i(x)}{dx} = -j\kappa_i^2 U_i(x) \tag{4.73}$$

Eqs. (4.72) and (4.73) can be written in a matrix form as

$$\frac{d}{dx}\begin{bmatrix} U_i(x) \\ V_i(x) \end{bmatrix} = \begin{bmatrix} 0 & -j \\ -j\kappa_i^2 & 0 \end{bmatrix}\begin{bmatrix} U_i(x) \\ V_i(x) \end{bmatrix} \tag{4.74}$$

From Maxwell's equations one can verify the following relation:

$$j\frac{dU_i(x)}{dx} = j\frac{\partial E_{yi}(x)}{\partial x} = \omega\mu_0 H_{zi}(x) = V_i(x)$$

Using the above relation, the solutions of Eqs. (4.74) are

$$U_i(x) = A_i e^{-j\kappa_i(x-x_i)} + B_i e^{j\kappa_i(x-x_i)}$$
$$V_i(x) = j\frac{dU_i(x)}{dx} = \kappa_i\left\{A_i e^{-j\kappa_i(x-x_i)} - B_i e^{j\kappa_i(x-x_i)}\right\}$$

They can be written in a matrix form as

$$\begin{bmatrix} U_i(x) \\ V_i(x) \end{bmatrix} = \begin{bmatrix} e^{-j\kappa_i(x-x_i)} & e^{j\kappa_i(x-x_i)} \\ \kappa_i e^{-j\kappa_i(x-x_i)} & -\kappa_i e^{j\kappa_i(x-x_i)} \end{bmatrix}\begin{bmatrix} A_i \\ B_i \end{bmatrix} \tag{4.75}$$

From the above, for x_i one finds

$$\begin{bmatrix} U_i(x_i) \\ V_i(x_i) \end{bmatrix} = \begin{bmatrix} 1 & 1 \\ \kappa_i & -\kappa_i \end{bmatrix}\begin{bmatrix} A_i \\ B_i \end{bmatrix}$$

Inverting the last equation gives the expression for coefficients in terms of field and its derivative

$$\begin{bmatrix} A_i \\ B_i \end{bmatrix} = \frac{1}{2}\begin{bmatrix} 1 & \frac{1}{j\kappa_i} \\ 1 & -\frac{1}{j\kappa_i} \end{bmatrix}\begin{bmatrix} U_i(x_i) \\ V_i(x_i) \end{bmatrix}$$

Substituting the last equation into (4.75) gives

$$\begin{bmatrix} U_i(x) \\ V_i(x) \end{bmatrix} = \begin{bmatrix} e^{-j\kappa_i(x-x_i)} & e^{j\kappa_i(x-x_i)} \\ \kappa_i e^{-j\kappa_i(x-x_i)} & -\kappa_i e^{j\kappa_i(x-x_i)} \end{bmatrix}\frac{1}{2}\begin{bmatrix} 1 & \frac{1}{j\kappa_i} \\ 1 & -\frac{1}{j\kappa_i} \end{bmatrix}\begin{bmatrix} U_i(x_i) \\ V_i(x_i) \end{bmatrix} \tag{4.76}$$
$$= \overleftrightarrow{T_i}(x_i)\begin{bmatrix} U_i(x_i) \\ V_i(x_i) \end{bmatrix}$$

Explicitly, propagation matrix $\overleftrightarrow{T_i}(x_i)$ at point x_i is

$$\overleftrightarrow{T_i}(x_i) = \begin{bmatrix} \cos \kappa_i (x - x_i) & -\frac{j}{\kappa_i} \sin \kappa_i (x - x_i) \\ -j\kappa_i \sin \kappa_i (x - x_i) & \cos \kappa_i (x - x_i) \end{bmatrix} \quad (4.77)$$

Equation (4.76) 'propagates' values of the fields from point x_i to an arbitrary point x within layer i.

4.6.2 Propagation constant

A propagation constant is obtained if we propagate fields across all interfaces. First consider propagation between two consecutive interfaces, that is propagation from x_i location to x_{i+1}. One finds from Eq. (4.76)

$$\begin{bmatrix} U_{i+1}(x_{i+1}) \\ V_{i+1}(x_{i+1}) \end{bmatrix} = \begin{bmatrix} \cos \kappa_i (x_{i+1} - x_i) & -\frac{j}{\kappa_i} \sin \kappa_i (x_{i+1} - x_i) \\ -j\kappa_i \sin \kappa_i (x_{i+1} - x_i) & \cos \kappa_i (x_{i+1} - x_i) \end{bmatrix} \begin{bmatrix} U_i(x_i) \\ V_i(x_i) \end{bmatrix}$$

Let us define the thickness of layer i as

$$d_i = x_{i+1} - x_i$$

The propagation matrix which propagates fields from x_i to x_{i+1} is called $\overleftrightarrow{M_i}$ and has the form

$$\overleftrightarrow{M_i} = \begin{bmatrix} \cos \kappa_i d_i & -\frac{j}{\kappa_i} \sin \kappa_i d_i \\ -j\kappa_i \sin \kappa_i d_i & \cos \kappa_i d_i \end{bmatrix} \quad (4.78)$$

Using the above formalism, we can now propagate fields through the entire structure from substrate to a cladding region as follows:

$$\begin{bmatrix} U_c \\ V_c \end{bmatrix} = \prod_{i=1}^{N} \overleftrightarrow{M_i} \begin{bmatrix} U_s \\ V_s \end{bmatrix}$$

$$= \overleftrightarrow{M} \begin{bmatrix} U_s \\ V_s \end{bmatrix}$$

$$\equiv \begin{bmatrix} m_{11} & m_{12} \\ m_{21} & m_{22} \end{bmatrix} \begin{bmatrix} U_s \\ V_s \end{bmatrix} \quad (4.79)$$

In the above we have introduced the 2×2 matrix $\overleftrightarrow{M}$ with elements $m_{11}, m_{12}, m_{21}, m_{22}$ which plays central role in the numerical process of determining propagation constants. From the above equation one obtains

$$U_c = m_{11} U_s + m_{12} V_s \quad (4.80)$$
$$V_c = m_{21} U_s + m_{22} V_s \quad (4.81)$$

In the substrate and cladding, the fields decay exponentially (Fig. 4.10). Therefore, the expressions for fields in the cladding and substrate regions are

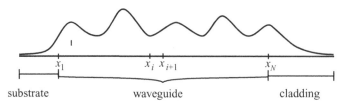

Fig. 4.10 Schematic of the field distribution. Exponential decay in the substrate and cladding regions is shown.

- for $x < x_1$ (substrate)

$$U_s(x) = A_s \exp[\gamma_s(x - x_1)] \tag{4.82}$$
$$V_s(x) = j\gamma_s A_s \exp[\gamma_s(x - x_1)] \tag{4.83}$$

- for $x > x_N$ (cladding)

$$U_c(x) = A_c \exp[-\gamma_c(x - x_N)] \tag{4.84}$$
$$V_c(x) = -j\gamma_c A_s \exp[-\gamma_c(x - x_N)] \tag{4.85}$$

In the above expressions we defined

$$\gamma_i^2 = \beta^2 - k_0^2 \bar{n}_i^2 \tag{4.86}$$
$$\gamma_i = j\kappa_i$$

From expressions (4.82) and (4.83) one obtains at point x_1

$$U_s(x_1) = A_s \tag{4.87}$$
$$V_s(x_1) = j\gamma_s A_s \tag{4.88}$$

In a similar way for point x_N one has

$$U_c(x_N) = A_c \tag{4.89}$$
$$V_c(x_N) = -j\gamma_c A_c \tag{4.90}$$

Substituting the above results into Eqs. (4.80) and (4.81), one obtains the expressions which are used to find propagation constants employing numerical procedure

$$\gamma_c \gamma_s m_{12} - m_{21} = j(\gamma_c m_{11} + \gamma_s m_{22}) \tag{4.91}$$

Values of matrix elements $m_{11}, m_{12}, m_{21}, m_{22}$ are obtained numerically for a particular guiding structure.

4.6.3 Electric field

Once propagation constants are found, one can obtain profile of electric field for each mode. To do this one uses Eq. (4.76), which after multiplication of matrices is

$$\begin{bmatrix} U_i(x) \\ V_i(x) \end{bmatrix} = \begin{bmatrix} \cos \kappa_i(x - x_i) & -\frac{j}{\kappa_i} \sin \kappa_i(x - x_i) \\ -j\kappa_i \sin \kappa_i(x - x_i) & \cos \kappa_i(x - x_i) \end{bmatrix} \begin{bmatrix} U_i(x_i) \\ V_i(x_i) \end{bmatrix} \tag{4.92}$$

where for layer i $x_i \leq x \leq x_{i+1}$. The above determines electric field at an arbitrary location within layer i once the values of fields are known at point x_i. In practice one assumes some value of electric field at point x_1 and using expression (4.92) determines field within all layers. The above procedure will be applied to analyse several guiding structures in the next section. Before this, we will formulate basic equations for TM modes.

4.6.4 TM modes

The analysis for TM modes follows exactly the same path as for TE modes. Here we outline basic equations only. For TM modes one has the following relations:

$$E_y = H_z = H_x = 0 \tag{4.93}$$

For each layer one has

$$\begin{aligned} \beta H_y &= \omega \varepsilon E_x \\ \frac{\partial H_y}{\partial x} &= j\omega \varepsilon E_z \\ \frac{\partial E_z}{\partial x} + j\beta E_x &= j\omega \mu H_y \end{aligned} \tag{4.94}$$

Define variables

$$\begin{aligned} U &= H_y \\ V &= \omega \varepsilon_0 E_z \end{aligned} \tag{4.95}$$

They fulfill the following equations:

$$\begin{aligned} \frac{dU}{dx} &= j\bar{n}^2 V \\ \frac{dV}{dx} &= j\left(k^2 - \frac{\beta^2}{\bar{n}^2}\right) U \end{aligned} \tag{4.96}$$

From this point one follows a similar approach as for TE modes. The remaining analysis is left as a project.

4.7 Examples: 1D approach

In this section we will consider several waveguiding structures described in literature and apply our formalism to analyse those structures. The tests were performed in searching for propagation constant β and then determining the profile of electric field. Our tests were performed for several structures analysed in References [2], [3], [4] and [5].

We attempted to find all modes. Excellent agreement was found as compared with the published results.

Table 4.3 Geometry of a four-layer lossless dielectric waveguide [3].

Layer	Refractive index	Thickness	Description
D_6	$\bar{n}_6 = 1.00$	1 μm	Cladding
D_5	$\bar{n}_5 = 1.66$	0.50 μm	
D_4	$\bar{n}_4 = 1.53$	0.50 μm	
D_3	$\bar{n}_3 = 1.60$	0.50 μm	
D_2	$\bar{n}_2 = 1.66$	0.50 μm	
D_1	$\bar{n}_1 = 1.50$	1 μm	Substrate

Table 4.4 Propagation constants for a four-layer lossless dielectric waveguide defined in Table 4.3.

Mode	Propagation constant
TE_0	$1.622728682325434 + 0.000000000000441i$
TE_1	$1.605275698094530 + 0.000000000000014i$
TE_2	$1.557136152293222 + 0.000000000000436i$
TE_3	1.503587112022723

The list of MATLAB files is provided in Table 4.12[1]. All MATLAB files are listed in Appendix 4A.3.

4.7.1 Four-layer lossless waveguide

The first tested structure is the one considered by Chilwell and Hodgkinson [3] and also by Chen *et al.* [4]. It is a four-layer lossless dielectric structure shown in Table 4.3. Wavelength is $\lambda = 0.6328$ μm. Determined propagation constants for all four modes are summarized in Table 4.4. They show an excellent agreement with the results published in Ref. [3] and [4].

Electric field for fundamental TE-mode has been calculated using mesh points generated by function *mesh_x.m*. The resulting profile of TE fundamental mode is shown in Fig. 4.11.

4.7.2 Six-layer lossy waveguide

The second structure tested by our program is defined in Table 4.5, from [4]. It is a six-layer lossy waveguide structure operating at the wavelength $\lambda = 1.523$ μm. Propagation constants for this structure are summarized in Table 4.6. Obtained results show an excellent agreement with the published data.

The TE electric field distribution of the fundamental mode for this structure is shown in Fig. 4.12.

---

[1] I acknowledge the help of my former student, Mr L. Glowacki, in developing some of those routines.

Table 4.5 Geometry of a six-layer lossy dielectric waveguide.

Layer	Refractive index	Thickness (μm)	Description
D_8	$\bar{n}_8 = 1.00$	1	Cladding
D_7	$\bar{n}_7 = 3.38327$	0.10	
D_6	$\bar{n}_6 = 3.39614$	0.20	
D_5	$\bar{n}_5 = 3.5321 - j0.08817$	0.60	
D_4	$\bar{n}_4 = 3.39583$	0.518	
D_3	$\bar{n}_3 = 3.22534$	1.60	
D_2	$\bar{n}_2 = 3.16455$	0.60	
D_1	$\bar{n}_1 = 3.172951$	1	Substrate

Table 4.6 Results of propagation constants for six-layer lossy dielectric waveguide.

Mode	Propagation constant
TE_0	$3.460829693510364 + 0.072663342917385i$
TE_1	$3.316707802046375 + 0.023275817588121i$
TE_2	$3.208555428734457 + 0.012782067986633i$
TE_3	$3.195490593396514 + 0.012585955654404i$

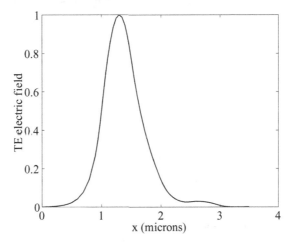

Fig. 4.11 Field profile of fundamental TE mode for four-layer lossless structure.

4.7.3 Structure by Visser

This structure was described by Visser et al. [5]. It is summarized in Table 4.7.

The results of calculations of propagation constants for TE modes are summarized in Table 4.8. They are identical to the values published by Visser et al. [5]. The profile of the electric field for mode TE_1 is shown in Fig. 4.13.

Table 4.7 Geometry of a six-layer lossy dielectric waveguide described by Visser *et al.* [5].

Layer	Thickness (μm)	Refractive index
clad	1.0	
3	0.6	3.40 − 0.002i
2	0.4	3.60 + 0.010i
1	0.6	3.40 − 0.002i
substrate	1.0	

Table 4.8 Propagation constants obtained for the structure shown in Table 4.7.

Layer	Propagation constant
TE_0	3.503443332950034 − 0.007103000978683i
TE_1	3.337286858209064 + 0.000229491103731i
TE_2	3.251685206983383 + 0.000530514779912i
TE_3	3.104251421414573 − 0.001337986339752i
TE_4	2.878636779881233 + 0.000173729890361i
TE_5	2.628139320470185 − 0.001548644353782i
TE_6	2.243951362601185 − 0.000708377958009i
TE_7	1.768190960412424 − 0.001353217183860i
TE_8	1.074262026525778 − 0.002457891473570i

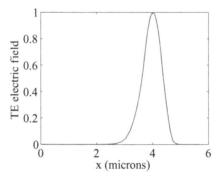

Fig. 4.12 Field profile of fundamental TE mode for a six-layer lossy waveguide.

4.8 Two-dimensional (2D) structures

In the previous sections we have discussed one-dimensional guiding structures, also known as planar waveguides. However, in practical devices the waveguides are essentially two-dimensional (2D) In those structures, refractive index $\bar{n}(x, y)$ depends on transverse coordinate x and lateral coordinate y. Two examples of popular structures are shown in

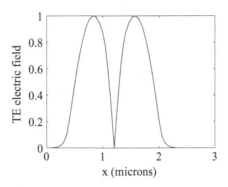

Fig. 4.13 Field profile of TE_1 mode for structure discussed by Visser et al. [5].

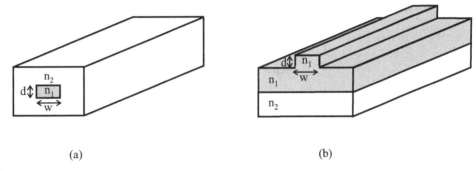

Fig. 4.14 Popular types of 2D waveguides. (a) Buried-channel waveguide. (b) Rib waveguide.

Fig. 4.14. They exhibit distinctive features of their refractive index. The structures shown are known as the buried channel waveguide and the rib waveguide.

In what follows we will discuss one of the methods of obtaining approximate solution for propagating mode, known as the effective index method.

4.8.1 Effective index method

To illustrate the effective index method [6], let us start with a scalar wave equation in 2D which is

$$\frac{\partial^2 \phi(x,y)}{\partial x^2} + \frac{\partial^2 \phi(x,y)}{\partial y^2} + k_0^2 \left[\varepsilon_r(x,y) - N_{eff}^2 \right] \phi(x,y) = 0 \qquad (4.97)$$

Here N_{eff}^2 is the effective index which we want to determine. Assume that there is no 'interaction' between variables x and y. This allows us to separate fields as [7]

$$\phi(x,y) = X(x) \cdot Y(y)$$

Substitute postulated solution into wave equation, evaluate the relevant derivatives and divide both sides of the resulting equation by $X(x) \cdot Y(y)$. One obtains

$$\frac{1}{X(x)} X''(x) + \frac{1}{Y(y)} Y''(y) + k_0^2 \left[\varepsilon_r(x,y) - N_{eff}^2 \right] = 0 \qquad (4.98)$$

Step 1

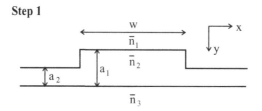

The above is equivalent to

Step 2

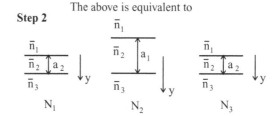

Step 3

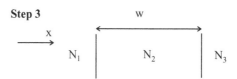

Fig. 4.15 Illustration of the effective index method.

Introduce separation constant $k_0^2 \lambda^2$ as

$$\frac{1}{Y(y)} Y''(y) + k_0^2 \varepsilon_r(x, y) = k_0^2 \lambda^2 \qquad (4.99)$$

The remaining terms are

$$\frac{1}{X(x)} X''(x) - k_0^2 N_{eff}^2 = -k_0^2 \lambda^2 \qquad (4.100)$$

Based on the above theory it is possible to design a method, known as effective index method to determine effective index N_{eff}^2. The method (see Fig. 4.15) consists of several steps:

1. Replace the 2D waveguide with a combination of 1D waveguides along x-axis.
2. Solve each of the 1D problems separately by calculating the effective index along y axis and obtain N_{eff} in each case (for our structure three cases).
3. Construct a new 'effective' 1D waveguide (along x-axis) which will model the original 2D waveguide. The new effective waveguide has the values of refractive indices as indicated.
4. Determine an effective index by solving the 1D waveguide constructed in Step 3 along the x-axis.

Two-dimensional (2D) structures

Table 4.9 Numerical parameters used in an example of the effective index method.

Wavelength	Refractive indices	Geometry
$\lambda = 0.8\,\mu m$	$\bar{n}_c = 1$	$h = 1.8\,\mu m$
	$\bar{n}_f = 2.234$	$d = 1\,\mu m$
	$\bar{n}_s = 2.214$	$w = 2\,\mu m$

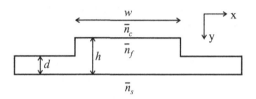

Fig. 4.16 Structure used as an example of the application of the effective index method.

An example

We illustrate the effective index method with an example can be found in [8]. The structure is shown in Fig. 4.16 and the numerical parameters in Table 4.9.

The method proceeds with the following four steps:

1. Determine normalized thicknesses using the definition of V

$$V_f = k \cdot h \sqrt{n_f^2 - \bar{n}_s^2}, \quad V_d = k \cdot d \sqrt{n_f^2 - \bar{n}_s^2}$$

Using the values from Table 4.9, one finds

$$V_f = 4.2, \quad V_d = 2.3$$

2. Obtain values of b using $b - V$ diagram. From the diagram, for TE modes we obtain (for $a = \infty$)

$$b_f = 0.65, \quad b_d = 0.2$$

Effective indices N_f and N_d are determined using

$$N_{f,d}^2 = \bar{n}_s^2 + b_{f,d}\left(\bar{n}_f^2 - \bar{n}_s^2\right)$$

One obtains

$$N_f = 2.227, \quad N_d = 2.218$$

3. The same structure along the y-direction is approximated as the one-dimensional, using the previous values. The V number is determined using

$$V_{eg} = k \cdot w \sqrt{N_f^2 - N_d^2} = kw\sqrt{\left(\bar{n}_f^2 - \bar{n}_s^2\right)(b_f - b_d)} \qquad (4.101)$$

One obtains

$$V_{eg} = 3.14, \quad b_{eg} = 0.64 \qquad (4.102)$$

4. Finally, the approximate effective index $N = N_{eq}$ of the equivalent guide is determined using

$$N \equiv N_{eg} = N_d^2 + b_{eq}\left(N_f^2 - N_d^2\right) \qquad (4.103)$$

For this example, we obtain $N = 2.224$.

More examples of the applications of effective index method can be found in books by Coldren and Corzine [9] and Liu [10].

4.9 Problems

1. Find the effective refractive indices and the number of TE modes in an asymmetric waveguide with an asymmetry parameter $a = 10$ and thickness $h = 1.50$ μm. Assume $\lambda_0 = 0.82$ μm.
2. Compute the largest thickness that will guarantee single TE mode operation of the above waveguide.
3. Analyse the three-layer, asymmetric guiding structure. Derive a transcendental equation for the propagation constant.
4. Conduct analysis of a four-layer waveguiding structure [8]. Derive equations for all elements of matrix $\overline{M}$. Obtain an analytical formula for dispersion relation.

4.10 Projects

1. Based on theoretical results developed for the three-layer symmetrical guiding structure, write a MATLAB program which will solve transcendental equations using functions given by (4.46) and (4.47). Select from the literature an appropriate waveguiding structure. Use your program to determine propagation constants and profiles of electric fields. Compare your results with those from the literature.
2. Analyse TM modes for the three-layer asymmetric waveguide defined in Table 4.1. Obtain analytical solutions in all regions and derive an appropriate transcendental equation. Write a MATLAB program to determine propagation constants.
3. Plasmonic waveguide contains a metallic layer. The simplest cases are the three-layer structures: metal-dielectric-metal (MDM) and dielectric-metal-dielectric (DMD). Derive a transcendental equation describing plasmonic waveguide (similar to Eq. 4.45 for a symmetrical dielectric guiding structure). Consult the paper by Kekatpure *et al.* [11]. With the equation at hand, write a MATLAB program to determine propagation constants

for TM modes for the MDM structure defined in Table 4.10 [11]. Analyse the obtained results.

Table 4.10 Plasmonic structure.

Material	Thickness (μm)	Relative permittivity
gold	3	$-95.92 - i10.97$
silica	50	2.1025
silver	3	$-143.49 - i9.52$

4. Conduct analysis of the multilayered structure for TM modes. Derive all necessary equations. Write a MATLAB program to determine propagation constants and field profiles.

Appendix 4A: MATLAB listings

In Tables 4.11 and 4.12 we provide a list of MATLAB files created for Chapter 4 and a short description of each function. In Table 4.12 we summarize a list of files used to analyse planar waveguides.

Table 4.11 List of MATLAB functions for Chapter 4.

Listing	Function name	Description
4A.1	$b\_V\_diagram.m$	Plots b-V diagram for planar waveguide
4A.2.1	$a3L.m$	Analysis for three-layer planar waveguide
4A.2.2	$func\_asym.m$	Function used by $a3L.m$

Table 4.12 Functions used to analyse a 1D planar waveguide.

Listing	Function name	Description
4A.3.1	$slab.m$	*driver function (determines prop. constants and field profile)*
4A.3.2	$lossless.m$	*data for waveguide structure without losses*
4A.3.3	$lossy.m$	*data for waveguide structure with losses*
4A.3.4	$visser.m$	*data for structure described by Visser*
4A.3.5	$muller.m$	*implements Muller's method*
4A.3.6	$f\_TE.m$	*creates transcendental equation for TE field*
4A.3.7	$mesh\_x.m$	*generates 1D mesh*
4A.3.8	$refindex.m$	*assigns the value of ref. index to each mesh point*
4A.3.9	$TE\_field.m$	*determines TE field for all mesh points*

Listing 4A.1 Program b_V_diagram.m. MATLAB program to analyse the transverse resonance condition (4.12) and to plot a normalized guide index b versus variable V. Allows to input different values of the asymmetry parameter a.

```
% File name: b_V_diagram.m
% function which plots b-V diagram of a planar slab waveguide
clear all
N_max = 400;                        % number of points for plot
b = linspace(0,1.0,N_max);
hold on
for nu = [0, 1, 2]
for a = [0.0, 8.0,50.0];            % asymmetry coefficient
% determine V
V1 = atan(sqrt(b./(1-b)) );
V2 = atan(sqrt((b+a)./(1.0-b)));
V3 = 1./sqrt(1.0-b);
V = (nu*pi + V1 + V2).*V3;
%
plot(V,b,'LineWidth',1.2)
grid on
axis([0.0 20.0 0.0 1.0])            % change axis limit
end
end
box on    % makes frame around plot
xlabel ('V','FontSize',14);
ylabel('b','FontSize',14);
set(gca,'FontSize',14);             % size of tick marks on both axes
text(10, 0.96, '\nu=0','Fontsize',14)
text(10, 0.8, '\nu=1','Fontsize',14)
text(10, 0.55, '\nu=2','Fontsize',14)
%
text(0.1, 0.3, 'a=0','Fontsize',14)
text(2.5, 0.3, 'a=50','Fontsize',14)
%
text(3.3, 0.2, 'a=0','Fontsize',14)
text(5.8, 0.2, 'a=50','Fontsize',14)
%
text(6, 0.1, 'a=0','Fontsize',14)
text(8.8, 0.1, 'a=50','Fontsize',14)
pause
close all
```

Listing 4A.2.1 Program a3L.m. Determines propagation constants and plots field profiles for an asymmetric three-layer planar waveguide. Uses *func_asym.m*.

```
% File name: a3L.m
% Analysis for TE modes
% First, we conduct test plots to find ranges of beta where possible
% solutions exists. This is done in several steps:
% 1. Plots are done treating kappa_f as an independent variable
% 2. Ranges of kappa_f are determined where there are zeros of functions
% 3. Corresponding ranges of beta are determined
% 4. Searches are performed to find propagation constants
%
clear all
% Definition of structure
n_f    = 1.50;              % ref. index of film layer
n_s    = 1.45;              % ref. index of substrate
n_c    = 1.40;              % ref. index of cladding
lambda = 1.0;               % wavelength in microns
h      = 5.0;               % thickness of film layer in microns
a_c    = 2*h;               % thickness of cladding region
a_s    = 2*h;               % thickness of substrate region
%
k = 2*pi/lambda;            % wave number
kappa_f = 0:0.01:3.0;       % establish range of kappa_f
beta_temp = sqrt((n_f*k)^2 - kappa_f.^2);
beta_min = min(beta_temp);
beta_max = max(beta_temp);
% Before searches, we plot search function versus beta
beta = beta_min:0.001:beta_max; % establish range of beta
N = beta./k;
ff = func_asym(beta,n_c,n_s,n_f,k,h);
plot(beta,ff)
xlabel('\beta','FontSize',22);
ylabel('Search function','FontSize',22);
ylim([-10.0 10.0])
grid on
pause
close all
% From the above plot, one must choose proper search range for each mode.
% Search numbers provided below are only for the waveguide defined above.
% For different waveguide, one must choose different ranges
% for searches.
beta0 = fzero(@(beta) func_asym(beta,n_c,n_s,n_f,k,h),[9.40 9.41])
beta1 = fzero(@(beta) func_asym(beta,n_c,n_s,n_f,k,h),[9.35 9.37])
beta2 = fzero(@(beta) func_asym(beta,n_c,n_s,n_f,k,h),[9.27 9.29])
beta3 = fzero(@(beta) func_asym(beta,n_c,n_s,n_f,k,h),[9.17 9.18])
% Plot of field profiles
```

```
A_s = 1.0;
thickness = h + a_c + a_s;
beta_field = beta0; % Select appropriate propagation constant for plotting
gamma_s = sqrt(beta_field^2 - (n_s*k)^2);
gamma_c = sqrt(beta_field^2 - (n_c*k)^2);
kappa_f = sqrt((n_f*k)^2 - beta_field^2);
%
% In the formulas below for electric field E_y we have shifted
% x-coordinate by a_s
% We also 'reversed' direction of plot in the substrate region
NN = 100;
delta = thickness/NN;
x = 0.0:delta:thickness;   % coordinates of plot points
x_t = 0;
for i=1:NN+1
    x_t(i+1)= x_t(i) + delta;
    if (x_t(i)<=a_s);
        E_y(i) = A_s*exp(gamma_s*(x_t(i)-a_s));
    elseif (a_s<=x_t(i)) && (x_t(i)<=a_s+h);
        E_y(i) = A_s*(cos(kappa_f*(x_t(i)-a_s))+...
            gamma_s*sin(kappa_f*(x_t(i)-a_s))/kappa_f);
    else (a_s+h<=x_t(i)) & (x_t(i)<=thickness);
        E_y(i) = A_s*(cos(kappa_f*h)+gamma_s*sin(kappa_f*h)/kappa_f)...
            *exp(-gamma_c*(x_t(i)-h-a_s));
    end
end
%
h=plot(x,E_y);
% add text on x-axix and y-axis and size of x and y labels
xlabel('x (microns)','FontSize',22);
ylabel('TE electric field','FontSize',22);
set(h,'LineWidth',1.5);             % new thickness of plotting lines
set(gca,'FontSize',22);             % new size of tick marks on both axes
grid on
pause
close all
```

Listing 4A.2.2 Function func_asym.m. Search function for the asymmetric three-layer waveguide. Used by program a3L.m.

```
function f = func_asym(beta,n_c,n_s,n_f,k,h)
% Construction of search function for asymmetric 3-layers waveguide
%
gamma_c = sqrt(beta.^2 - (n_c*k)^2);
gamma_s = sqrt(beta.^2 - (n_s*k)^2);
kappa_f = sqrt((n_f*k)^2 - beta.^2);
```

```
%
denom = kappa_f - (gamma_c.*gamma_s)./kappa_f;
f = tan(kappa_f*h) - (gamma_s+gamma_c)./denom;
```

Listing 4A.3.1 Program slab.m. Driver function which determines propagation constant for fundamental TE mode and plots electric field profiles. One-dimensional planar waveguide of an arbitrary number of layers is assumed. The list of files is shown in Table 4.12.

```
% File name: slab.m
% Driver function which determines propagation constants and
% electric field profiles (TE mode) for multilayered slab structure
clear all;
format long
% Input structure for analysis (select appropriate input)
%lossless;
%lossy;
visser
%
epsilon = 1e-6;                  % numerical parameter
TE_mode = [];
n_max = max(n_layer);
z1 = n_max;                      % max value of refractive index
n_min = max(n_s,n_c) + 0.001;    % min value of refractive index
dz = 0.005;                      % iteration step
mode_control = 0;
%
while(z1 > n_min)
    z0 = z1 - dz;                % starting point for Muller method
    z2 = 0.5*(z1 + z0);          % starting point for Muller method
    z_new = muller(@f_TE , z0, z1, z2);
    if (z_new ~= 0)
                                 % verifying for mode existence
    for u=1 : length(TE_mode)
        if(abs(TE_mode(u) - z_new) < epsilon)
            mode_control = 1; break; % mode found
        end
    end
    if (mode_control == 1)
        mode_control = 0;
    else
        TE_mode(length(TE_mode) + 1) = z_new;
    end
    end
    z1 = z0;
end
%
```

```
TE_mode = sort(TE_mode, 'descend');
%TE_mode'                    % outputs all calculated modes
beta = TE_mode(2);           % selects mode for plotting field profile
x = mesh_x(d_s,d_layer,d_c,NumberMesh);
n_total = [n_s,n_layer,n_c]; % ref index for all layers
n_mesh = refindex(x,NumberMesh,n_total);
TE_mode_field = TE_field(beta,n_mesh,x,k_0);
```

Listing 4A.3.2 Function lossless.m. Contains data for a lossless waveguide.

```
% File name: lossless.m
% Contains data for lossless waveguide
% Reference:
% J. Chilwell and I. Hodgkinson,
% "Thin-films field-transfer matrix theory of planar multilayer
% waveguides and reflection from prism-loaded waveguide",
% J. Opt. Soc. Amer.A, vol.1, pp. 742-753 (1984).
% Fig.3
% Global variables to be transferred to function f_TE.m
%
global n_c              % ref. index cladding
global n_layer          % ref. index of internal layers
global n_s              % ref. index substrate
global d_c              % thickness of cladding (microns)
global d_layer          % thicknesses of internal layers (microns)
global d_s              % thickness of substrate (microns)
global k_0              % wavenumber
global NumberMesh       % number of mesh points in each layer
                        % (including substrate and cladding)
n_c       = 1.0;
n_layer   = [1.66 1.60 1.53 1.66];
n_s       = 1.5;
d_c       = 0.5;
d_layer   = [0.5 0.5 0.5 0.5];
d_s       = 1.0;
NumberMesh = [10 10 10 10 10 10];
lambda    = 0.6328; % wavelength in microns
k_0       = 2*pi/lambda;
```

Listing 4A.3.3 Function lossy.m. Contains data for a lossy waveguide.

```
% File name: lossy.m
% Contains data for lossy waveguide.
% Reference:
```

```
% C. Chen et al, Proc. SPIE, v.3795 (1999)
% Global variables to be transferred to function f_TE.m
global n_c           % ref. index cladding
global n_layer       % ref. index of internal layers
global n_s           % ref. index substrate
global d_c           % thickness of cladding (microns)
global d_layer       % thicknesses of internal layers (microns)
global d_s           % thickness of substrate (microns)
global k_0           % wavenumber
global NumberMesh    % number of mesh points in each layer
                     % (including substrate and cladding)
n_c       = 1.0;
n_layer   = [3.16455 3.22534 3.39583 3.5321-1j*0.08817 3.39614 3.38327];
n_s       = 3.172951;
d_c       = 1.0;
d_layer   = [0.6 1.6 0.518 0.6 0.2 0.1];
d_s       = 1.0;
NumberMesh = [10 10 10 10 10 10 10 10];
lambda    = 1.523;   % wavelength in microns
k_0       = 2*pi/lambda;
```

Listing 4A.3.4 Function visser.m. Contains data for a five-layer waveguide described by Visser *et al*.

```
% File name: visser.m
% Contains data for five-layer waveguide with gain and losses
% Reference:
% T.D. Visser et al, JQE v.31, p.1803 (1995)
% Fig.6
% Global variables to be transferred to function f_TE.m
%
global n_c           % ref. index cladding
global n_layer       % ref. index of internal layers
global n_s           % ref. index substrate
global d_c           % thickness of cladding (microns)
global d_layer       % thicknesses of internal layers (microns)
global d_s           % thickness of substrate (microns)
global k_0           % wavenumber
global NumberMesh    % number of mesh points in each layer
                     % (including substrate and cladding)
n_c       = 1.0;
n_layer   = [3.40-1i*0.002 3.60+1i*0.010 3.40-1i*0.002];
n_s       = 1.0;
d_s       = 0.4;
d_layer   = [0.6 0.4 0.6];
```

```
d_c        = 0.5;
NumberMesh = [10 10 10 10 10];
lambda     = 1.3;   % wavelength in microns
k_0        = 2*pi/lambda;
```

Listing 4A.3.5 Function muller.m. This function implements Muller's method. It is good for finding complex roots.

```
function f_val = muller (f, x0, x1, x2)
% Function implements Muller's method
iter_max = 100;      % max number of steps in Muller method
f_tol    = 1e-6;     % numerical parameters
x_tol = 1e-6;
y0 = f(x0);
y1 = f(x1);
y2 = f(x2);
iter = 0;
while(iter <= iter_max)
    iter = iter + 1;
    a =( (x1 - x2)*(y0 - y2) - (x0 - x2)*(y1 - y2)) / ...
        ( (x0 - x2)*(x1 - x2)*(x0 - x1) );
    %
    b = ( ( x0 - x2 )^2 *( y1 - y2 ) - ( x1 - x2 )^2 *( y0 - y2 )) / ...
        ( (x0 - x2)*(x1 - x2)*(x0 - x1) );
    %
    c = y2;
    %
    if (a~=0)
        D = sqrt(b*b - 4*a*c);
        q1 = b + D;
        q2 = b - D;
        if (abs(q1) < abs(q2))
            dx = - 2*c/q2;
        else
            dx = - 2*c/q1;
        end
        elseif (b~=0)
            dx = -c/b;
    else
        warning('Muller method failed to find a root')
        break;
    end
    x3 = x2 + dx;
    x0 = x1;
    x1 = x2;
```

```
        x2 = x3;
        y0 = y1;
        y1 = y2;
        y2 = feval(f, x2);
        if (abs(dx) < x_tol && abs (y2) < f_tol)
        break;
        end
end
% Lines below ensure that only proper values are calculated
if (abs(y2) < f_tol)
    f_val = x2;
    return;
else
    f_val = 0;
end
```

Listing 4A.3.6 Function f_TE.m. Using the transfer matrix approach, a transcendental equation is created which is then used to determine a propagation constant.

```
function result = f_TE(z)
% Creates function used to determine propagation constant
% Variable description:
% result - expression used in search for propagation  constant
% z      - actual value of propagation constant
%
% Global variables:
% Global variables are used to transfer values from data functions
global n_s           % ref. index substrate
global n_c           % ref. index cladding
global n_layer       % ref. index of internal layers
global d_layer       % thicknesses of internal layers (microns)
global k_0           % wavenumber
%
zz=z*k_0;
NumLayers = length(d_layer);
%
% Creation for substrate and cladding
gamma_sub=sqrt(zz^2-(k_0*n_s)^2);
gamma_clad=sqrt(zz^2-(k_0*n_c)^2);
%
% Creation of kappa for internal layers
kappa=sqrt(k_0^2*n_layer.^2-zz.^2);
temp = kappa.*d_layer;
%
```

```
% Construction of transfer matrix for first layer
cc =  cos(temp);
ss =  sin(temp);
m(1,1) = cc(1);
m(1,2) = -1j*ss(1)/kappa(1);
m(2,1) = -1j*kappa(1)*ss(1);
m(2,2) = cc(1);
%
% Construction of transfer matrices for remaining layers
% and multiplication of matrices
for i=2:NumLayers
    mt(1,1) = cc(i);
    mt(1,2) = -1j*ss(i)/kappa(i);
    mt(2,1) = -1j*ss(i)*kappa(i);
    mt(2,2) = cc(i);
    m = mt*m;
end
%
result = 1j*(gamma_clad*m(1,1)+gamma_sub*m(2,2))...
     + m(2,1) - gamma_sub*gamma_clad*m(1,2);
```

Listing 4A.3.7 Function mesh_x.m. Generates a one-dimensional mesh along the transversal direction (which is taken as the *x*-axis) of a dielectric waveguide.

```
function x = mesh_x(d_s,d_layer,d_c,NumberMesh)
% Generates one-dimensional mesh along x-axis
% Variable description:
% Input
% d_layer     - thicknesses of each layer
% NumberMesh  - number of mesh points in each layer
% Output
% x           - mesh point coordinates
%
d_total = [d_s,d_layer,d_c];      % thicknesses of all layers
NumberOfLayers = length(d_total); % determine number of layers
delta = d_total./NumberMesh;      % separation of points for all layers
%
x(1) = 0.0;                       % coordinate of first mesh point
i_mesh = 1;
for k = 1:NumberOfLayers          % loop over all layers
    for i = 1:NumberMesh(k)       % loop within layer
        x(i_mesh+1) = x(i_mesh) + delta(k);
        i_mesh = i_mesh + 1;
    end
end
```

Listing 4A.3.8 Function refindex.m. Determines the refractive index at all mesh points.

```
function n_mesh = refindex(x,interface,index_layer)
% Assigns the values of refractive indices to mesh points
% in all layers
% Input
% x()            -- mesh points coordinates
% interface(n)   -- number of mesh points in layers
% index_layer()  -- refrective index in layers
% Output
% index_mesh     -- refractive index for each mesh point
%
% Within a given layer, refractive index is assigned the same value.
% Loop scans over all mesh points.
% For all mesh points selected for a given layer, the same
% value of refractive index is assigned.
%
N_mesh = length(x);
NumberOfLayers = length(index_layer);
%
i_mesh = 1;
for k = 1:NumberOfLayers        % loop over all layers
    for i = 1:interface(k)      % loop within layer
        n_mesh(i_mesh+1) = index_layer(k);
        i_mesh = i_mesh + 1;
    end
end
```

Listing 4A.3.9 Function TE_field.m. Determines TE optical field for all layers.

```
function TE_mode_field = TE_field(beta,index_mesh,x,k_zero)
% Determines TE optical field for all layers
%
% x - grid created in mesh_x.m
TotalMesh = length(x);  % total number of mesh points
%
zz=beta*k_zero;
%
% Creation of constants at each mesh point
kappa = 0;
for n = 1:(TotalMesh)
   kappa(n)=sqrt((k_zero*index_mesh(n))^2-zz^2);
end
%
% Establish boundary conditions in first layer (substrate).
```

```
% Values of the fields U and V are numbered by index not by
% location along x-axis.
% For visualization purposes boundary conditions are set at first point.
U(1) = 1.0;
temp = imag(kappa(1));
if(temp<0), kappa(1) = - kappa(1);
end
% The above ensures that we get a field decaying in the substrate
V(1) = kappa(1);
%
for n=2:(TotalMesh)
   cc=cos( kappa(n)*(x(n)-x(n-1)) );
   ss=sin( kappa(n)*(x(n)-x(n-1)) );
   m(1,1)=cc;
   m(1,2)=-1i/kappa(n)*ss;
   m(2,1)=-1i*kappa(n)*ss;
   m(2,2)=cc;
   %
   U(n)=m(1,1)*U(n-1)+m(1,2)*V(n-1);
   V(n)=m(2,1)*U(n-1)+m(2,2)*V(n-1);
end
%
TE_mode_field = abs(U);              % Finds Abs(E)
max_value = max(TE_mode_field);
h = plot(x,TE_mode_field/max_value); % plot normalized value of TE field
% adds text on x-axix and size of x label
xlabel('x (microns)','FontSize',22);
% adds text on y-axix and size of y label
ylabel('TE electric field','FontSize',22);
set(h,'LineWidth',1.5);              % new thickness of plotting lines
set(gca,'FontSize',22);              % new size of tick marks on both axes
pause
close all
```

References

[1] C. R. Pollock and M. Lipson. *Integrated Photonics*. Kluwer Academic Publishers, Boston, 2003.

[2] C. Chen, P. Berini, D. Feng, S. Tanev, and V. P. Tzolov. Efficient and accurate numerical analysis of multilayer planar optical waveguides in lossy anisotropic media. *Optics Express*, **7**:260–72, 2000.

[3] J. Chilwell and I. Hodgkinson. Thin-films field-transfer matrix theory of planar multilayer waveguides and reflection from prism-loaded waveguide. *J. Opt. Soc. Amer. A*, **1**:742–53, 1984.

[4] C. Chen, P. Berini, D. Feng, S. Tanev, and V. P. Tzolov. Efficient and accurate numerical analysis of multilayer planar optical waveguides. *Proc. SPIE*, **3795**:676, 1999.

[5] T. D. Visser, H. Blok, and D. Lenstra. Modal analysis of a planar waveguide with gain and losses. *IEEE J. Quantum Electron.*, **31**:1803–18, 1995.

[6] R. Maerz. *Integrated Optics. Design and Modeling.* Artech House, Boston, 1995.

[7] M. N. O. Sadiku. *Elements of Electromagnetics. Second Edition.* Oxford University Press, New York, Oxford, 1995.

[8] H. Kogelnik. Theory of optical waveguides. In T. Tamir, ed., *Springer Series in Electronics and Photonics. Guided-Wave Optoelectronics*, volume 26, pp. 7–88. Springer-Verlag, Berlin, 1988.

[9] L. A. Coldren and S. W. Corzine. *Diode Lasers and Photonic Integrated Circuits.* Wiley, New York, 1995.

[10] J.-M. Liu. *Photonic Devices.* Cambridge University Press, Cambridge, 2005.

[11] R. D. Kekatpure, A. C. Hryciw, E. S. Barnard, and M. L. Brongersma. Solving dielectric and plasmonic waveguide dispersion relations on a pocket calculator. *Optics Express*, **17**:24112–29, 2009.

5 Linear optical fibre and signal degradation

In this chapter we will discuss properties of the linear fibre, where refractive index does not depend on field intensity. We will concentrate on the following topics:

- geometrical-optics description
- wave propagation
- dispersion in single-mode fibres
- dispersion-induced limitations.

Due to dispersion, the width of propagating pulse will increase. In multimode fibres, broadening of optical pulses is significant, about 10 ns km^{-1}. In the geometrical-optics description, such broadening is attributed to different paths followed by different rays. In a single mode fibre, intermodal dispersion is absent; all energy is transported by a single mode. However, pulse broadening exists. Different spectral components of the pulse travel at slightly different group velocities. This effect is known as group velocity dispersion (GVD).

5.1 Geometrical-optics description

Consider an optical fibre whose cross-section is shown in Fig. 5.1. The corresponding change in refractive index profiles is shown in Fig. 5.2. We illustrate two profiles: step index and graded index.

In the following discussion we will assume validity of a geometrical optics description which holds in the limit $\lambda \ll a$, where λ is the light wavelength, and a is the core radius [1], [2].

5.1.1 Numerical aperture (NA)

To start our discussion, consider the propagation of rays entering a cylindrical fibre at different angles in the plane passing through the core centre, see Fig. 5.3. Ray B will travel in the cladding region and it is known as an unguided ray, whereas ray A will stay within core region, and it will form a guided ray.

There exists an angle θ_a (known as the acceptance angle) such that all rays incident from a region with refractive index $\bar{n}_0$ at angles fulfilling the relation

$$\theta < \theta_a$$

will stay within a core region (such as ray A). All rays incident at the acceptance angle θ_a will travel at the boundary between core and cladding.

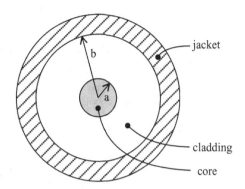

Fig. 5.1 Cross-section of a typical optical fibre.

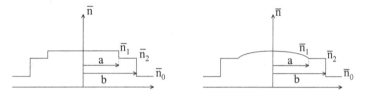

Fig. 5.2 Typical profiles of optical fibre. Linear step (left) and graded (right).

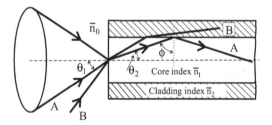

Fig. 5.3 Acceptance angle.

The numerical aperture (NA) is defined as

$$NA \equiv \bar{n}_0 \sin \theta_a$$

It is a dimensionless quantity. Typical values of NA are within the 0.14–0.50 range. Evaluation of NA follows exactly the same procedure as described in Chapter 4. The final expression is

$$NA = \bar{n}_1 (2\Delta)^{1/2} \tag{5.1}$$

In practice $n_1 \approx n_2$ and the relative refractive index difference Δ is

$$\Delta = \frac{(\bar{n}_1 + \bar{n}_2)(\bar{n}_1 - \bar{n}_2)}{2\bar{n}_1^2} \simeq \frac{2\bar{n}_1(\bar{n}_1 - \bar{n}_2)}{2\bar{n}_1^2} = \frac{\bar{n}_1 - \bar{n}_2}{\bar{n}_1} \tag{5.2}$$

A typical value of parameter Δ for a single-mode fibre is around 0.01.

5.1.2 Multipath dispersion

A very important physical effect originates from the fact that different rays travel along paths of different lengths. The shortest path, for incident angle $\theta_1 = 0$ is equal to L, fibre length; the longest path, for $\theta_1 = \theta_{\max}$ (which corresponds to ϕ_c) is for rays travelling the distance $\frac{L}{\sin \phi_c}$. Time delay between both paths is ΔT. It can be estimated as follows:

$$\Delta T = \frac{\Delta \text{distance}}{v}$$
$$= \frac{\frac{L}{\sin \phi_c} - L}{\frac{c}{\bar{n}_1}} = \frac{\bar{n}_1}{c}\left(\frac{L}{\sin \phi_c} - L\right)$$
$$= \frac{\bar{n}_1}{c}\left(\frac{L}{\frac{\bar{n}_2}{\bar{n}_1}} - L\right) = \frac{L\bar{n}_1}{c}\left(\frac{\bar{n}_1}{\bar{n}_2} - 1\right) = \frac{L\bar{n}_1}{c\bar{n}_2}\frac{\bar{n}_1 - \bar{n}_2}{\bar{n}_1}$$
$$= \frac{L\bar{n}_1^2}{c\bar{n}_2}\Delta \quad (5.3)$$

where we have used $v = \frac{c}{\bar{n}_1}$ for speed of light in fibre. Time ΔT is a measure of broadening of an impulse.

5.1.3 Information-carrying capacity of the fibre

Next, we will estimate the information-carrying capacity of the fibre. If we introduce B as a bit rate, and L fibre length, then the measure of the information-capacity of the fibre is determined by a product $B \cdot L$.

The estimation of this product is as follows. Let T_B where $T_B = \frac{1}{B}$ be the time allocated for each bit slot. We use the condition

$$\Delta T < T_B \quad (5.4)$$

which requires that pulse broadening is smaller than time slot allocated for each bit. Let us substitute the previous expression for ΔT, and we obtain

$$\frac{L\bar{n}_1^2}{c\bar{n}_2}\Delta \cdot B < 1 \text{ or}$$
$$L \cdot B < \frac{\bar{n}_2}{\bar{n}_1^2}\frac{c}{\Delta} \quad (5.5)$$

Example Assume $\bar{n}_1 = 1.5, \bar{n}_2 = 1.0$. From the above, one finds $L \cdot B < 0.4 \frac{Mb}{s} \cdot$ km. On the other hand, for $\Delta = 2 \times 10^{-3} \Rightarrow B \cdot L < 100 \frac{Mb}{s} \cdot$ km. This means that one can carry data of 10 Mb s^{-1} over 10 km.

5.1.4 Loss mechanisms in silica fibre

Attenuation (losses) of optical signal is a major factor in the design of optical communication systems. Losses can occur at the input coupler, splices and connectors but also within the

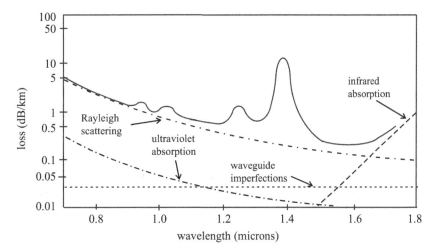

Fig. 5.4 Spectrum of losses in glass. Adapted from [3] with permission from the Institution of Engineering and Technology.

fibre itself. In this section we will concentrate on losses in the fibre. A typical spectrum of losses in modern fibre is shown in Fig. 5.4. Losses are usually expressed in decibels per unit of distance (as was mentioned in Chapter 2) according to formula

$$\alpha_{loss} = \frac{10}{L} \log_{10} \frac{P_{out}}{P_{in}} \qquad (5.6)$$

where P_{in} and P_{out} are, respectively, the input and output optical powers for a fibre of length L.

Loss mechanisms are classified as [4]:

- *intrinsic* which arise from the fundamental material properties of the glass. These can be reduced, e.g. by appropriate choice of wavelength
- *extrinsic* which result due to imperfections in the fabrication process. Sources of extrinsic loss include impurities or structural imperfections.

5.1.5 Intrinsic loss

In pure silica, the three sources of loss important at visible and near-infrared wavelengths are:

- ultraviolet absorption
- infrared absorption
- Rayleigh scattering.

All of the three mechanisms are wavelength dependent. The first two involve two resonances centred in the ultraviolet and mid-infrared, see Fig. 5.4. They originate due to strong electronic and molecular transition bands. The ultraviolet resonance is of electronic origin and is centred near $\lambda = 0.1$ μm [4]. The infrared absorption originates from lattice vibrational modes of silica and dopants. The vibrational modes produce absorptive resonances

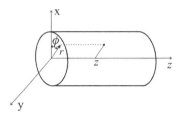

Fig. 5.5 Cylindrical coordinate system.

centered between 7 μm and 11 μm. The resonances for silica and germania occur at 9 μm and 11 μm, respectively [4]. Intraband absorption results in a power loss approximately given by

$$\alpha_{ir} = A \exp\left(-\frac{a_{ir}}{\lambda}\right) \quad [\text{dB km}^{-1}] \tag{5.7}$$

where λ is in micrometres. For $GeO_2 - SiO_2$ glass, the values of A and a_{ir} are $A = 7.81 \times 10^{11}$ dB km^{-1} and $a_{ir} = 48.48$ μm [4].

Rayleigh scattering in pure fused silica can be approximated by the expression

$$\alpha_R = 1.7 \left(\frac{0.85}{\lambda}\right)^4 \tag{5.8}$$

where λ is in micrometres and loss α_R is in dB km^{-1}.

Rayleigh scattering appears when a wave travels through a medium having scattering objects smaller than a wavelength [5]. In a glass those small objects are created as local variations in refractive index in otherwise homogeneous material. Local variations of refractive index in turn can be due to localized variations in a material's density.

5.1.6 Extrinsic loss

These losses are not associated with the fundamental properties of material. There are many sources of extrinsic losses, for example presence of additional impurities and surface irregularities at the core-cladding boundary. Some other related losses are due to bending and source-fibre coupling.

5.2 Fibre modes in cylindrical coordinates

Fibre exhibits cylindrical symmetry, see Fig. 5.5. In the cylindrical coordinate system we use the following variables: r, ϕ, z. Refractive index is expressed as (see Fig. 5.1)

$$\overline{n} = \begin{Bmatrix} \overline{n}_1, & r \leq a \\ \overline{n}_2, & r > a \end{Bmatrix} \tag{5.9}$$

where a is the radius of the central region known as core.

5.2.1 Maxwell's equations in cylindrical coordinates

In cylindrical coordinates, one has the following general relation for arbitrary vector **A** [6]:

$$\nabla \times \mathbf{A} = \frac{1}{r} \begin{vmatrix} \hat{\mathbf{a}}_r & r\hat{\mathbf{a}}_\phi & \hat{\mathbf{a}}_z \\ \frac{\partial}{\partial r} & \frac{\partial}{\partial \phi} & \frac{\partial}{\partial z} \\ A_r & rA_\phi & A_z \end{vmatrix}$$

$$= \hat{\mathbf{a}}_r \left(\frac{\partial A_z}{r \partial \phi} - \frac{\partial A_\phi}{\partial z} \right) + \hat{\mathbf{a}}_\phi \left(\frac{\partial A_r}{\partial z} - \frac{\partial A_z}{\partial r} \right) + \hat{\mathbf{a}}_z \frac{1}{r} \left[\frac{\partial}{\partial r}(rA_\phi) - \frac{\partial A_r}{\partial \phi} \right]$$

where $\hat{\mathbf{a}}_r, \hat{\mathbf{a}}_\phi, \hat{\mathbf{a}}_z$ are unit vectors in the corresponding directions. Vector **A** is expressed as

$$\mathbf{A} = \hat{\mathbf{a}}_r A_r + \hat{\mathbf{a}}_\phi A_\phi + \hat{\mathbf{a}}_z A_z$$

Assuming $e^{j\omega t}$ dependence for fields, the first Maxwell equation in cylindrical coordinates is

$$\frac{1}{r} \frac{\partial E_z}{\partial \phi} - \frac{\partial E_\phi}{\partial z} = -j\omega\mu H_r$$

$$\frac{\partial E_r}{\partial z} - \frac{\partial E_z}{\partial r} = -j\omega\mu H_\phi$$

$$\frac{1}{r} \left[\frac{\partial}{\partial r}(rE_\phi) - \frac{\partial E_r}{\partial \phi} \right] = -j\omega\mu H_z$$

Assuming additionally $e^{j\beta z}$ dependence, one finds

$$\frac{1}{r} \frac{\partial E_z}{\partial \phi} + j\beta E_\phi = -j\omega\mu H_r \tag{5.10}$$

$$j\beta E_r + \frac{\partial E_z}{\partial r} = j\omega\mu H_\phi \tag{5.11}$$

$$\frac{1}{r} \left[\frac{\partial}{\partial r}(rE_\phi) - \frac{\partial E_r}{\partial \phi} \right] = -j\omega\mu H_z \tag{5.12}$$

Similarly, from the second Maxwell equation we obtain

$$\frac{1}{r} \frac{\partial H_z}{\partial \phi} + j\beta H_\phi = j\omega\varepsilon E_r \tag{5.13}$$

$$j\beta H_r + \frac{\partial H_z}{\partial r} = -j\omega\varepsilon E_\phi \tag{5.14}$$

$$\frac{1}{r} \left[\frac{\partial}{\partial r}(rH_\phi) - \frac{\partial H_r}{\partial \phi} \right] = j\omega\varepsilon E_z \tag{5.15}$$

Expressing H_r from (5.10) and (5.14) and comparing both expressions, one obtains for E_ϕ

$$E_\phi = -\frac{j}{q^2} \left(\frac{\beta}{r} \frac{\partial E_z}{\partial \phi} - \omega\mu \frac{\partial H_z}{\partial r} \right) \tag{5.16}$$

where

$$q^2 = \omega^2 \varepsilon \mu - \beta^2 \tag{5.17}$$
$$= \bar{n}^2 k_0^2 - \beta^2$$

Similarly, using the same procedure, from (5.11) and (5.13) one obtains for E_r

$$E_r = -\frac{j}{q^2}\left(\beta\frac{\partial E_z}{\partial r} + \frac{\omega\mu}{r}\frac{\partial H_z}{\partial \phi}\right) \tag{5.18}$$

In an analogous way one finds expressions for H_ϕ and H_r:

$$H_\phi = -\frac{j}{q^2}\left(\frac{\beta}{r}\frac{\partial H_z}{\partial \phi} + \varepsilon\omega\frac{\partial E_z}{\partial r}\right) \tag{5.19}$$

$$H_r = -\frac{j}{q^2}\left(\beta\frac{\partial H_z}{\partial r} - \varepsilon\frac{\omega}{r}\frac{\partial E_z}{\partial \phi}\right) \tag{5.20}$$

Equations (5.16), (5.18), (5.19) and (5.20) express relevant components of **E** and **H** fields in terms of E_z and H_z components. In the next step, we derive equations for those components.

5.2.2 Wave equations in cylindrical coordinates

From the analysis above we can observe that the z-component of the electric field E_z (and same applies also to magnetic z-component H_z) can be considered an independent variable and that the remaining components can be expressed in terms of E_z. (The same is true for H_z.) Hence, we only need to work with E_z (or H_z). For that purpose we will derive the appropriate wave equation from which we can obtain a solution for E_z. Once we have a solution for E_z we will use it to find E_r and E_ϕ. In this way, all components of the electric field within a circular waveguide are derived.

To derive the equation for E_z we start with Eq. (5.15). Replacing derivatives on the left hand side by expressions (5.19) and (5.20) and rearranging terms, one finds

$$\frac{\partial^2 E_z}{\partial r^2} + \frac{1}{r}\frac{\partial E_z}{\partial r} + \frac{1}{r^2}\frac{\partial^2 E_z}{\partial \phi^2} + q^2 E_z = 0 \tag{5.21}$$

where q^2 is given by Eq. (5.17).

Similarly, we can derive an equation for H_z. For the last step we will use Eq. (5.12). We replace E_ϕ and E_r and obtain a wave equation for H_z in cylindrical coordinates

$$\frac{\partial^2 H_z}{\partial r^2} + \frac{1}{r}\frac{\partial H_z}{\partial r} + \frac{1}{r^2}\frac{\partial^2 H_z}{\partial \phi^2} + n^2 k_0^2 H_z = 0 \tag{5.22}$$

5.2.3 Solution of the wave equation in cylindrical coordinates

Here, we will analyse the equation for E_z. An equation for H_z can be solved in a similar way. Assume that the following separation of variables is possible:

$$E_z(r,\phi) = R(r)\Phi(\phi) \tag{5.23}$$

After separation, one obtains two equations

$$\frac{d^2\Phi(\phi)}{d\phi^2} + m^2\Phi(\phi) = 0 \tag{5.24}$$

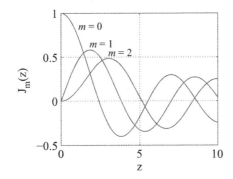 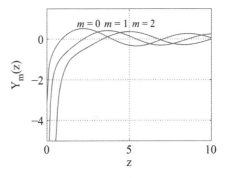

Fig. 5.6 Ordinary Bessel functions. (left) Functions J_m, (right) functions Y_m.

and

$$r^2 \frac{d^2R}{dr^2} + r\frac{dR}{dr} + (r^2 q^2 - m^2) R = 0 \quad (5.25)$$

where m^2 is a separation constant. Eq. (5.25) is known as the Bessel equation and its solutions are known as Bessel functions [7]. The solution of Eq. (5.24) is

$$\Phi(\phi) = e^{im\phi} \quad (5.26)$$

Here, m takes integer values because $\Phi(\phi + 2\pi) = \Phi(\phi)$.

The Bessel equation will be written and solved separately for core and cladding regions. In the core region it reads

$$r^2 \frac{d^2R}{dr^2} + r\frac{dR}{dr} + (r^2 \kappa^2 - m^2) R = 0 \quad (5.27)$$

where

$$\kappa^2 = \bar{n}_1^2 k_0^2 - \beta^2 > 0 \quad (5.28)$$

The solutions of this equation are known as ordinary Bessel functions J_m and N_m of the first order and are shown in Fig. 5.6.

The Bessel equation for the cladding region is

$$r^2 \frac{d^2R}{dr^2} + r\frac{dR}{dr} + (r^2 \gamma^2 + m^2) R = 0 \quad (5.29)$$

where

$$\gamma^2 = \beta^2 - \bar{n}_2^2 k_0^2 > 0 \quad (5.30)$$

The solutions of this equation are known as modified Bessel functions, K_m, I_m of the second kind and of order m. Those functions are plotted in Fig. 5.7.

MATLAB codes for obtaining plots of Bessel functions are provided in Appendix 5C.1. Using the above results for solutions of Bessel equations, one can obtain the solutions for

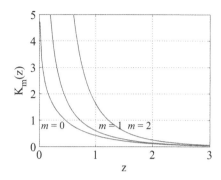 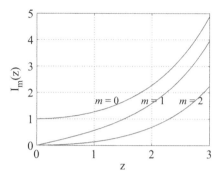

Fig. 5.7 Modified Bessel functions. (left) Functions K_m, (right) functions I_m.

core and cladding regions as [4]

$$R(r) = \begin{cases} AJ_m(\kappa r) + A'Y_m(\kappa r), & r \leq a \\ CK_m(\gamma r) + C'I_m(\gamma r), & r > a \end{cases} \quad (5.31)$$

where A, A', C, C' are constants.

We expect the following properties of guided modes:

1. the solution in the core will be oscillatory and finite at $r = 0$
2. the solution in the cladding will decrease monotonically as radius increases, i.e. field $\to 0$ for $r \to \infty$.

The above properties imply that for core the J_m function should be used and since Y_m has a singularity (see Fig. 5.6), for $r = 0$ we set $A' = 0$. Similarly for cladding the function I_m increases (see Fig. 5.7) and we set $C' = 0$. In cladding we use only K_m.

In this way one finds the solutions

$$E_z = \begin{cases} AJ_m(\kappa r) e^{jm\phi}, & r \leq a \\ CK_m(\gamma r) e^{jm\phi}, & r > a \end{cases} \quad (5.32)$$

In exactly the same way

$$H_z = \begin{cases} BJ_m(\kappa r) e^{jm\phi}, & r \leq a \\ DK_m(\gamma r) e^{jm\phi}, & r > a \end{cases} \quad (5.33)$$

Employing Eqs. (5.16), (5.18), (5.19) and (5.20), one can express the other field components in terms of E_z and H_z. After a little algebra, the full solutions are:

Case: $r < a$ (core)

$$E_\phi = -\frac{j}{\kappa^2} \left(j\frac{\beta}{r} mAJ_m(\kappa r) - \omega\mu\kappa BJ'_m(\kappa r) \right) e^{jm\phi} \quad (5.34)$$

$$E_r = -\frac{j}{\kappa^2} \left(\beta\kappa AJ'_m(\kappa r) + j\frac{\omega\mu}{r} mBJ_m(\kappa r) \right) e^{jm\phi} \quad (5.35)$$

$$H_\phi = -\frac{j}{\kappa^2} \left(\omega\varepsilon_1 \kappa AJ'_m(\kappa r) + j\frac{\beta}{r} mBJ_m(\kappa r) \right) e^{jm\phi} \quad (5.36)$$

$$H_r = -\frac{j}{\kappa^2}\left(\beta\kappa B J'_m(\kappa r) - j\frac{\omega\varepsilon_1}{r}m A J_m(\kappa r)\right)e^{jm\phi} \quad (5.37)$$

Case: $r > a$ (cladding)

$$E_\phi = \frac{j}{\gamma^2}\left(j\frac{\beta}{r}m C K_m(\gamma r) - \omega\mu\gamma D K'_m(\gamma r)\right)e^{jm\phi} \quad (5.38)$$

$$E_r = \frac{j}{\gamma^2}\left(\beta\gamma C K'_m(\gamma r) + j\frac{\omega\mu}{r}m D K_m(\gamma r)\right)e^{jm\phi} \quad (5.39)$$

$$H_\phi = \frac{j}{\gamma^2}\left(\omega\varepsilon_2\gamma C K'_m(\gamma r) + j\frac{\beta}{r}m D K_m(\gamma r)\right)e^{jm\phi} \quad (5.40)$$

$$H_r = \frac{j}{\gamma^2}\left(\beta\gamma D K'_m(\gamma r) - j\frac{\omega\varepsilon_2}{r}m C K_m(\gamma r)\right)e^{jm\phi} \quad (5.41)$$

where we have introduced the following definitions

$$J'_m(\kappa r) = \frac{dJ_m(\kappa r)}{d(\kappa r)}$$

and

$$K'_m(\gamma r) = \frac{K_m(\gamma r)}{d(\gamma r)}$$

5.2.4 Boundary conditions and modal equation

We use boundary conditions derived previously which require that all field components tangent to the core-cladding boundary at $r = a$ must be continuous across it. We therefore have [8]

$$E_z, H_z, E_\phi, H_\phi \text{ are continuous at } r = a \quad (5.42)$$

Applying the above requirement, we obtain the following equations.

1. from continuity of E_z

$$A J_m(\kappa a) = C K_m(\gamma a) \quad (5.43)$$

2. from continuity of H_z

$$B J_m(\kappa a) = D K_m(\gamma a) \quad (5.44)$$

3. from continuity of E_ϕ

$$\frac{\beta m}{\kappa^2 a}J_m(\kappa a)A + j\frac{\omega\mu}{\kappa}J'_m(\kappa a)B + \frac{\beta m}{\gamma^2 a}K_m(\gamma a)C + j\frac{\omega\mu}{\gamma}K'_m(\gamma a)D = 0 \quad (5.45)$$

4. from continuity of H_ϕ

$$-j\frac{\omega\varepsilon_1}{\kappa}J'_m(\kappa a)A + \frac{\beta m}{a\kappa^2}J_m(\kappa a)B - j\frac{\omega\varepsilon_2}{\gamma}K'_m(\gamma a)C + \frac{\beta m}{a\gamma^2}K_m(\gamma a)D = 0 \quad (5.46)$$

The above form a set of simultaneous equations for unknown coefficients A, B, C, D. The nontrivial solution is obtained if the determinant vanishes, i.e.

$$\begin{vmatrix} J_m(\kappa a) & 0 & -K_m(\gamma a) & 0 \\ 0 & J_m(\kappa a) & 0 & -K_m(\gamma a) \\ \dfrac{\beta m}{\kappa^2 a} J_m(\kappa a) & j\dfrac{\omega\mu}{\kappa} J'_m(\kappa a) & \dfrac{\beta m}{\gamma^2 a} K_m(\gamma a) & j\dfrac{\omega\mu}{\gamma} K'_m(\gamma a) \\ -j\dfrac{\omega\varepsilon_1}{\kappa} J'_m(\kappa a) & \dfrac{\beta m}{a\kappa^2} J_m(\kappa a) & -j\dfrac{\omega\varepsilon_2}{\gamma} K'_m(\gamma a) & \dfrac{\beta m}{a\gamma^2} K_m(\gamma a) \end{vmatrix} = 0 \quad (5.47)$$

Expansion and evaluation of this determinant is described in Appendix 5B. The resulting characteristic equation is

$$\left[\frac{J'_m(\kappa a)}{\kappa J_m(\kappa a)} + \frac{K'_m(\gamma a)}{\gamma K_m(\gamma a)} \right] \left[\frac{k_0^2 \bar{n}_1^2 J'_m(\kappa a)}{\kappa J_m(\kappa a)} + \frac{k_0^2 \bar{n}_2^2 K'_m(\gamma a)}{\gamma K_m(\gamma a)} \right] = \frac{\beta^2 m^2}{a^2} \left(\frac{1}{\kappa^2} + \frac{1}{\gamma^2} \right)$$

(5.48)

This equation forms the basis for obtaining propagation constants for guided modes propagating in cylindrical fibre. Before we start a detailed discussion, let us look at the general classification of modes.

5.2.5 Mode classification

Variable m is the main parameter used in mode classification. It is known as azimuthal mode number. Because of the oscillatory behavior of $J_m(\kappa a)$ for a given value of m as seen in Figs. 5.6, multiple solutions exist for each integer value m. We denote them as β_{mn}, where $n = 1, 2, \ldots$ Index n is called the radial mode number and it represents the number of radial modes that exist in the field distribution [4].

In general, electromagnetic waves in a step index optical cylindrical fibre can be divided into three distinct categories [9]:

1. Transverse electric (TE); sometimes designated as magnetic (H) waves. They are characterized by $E_z = 0$ and $H_z \neq 0$.
2. Transverse magnetic (TM); sometimes designated as electric (E) waves. They are characterized by $E_z \neq 0$ and $H_z = 0$.
3. Hybrid waves. Characterized by $E_z \neq 0$ and $H_z \neq 0$.

In more detail, the following cases are treated separately [8]:

1. $m = 0$. Still, solutions are separated into two groups:
   a. TM_{0n} (transverse magnetic) because $E_\phi = H_r = H_z = 0$. In this case coefficients $B = D = 0$.
   b. TE_{0n} (transverse electric) because $H_\phi = E_r = E_z = 0$. In this case coefficients $A = C = 0$.
2. $m \neq 0$. Hybrid modes.

The modes are called EH or HE modes (terminology has historical roots traced to microwaves).

Values of β will correspond to modes which have finite components of both E_z and H_z. There are neither TE nor TM modes. General classification is [10], [11], [12].

if $A = 0$ the mode is called TE mode
if $B = 0$ the mode is called TM mode
if $A > B = 0$ the mode is called HE mode (E_z dominates H_z)
if $A < B = 0$ the mode is called EH mode (H_z dominates E_z).

Now, we will discuss some of those cases in more detail.

5.2.6 Modes with $m = 0$

The simplest solutions of characteristic Eq. (5.48) are obtained for $m = 0$. In this case there is no angular dependence on ϕ, and therefore field components E_z and H_z are rotationally invariant. The characteristic equation is reduced into two equations which describe TE and TM modes:

TE modes
$$\frac{J_0'(\kappa a)}{\kappa J_0(\kappa a)} + \frac{K_0'(\gamma a)}{\gamma K_0(\gamma a)} = 0 \qquad (5.49)$$

TM modes
$$\frac{J_0'(\kappa a)}{\kappa J_0(\kappa a)} + \frac{\overline{n}_2^2}{\overline{n}_1^2} \frac{K_0'(\gamma a)}{\gamma K_0(\gamma a)} = 0 \qquad (5.50)$$

Using the following relations for Bessel functions, (see Appendix 5A)
$$\frac{J_0'(u)}{J_0(u)} = \frac{J_{-1}(u)}{J_0(u)} \quad \text{and} \quad \frac{K_0'(w)}{wK_0(w)} = -\frac{K_{-1}(w)}{wK_0(w)}$$

and $J_{-1} = (-1)^1 J_1$, $K_{-1} = K_1$, one obtains
$$\frac{J_1(u)}{J_0(u)} + \frac{u}{w} \frac{K_1(w)}{K_0(w)} = 0 \qquad (5.51)$$

and
$$\frac{J_1(u)}{J_0(u)} + \frac{\overline{n}_2^2}{\overline{n}_1^2} \frac{u}{w} \frac{K_1(w)}{K_0(w)} = 0 \qquad (5.52)$$

where we used definitions $u = a\kappa$ and $w = a\gamma$.

The interpretation of relations (5.51) and (5.52) as describing TE and TM modes comes from the following observations.

From continuity equations for E_z and H_z (established in the previous section), one can express constants C and D in terms of A and B. Substitute these results into the continuity equation for E_ϕ and obtain

$$A\frac{\beta m}{a}\left(\frac{1}{\kappa^2} + \frac{1}{\gamma^2}\right) + j\omega\mu \left[\frac{J_m'(\kappa a)}{\kappa J_m(\kappa a)} + \frac{K_m'(\gamma a)}{\gamma K_m(\gamma a)}\right] B = 0$$

When the term in the square bracket is zero, then $A = 0$ and because of Eq. (5.32), $E_z = 0$ which means that electric field is transverse, i.e. TE mode. Similarly, substituting expressions for constants C and D into continuity equation for H_ϕ one obtains

$$-j\left[n_1^2\frac{J'_m(\kappa a)}{\kappa J_m(\kappa a)} + n_2^2\frac{K'_m(\gamma a)}{\gamma K_m(\gamma a)}\right]A + \frac{\beta m}{a}\omega\mu\left(\frac{1}{\kappa^2} + \frac{1}{\gamma^2}\right)B = 0$$

When the term in the square bracket is zero, then $B = 0$ and because of Eq. (5.33), $H_z = 0$ which means that magnetic field is transverse, i.e. TM mode.

Interpretation

Here, we will provide an interpretation of relations (5.51) and (5.52) which describe TE and TM modes.

1. TM modes

Assuming $B = D = 0$ (and also $m = 0$), from Eqs. (5.34)–(5.37) in the core region one finds

$$E_\phi = 0, \quad E_r = -\frac{j}{\kappa}\beta A J'_0(\kappa r)$$

$$H_\phi = -\frac{j}{\kappa}\omega\varepsilon_1 A J'_0(\kappa r), \quad H_r = 0$$

Also, from (5.32) and (5.33) one has

$$E_z = A J_0(\kappa r), \quad H_z = 0$$

For such modes longitudinal magnetic component H_z is zero and it is therefore called TM mode (transverse magnetic). Designation of such modes is TM_{0n}.

2. TE modes

In a similar way, taking $A = C = 0$ (and also $m = 0$), from Eqs. (5.34)–(5.37) in the core region one obtains

$$E_\phi = \frac{j}{\kappa}\omega\mu B J'_0(\kappa r), \quad E_r = 0$$

$$H_\phi = 0, \quad H_r = -\frac{j}{\kappa}\beta B J'_0(\kappa r)$$

and also, from (5.32) and (5.33) one has

$$E_z = 0, \quad H_z = B J_0(\kappa r)$$

For such modes longitudinal electric component E_z is zero and it is therefore called TE mode (transverse electric). Designation of such modes is TE_{0n}.

A summary of mode properties for $m = 0$ is provided in Table 5.1.

5.2.7 Weakly guiding approximation (wga)

General solution of the dispersion Eq. (5.48) is very complicated. Gloge [13] developed an approximation known as the weakly guiding approximation (wga). It is based on the fact

Table 5.1 Classification of modes for $m = 0$.

Mode description	Equation	Values of coefficients	Nonzero field components	Zero field components
TE_{0n}	$\dfrac{J_0'(u)}{uJ_0(u)} + \dfrac{K_0'(w)}{\gamma K_0(w)} = 0$	$A = C = 0$	H_z, H_r, E_ϕ	$E_z = E_r = H_\phi = 0$
TM_{0n}	$\dfrac{J_0'(u)}{uJ_0(u)} + \dfrac{\bar{n}_2^2}{\bar{n}_1^2}\dfrac{K_0'(w)}{\gamma K_0(w)} = 0$	$B = D = 0$	E_z, E_r, H_ϕ	$H_z, H_r, E_\phi = 0$

that $\bar{n}_1 - \bar{n}_2 \ll 1$. Under such approximation $\bar{n}_1^2 \approx \bar{n}_2^2 = \bar{n}^2$ and also $\beta^2 \approx k_0^2 \bar{n}^2$. Introducing definitions $u = \kappa a$ and $w = \gamma a$, general Eq. (5.48) reduces to

$$\frac{J_m'(u)}{uJ_m(u)} + \frac{K_m'(w)}{wK_m(w)} = \pm m\left(\frac{1}{u^2} + \frac{1}{w^2}\right) \tag{5.53}$$

One can observe that two possible cases exist. We will consider them separately.

Case '+':
Applying properties of Bessel functions (see Appendix 5A) given by Eqs. (5.85) and (5.87), from Eq. (5.53) one obtains

$$\frac{J_{m+1}(u)}{uJ_m(u)} + \frac{K_{m+1}(w)}{wK_m(w)} = 0 \tag{5.54}$$

Case '−':
Applying properties of Bessel functions (see Appendix 5A) given by Eqs. (5.84) and (5.86), from Eq. (5.53) one has

$$\frac{J_{m-1}(u)}{uJ_m(u)} + \frac{K_{m-1}(w)}{wK_m(w)} = 0 \tag{5.55}$$

Eigenvalue Eq. (5.53) is solved numerically to find propagation constant β. In general, it has multiple solutions for each m. They are labelled by n ($n = 1, 2, \ldots$). Thus $\beta = \beta_{mn}$. Each value of β_{mn} corresponds to one possible mode. Fibre modes are referred to as hybrid modes and denoted by [4]

$$HE_{mn} \ (\text{dominates} H_z)$$

$$EH_{mn} \ (\text{dominates} E_z)$$

In special case $m = 0$

$$\left.\begin{array}{l} HE_{0n} \equiv TE_{0n} \\ EH_{0n} \equiv TM_{0n} \end{array}\right\} \text{correspond to transverse electric and transverse magnetic}$$

LP modes

LP modes originate in a weakly guiding approximation. Those approximate solutions are much simpler than the exact solutions. Solving characteristic equation leads to similar eigenvalues for both the $EH_{m-1,n}$ modes and $HE_{m+1,n}$ [9]. If the fields of these modes are combined, the resulting transversal field will be linearly polarized. The linearly polarized modes are labelled as $LP_{m,n}$. They are composed in the following way:

$$LP_{0n} = HE_{1n}$$
$$LP_{1n} = HE_{2n} + E_{0n} + H_{0n}$$
$$LP_{m,n} = HE_{m+1,n} + EH_{m-1,n}, \quad m > 2$$

For each n there are two LP_{0n} modes, polarized perpendicular to each other, while for each n and each $m > 0$ there are four $LP_{m,n}$. Here

$$LP_{mn} \text{ (LP stands for linearly polarized)} \qquad (5.56)$$

At the end, let us summarize some important definitions:

$$N \equiv \frac{\beta}{k_0}$$

which is known as mode index or effective index, and also

$$V = k_0 a \left(n\bar{n}_1^2 - n_2^2 \right)^{1/2} \approx \frac{2\pi}{\lambda} a \bar{n}_1 \sqrt{2\Delta}$$

where

$$\Delta = \frac{\bar{n}_1 - \bar{n}_2}{\bar{n}_1}$$

and finally

$$b = \frac{\frac{\beta}{k_0} - \bar{n}_2}{\bar{n}_1 - \bar{n}_2} \equiv \frac{\bar{n} - \bar{n}_2}{\bar{n}_1 - \bar{n}_2}$$

To finish this section, consider an example.

Example Typical values (for multimode fibres) are $a = 25$ μm, $\Delta = 5 \times 10^{-3}$, $V \approx 18$, at $\lambda = 1.3$ μm. Using approximate expression $M = V^2/2$, where M is the total number of modes (see Problem 1), one finds out that this particular fibre supports approximately 162 modes.

5.2.8 The unified expression

Equations derived so far for TE, TM, EH and HE modes in the weakly guiding approximation can be written in the form of a single equation. One should notice that in the wga, equations describing TE and TM modes degenerate into a single equation. The Eq. (5.55) for HE modes requires modification. First, write Eq. (5.55) as

$$\frac{uJ_m(u)}{J_{m-1}(u)} + \frac{wK_m(w)}{K_{m-1}(w)} = 0 \qquad (5.57)$$

In order to appropriately modify the above equations, consider general formulas for Bessel functions provided in the Appendix, Eqs. (5.79) and (5.80). By changing index as $m+1 \to m$

in those equations, one obtains

$$uJ_m(u) = 2(m-1)J_{m-1}(u) - uJ_{m-2}(u)$$
$$wK_m(w) = 2(m-1)K_{m-1}(w) + wK_{m-2}(u)$$

The above equations can now be substituted into (5.57) and finally obtain

$$-\frac{uJ_{m-2}(u)}{J_m(u)} = \frac{wK_{m-2}(w)}{K_m(w)} \qquad (5.58)$$

Equations (5.54) and (5.58) can now be written as a single equation if we define new index k as

$$k = \begin{cases} 1 & \text{for both TE and TM modes} \\ m+1 & \text{for EH modes} \\ m-1 & \text{for HE modes} \end{cases}$$

The universal equation reads

$$\frac{uJ_{k-1}(u)}{J_k(u)} = -\frac{wK_{k-1}(w)}{K_k(w)} \qquad (5.59)$$

5.2.9 Universal relation for fundamental mode HE_{11}

Fundamental mode HE_{11} (or LP_{01}) has the following dispersion relation (corresponding to $m = 1$):

$$\frac{J_1(u)}{J_0(u)} + \frac{u}{w}\frac{K_1(w)}{K_0(w)} = 0 \qquad (5.60)$$

where $u = a\sqrt{\bar{n}_1^2 k_0^2 - \beta^2}$ and $w = a\sqrt{\beta^2 - \bar{n}_2^2 k_0^2}$. Recall previous relations

$$b = \frac{\beta^2/k_0^2 - \bar{n}_2^2}{\bar{n}_1^2 - \bar{n}_2^2} = \frac{\beta^2 - \bar{n}_2^2 k_0^2}{\bar{n}_1^2 k_0^2 - \bar{n}_2^2 k_0^2}$$

and

$$V = k_0 a \left(\bar{n}_1^2 - \bar{n}_2^2\right)^{1/2}$$

From these definitions, one finds

$$b = \frac{w^2}{V^2}$$

After a little algebra

$$u^2 = a^2\left(\bar{n}_1^2 k_0^2 - \beta^2\right) = a^2 \bar{n}_1^2 k_0^2 - a^2 \bar{n}_2^2 k_0^2 + a^2 \bar{n}_2^2 k_0^2 - a^2 \beta^2$$
$$= V^2 - w^2 = V^2 - V^2 b$$

The arguments in Eq. (5.60) can therefore be expressed as

$$u = V\sqrt{1-b}$$

and

$$w = V\sqrt{b}$$

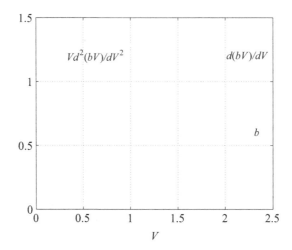

Fig. 5.8 Plot of the waveguide parameter b and its derivatives $d(bV)/dV$ and $Vd^2(bV)/dV^2$ as a function of the V number for HE_{11} mode.

Replacing arguments in Eq. (5.60) with the above relations, we finally obtain

$$\frac{J_1(V\sqrt{1-b})}{J_0(V\sqrt{1-b})} + \frac{\sqrt{1-b}}{\sqrt{b}}\frac{K_1(V\sqrt{b})}{K_0(V\sqrt{b})} = 0 \qquad (5.61)$$

The above formula holds for mode HE_{11}. It contains only two variables V and b. It is therefore called a universal relation. In Appendix 5C.2 we provided MATLAB program which plots $b = b(V)$ relation and also two other functions $\frac{d(bV)}{dV}$ and $\frac{Vd^2(bV)}{dV^2}$ which are important in determining dispersion relation for propagating mode. Those relations are shown in Fig. 5.8.

5.2.10 Single-mode fibres

The single-mode fibre supports only HE_{11} mode (fundamental mode which corresponds to $m = 0$). From Eq. (5.48) we have

$$\kappa J_0(\kappa a) K_0'(\gamma a) + \gamma J_0'(\gamma a) K_0(\gamma a) = 0$$
$$\kappa \bar{n}_2^2 J_0(\kappa a) K_0'(\gamma a) + \gamma \bar{n}_1^2 J_0'(\gamma a) K_0(\gamma a) = 0$$

At cutoff, $b = 0$ which gives that $\beta/\kappa_0 = N$ and that $\gamma = 0$. Also, $\kappa^2 a^2 = a^2 k_0^2(\bar{n}_1^2 - \bar{n}_2^2)$. Thus $V = \kappa a$ when $\gamma = 0$.

Thus cutoff condition is

$$J_0(V) = 0 \qquad (5.62)$$

which gives (for single mode)

$$V = 2.405 \qquad (5.63)$$

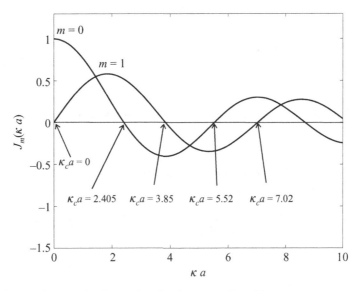

Fig. 5.9 Plot of Bessel functions for $m = 0$ and $m = 1$ used to illustrate cutoff conditions.

Example Estimates of a.

Assume $\lambda = 1.2\,\mu\text{m}$, $\bar{n}_1 = 1.45$ and $\Delta = 5 \times 10^{-3}$. For those values, the above condition for V gives $a < 3.2\,\mu\text{m}$.

5.2.11 Cutoff conditions

The condition $b = 0$, which results in $\beta^2 = k_0^2 \bar{n}_2^2$, defines the so-called cutoff of the mode [14]. A mode is cut off when its field in the cladding becomes evanescent and is detached from the guide. At cutoff $\beta = k_0 \bar{n}_2$ and also $b = 0$, $V = V_c$. The situation is illustrated in Fig. 5.9 for $m = 0$ and $m = 1$. Case $m = 0$ corresponds to modes TE_{0n} and TM_{0n}. The cutoff conditions for those modes are obtained from roots of the following equation:

$$J_0(\kappa_c a) = 0 \qquad (5.64)$$

Cutoff conditions are illustrated in Fig. 5.9. For $m = 0$ the few lowest roots are shown. They correspond to $\kappa_c a = 2.405$ and $\kappa_c a = 5.52$. For $m = 1$ the lowest cutoffs are: $\kappa_c a = 3.83$ and $\kappa_c a = 7.02$.

5.3 Dispersion

Dispersion is an effect responsible for spreading of a signal in time when it propagates. Signal is represented by pulse, so effectively we talk about pulse broadening.

Transmitted pulses broaden when they propagate in dispersive media. After long propagation it might be impossible to distinguish individual pulses, see Fig. 5.10.

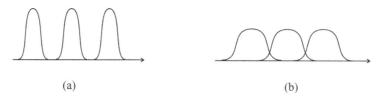

Fig. 5.10 The effect of dispersion on short temporal pulses. (a) Before entering dispersive medium. (b) After propagation in a dispersive medium.

The main types of dispersion are [10]:

1. Material dispersion.
   It exists when refractive index $\bar{n}(\lambda)$ depends on wavelength. As a result, different wavelengths travel with different velocities.
2. Modal dispersion.
   Exists in waveguides with more than one propagating mode. Each mode propagates with different velocity.
3. Waveguide dispersion.
   Exists because propagation constant β depends on the wavelength, so even for $\bar{n}(\lambda) = const$ different wavelengths will propagate with different speeds. Usually the smallest, but important close to the wavelength where dispersion is zero.

5.3.1 Group delay-general discussion

In a fibre, different spectral components of the pulse travel at slightly different group velocities. They produce an effect known as group-velocity dispersion (GVD), also known as intermodal dispersion or fibre dispersion. The name used depends on what description is used (geometrical optics or wave optics).

Detailed discussion is based on the following assumptions:

- optical signal excites all modes equally at the input of the fibre
- each mode contains all spectral components
- each spectral component travels independently (in the fibre)
- each spectral component undergoes a time delay or group delay.

Transit time of a pulse travelling through the fibre of length L is

$$\frac{\tau_g}{L} = \frac{1}{v_g} \tag{5.65}$$

where τ_g – time delay or group delay, L – distance, v_g – group velocity, β – propagation constant.

Group delay determines the transit time of a pulse travelling through the fibre of length L. Group velocity is defined as

$$\frac{1}{v_g} \equiv \frac{d\beta}{d\omega} \tag{5.66}$$

Typically, one can express τ_g in terms of derivatives with respect to k or λ or V. For that purpose we will summarize several relations:

$$\omega = 2\pi \nu, \quad \nu = \frac{c}{\lambda}, \quad k = \frac{2\pi}{\lambda}, \quad \omega = kc, \quad V = \left(\overline{n}_1^2 - \overline{n}_2^2\right)^{1/2} \cdot a \cdot k$$

From the above relations, one finds the following useful formulas:

$$\frac{d}{d\omega} = \frac{d}{cdk}, \quad dk = \frac{dk}{d\lambda}d\lambda, \quad \frac{dk}{d\lambda} = -\frac{2\pi}{\lambda^2}, \quad \frac{d}{dk} = -\frac{\lambda^2}{2\pi}\frac{d}{d\lambda} \tag{5.67}$$

To evaluate derivative with respect to V, one observes that $\beta = \beta(k(V))$. Therefore

$$\frac{d\beta}{dk} = \frac{d\beta}{dV}\frac{dV}{dk} = \frac{V}{k}\frac{d\beta}{dV}$$

where we applied the following result obtained using expression for V:

$$\frac{dV}{dk} = \frac{V}{k}$$

Using the above relations allows us to express τ_g in several forms as

$$\tau_g = L\frac{d\beta}{d\omega} = \frac{L}{c}\frac{d\beta}{dk} = \frac{L}{c}\frac{V}{k}\frac{d\beta}{dV} = -\frac{L\lambda^2}{2\pi c}\frac{d\beta}{d\lambda} \tag{5.68}$$

Because τ_g depends on λ, each spectral component of any particular mode travels the same distance in different time.

At this stage, a typical procedure is to explicitly consider two main physical mechanisms which contribute to group delay τ_g. One can therefore write

$$\tau_g = \frac{L}{c}\frac{d\beta}{dk}\bigg|_{\overline{n}\neq const} + \frac{L}{c}\frac{d\beta}{dk}\bigg|_{\overline{n}=const} \equiv \tau_{mat} + \tau_{wg} \tag{5.69}$$

where τ_{mat} is known as material dispersion whereas τ_{wg} is a waveguide dispersion.

Material dispersion accounts for the fact that refractive index depends on wavelength; i.e. $\overline{n} = \overline{n}(\lambda)$. If we neglect dependence of refractive index on wavelength, i.e. set $\overline{n} = const$, then due to the fact that group velocity depends on λ one obtains the contribution known as waveguide dispersion. We now consider both contributions separately.

5.3.2 Material dispersion: Sellmeier equation

Material dispersion originates from the fact that refractive index $\overline{n}(\lambda)$ is wavelength dependent. Propagation constant β is

$$\beta = \frac{2\pi}{\lambda}\overline{n}(\lambda)$$

Use the above and Eq. (5.68) to obtain

$$\tau_{mat} = \frac{L}{c}\left(\overline{n} - \lambda\frac{d\overline{n}}{d\lambda}\right) \tag{5.70}$$

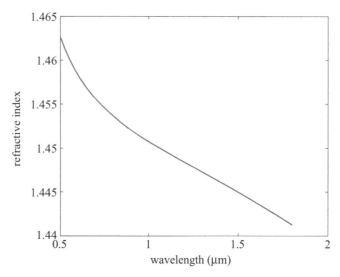

Fig. 5.11 Wavelength dependence of refractive index based on the Sellmeier equation for SiO_2.

Let us concentrate now on material dispersion in silica. In practice, very often an empirical expression for the refractive index of glass in terms of the wavelength λ is used. It is known as the Sellmeier equation and it was introduced in Chapter 2. Its form is

$$\overline{n}^2 = 1 + G_1 \frac{\lambda^2}{\lambda^2 - \lambda_1^2} + G_2 \frac{\lambda^2}{\lambda^2 - \lambda_2^2} + G_3 \frac{\lambda^2}{\lambda^2 - \lambda_3^2} \tag{5.71}$$

Here G_1, G_2, G_3 and $\lambda_1, \lambda_2, \lambda_3$ are constants (called Sellmeier coefficients) which are determined by fitting the above expression to the experimental data. Often in literature more elaborate expressions are used. Using data provided in Table 2.2, we plotted in Fig. 5.11a refractive index for SiO_2 as a function of wavelength. MATLAB code is provided in Appendix 5C.3.

5.3.3 Waveguide dispersion

Waveguide dispersion exists because group velocity v_g depends on the wavelength. The expression for waveguide dispersion τ_{wg} is expressed in terms of quantity b introduced earlier, so we start with the definition of parameter b

$$b = \frac{\beta^2/k^2 - \overline{n}_2^2}{\overline{n}_1^2 - \overline{n}_2^2} \approx \frac{\beta/k - \overline{n}_2}{\overline{n}_1 - \overline{n}_2}$$

where we used the fact that $\beta/k \approx \overline{n}_1 \approx \overline{n}_2$. We use the above equation to determine β as

$$\beta = k\overline{n}_2(1 + b\Delta) \tag{5.72}$$

Here $\Delta = \frac{\overline{n}_1 - \overline{n}_2}{\overline{n}_2}$. Waveguide dispersion is obtained by evaluatind derivative $d\beta/dk$ using Eq. (5.72) and assuming that $\overline{n}_2$ is k independent. One finds

$$\frac{d\beta}{dk} = \overline{n}_2 + \overline{n}_2 \Delta \frac{d(bV)}{dk}$$

Substitute the above result into general definition (5.69) and have

$$\tau_{wg} = \frac{L}{c}\left(\bar{n}_2 + \bar{n}_2\Delta\frac{d(Vb)}{dV}\right) \quad (5.73)$$

5.4 Pulse dispersion during propagation

Because τ_g depends on λ, each spectral component of any particular mode travels the same distance in different time. As a result, optical pulse spreads out with time. Let $\Delta\lambda$ be a spectral width of a source. Each wavelength component within $\Delta\lambda$ will propagate with different group velocity, resulting in a temporal broadening of a pulse. Pulse broadening $\Delta\tau$ is expressed as

$$\Delta\tau = \frac{d\tau}{d\lambda}\Delta\lambda \quad (5.74)$$

Assuming only material dispersion, i.e. $\tau = \tau_{mat}$ and using previous result (5.70) we have

$$\frac{d\tau_{mat}}{d\lambda} = \frac{L}{c}\left\{\frac{d\bar{n}}{d\lambda} - \frac{d\bar{n}}{d\lambda} - \lambda\frac{d^2\bar{n}}{d\lambda^2}\right\} = -\frac{L}{c}\lambda\frac{d^2\bar{n}}{d\lambda^2}$$

Therefore $\Delta\tau_{mat}$ can be written as

$$\Delta\tau_{mat} = -\frac{L}{c}\left(\lambda^2\frac{d^2\bar{n}}{d\lambda^2}\right)\left(\frac{\Delta\lambda}{\lambda}\right) \quad (5.75)$$

From the above, one can see that material dispersion can be specified in units of picoseconds per kilometre (length of fibre) per nanometre (spectral width of the source). One usually introduces

$$D_{mat} = \frac{\Delta\tau_{mat}}{L\Delta\lambda} = \frac{1}{L}\frac{d\tau_{mat}}{d\lambda} =$$
$$= -\frac{1}{c}\lambda\frac{d^2\bar{n}}{d\lambda^2} \quad (5.76)$$

which is known as material dispersion coefficient. In practical units [14]

$$D_{mat} = -\frac{1}{\lambda\cdot c}\left(\lambda^2\frac{d^2\bar{n}}{d\lambda^2}\right)\times 10^9 \frac{ps}{km\cdot nm} \quad (5.77)$$

where λ is in nm and velocity of light $c = 3\times 10^5$ km s^{-1}. Typical behaviour for $SiO_2-13.5\%$ GeO_2 is shown in Fig. 5.12. MATLAB code is provided in Appendix 5C.4.

For waveguide dispersion, we can also determine waveguide dispersion coefficient D_{wg} which is defined as

$$D_{wg} = \frac{1}{L}\frac{\Delta\tau_{wg}}{\Delta\lambda}$$

Here $\Delta\tau_{wg}$ is the pulse broadening due to waveguide. It can be expressed as

$$\Delta\tau_{wg} = \frac{d\tau_{wg}}{d\lambda}\Delta\lambda$$

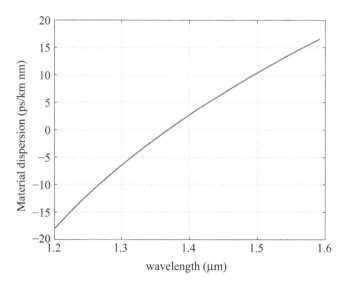

Fig. 5.12 Material dispersion as a function of optical wavelength.

Using Eq. (5.73), one can evaluate $\frac{d\tau_{wg}}{d\lambda}$. It is, however, typical to express it in terms of parameter V, which is $V = \frac{2\pi}{\lambda} a \bar{n}_2 \sqrt{2\Delta} \equiv \frac{A}{\lambda}$. Taking derivative

$$\frac{dV}{d\lambda} = -\frac{A}{\lambda^2}$$

allows us to express

$$\frac{d}{d\lambda} = -\frac{A}{\lambda^2} \frac{d}{dV}$$

Using the above, we have

$$\frac{d\tau_{wg}}{d\lambda} = -\frac{A}{\lambda^2} \frac{d\tau_{wg}}{dV} = -\frac{V}{\lambda} \frac{L}{c} \bar{n}_2 \Delta \frac{d^2(Vb)}{dV^2}$$

Finally

$$D_{wg} = -\frac{\bar{n}_2 \Delta}{c \cdot \lambda} V \frac{d^2(Vb)}{dV^2} \qquad (5.78)$$

Combining the above results for material and waveguide dispersion and using the Sellmeier equation with parameters for silica, one can determine total dispersion of fibre glass. The work is left as a project.

5.5 Problems

1. Based on ray theory, estimate the number of modes M in a multimode fibre when M is large.
2. Determine the cutoff wavelength for single-mode operation of an optical fibre. Assume the following parameters: $a = 5$ μm, $\bar{n}_2 = 1.450$, $\Delta = 0.002$.
3. Determine the value of waveguide dispersion for the above parameters.

4. A fibre of 10 km length has an attenuation coefficient of 0.6 dB km^{-1} at 1.3 μm and 0.3 dB km^{-1} at 1.55 μm. Assuming that 10 mW of optical power is launched into the fibre at each wavelength, determine the output power at each wavelength.

5.6 Projects

1. Write MATLAB code to analyse the combined effects of material and waveguide dispersion.
2. Using our results developed so far, establish methodologies of designing a single-mode fibre with predefined properties.
3. Based on the literature, analyse polarization-maintaining fibres.

Appendix 5A: Some properties of Bessel functions

Here, we will provide several properties of Bessel functions needed in the main text. For more information on Bessel functions consult books by Arfken [7] and Okoshi [8]. Some general formulas are

$$J_{m+1}(u) + J_{m-1}(u) = 2\frac{m}{u}J_m(u) \tag{5.79}$$

$$K_{m+1}(w) + K_{m-1}(w) = 2\frac{m}{u}K_m(w) \tag{5.80}$$

$$2J'_m = J_{m-1} - J_{m+1} \tag{5.81}$$

$$-2K'_m = K_{m-1} + K_{m+1} \tag{5.82}$$

also

$$J_{-m} = (-1)^m J_m, \quad K_{-m} = K_m \tag{5.83}$$

Start with (5.81) and replace J_{m+1} using (5.79). One finds

$$\frac{J'_m(u)}{uJ_m(u)} = \frac{J_{m-1}(u)}{uJ_m(u)} - \frac{m}{u^2} \tag{5.84}$$

Similarly, replacing J_{m-1} in Eq. (5.79) one finds

$$\frac{J'_m(u)}{uJ_m(u)} = \frac{m}{u^2} - \frac{J_{m+1}(u)}{uJ_m(u)} \tag{5.85}$$

Similar expressions can be derived for functions K_m. Using (5.82) and eliminating K_{m+1}, one has

$$\frac{K'_m(u)}{wK_m(u)} = -\frac{K_{m-1}(u)}{wK_m(u)} - \frac{m}{w^2} \tag{5.86}$$

Finally, eliminating K_{m-1} in (5.82) one has

$$\frac{K'_m(u)}{wK_m(u)} = -\frac{K_{m+1}(u)}{wK_m(u)} + \frac{m}{w^2} \tag{5.87}$$

Appendix 5B: Characteristic determinant

In this appendix we provide details of the evaluation of characteristic determinant. The determinant which describes guided modes has been obtained before and it is given by Eq. (5.47):

$$\begin{vmatrix} J_m(\kappa a) & 0 & -K_m(\gamma a) & 0 \\ \dfrac{\beta m}{\kappa^2 a} J_m(\kappa a) & i\dfrac{\omega \mu}{\kappa} J_m'(\kappa a) & \dfrac{\beta m}{\kappa^2 a} K_m(\gamma a) & i\dfrac{\omega \mu}{\gamma} K_m'(\gamma a) \\ 0 & J_m(\kappa a) & 0 & -K_m(\gamma a) \\ -i\dfrac{\omega \varepsilon_1}{\kappa} J_m'(\kappa a) & \dfrac{\beta m}{a\kappa^2} J_m(\kappa a) & -i\dfrac{\omega \varepsilon_2}{\gamma} K_m'(\gamma a) & \dfrac{\beta m}{a\gamma^2} K_m(\gamma a) \end{vmatrix} = 0$$

We will now evaluate it following the method of Cherin [15]. Introduce notation

$$A = J_m(\kappa a), \quad B = \frac{\beta m}{\kappa^2 a}, \quad C = i\frac{\omega}{\kappa} J_m'(\kappa a)$$

$$D = K_m(\gamma a), \quad E = \frac{\beta m}{a\gamma^2}, \quad F = i\frac{\omega}{\gamma} K_m'(\gamma a)$$

The determinant can be written as

$$\begin{vmatrix} A & 0 & -D & 0 \\ 0 & A & 0 & -D \\ AB & \mu C & DE & \mu F \\ -\varepsilon_1 C & AB & -\varepsilon_2 F & DE \end{vmatrix} = 0$$

Expand the determinant. In the first step we have

$$A \begin{vmatrix} A & 0 & -D \\ \mu C & DE & \mu F \\ AB & -\varepsilon_2 F & DE \end{vmatrix} - D \begin{vmatrix} 0 & A & -D \\ AB & \mu C & \mu F \\ -\varepsilon_1 C & AB & DE \end{vmatrix} = 0$$

and in the second step

$$A^2 \begin{vmatrix} DE & \mu F \\ -\varepsilon_2 F & DE \end{vmatrix} - AD \begin{vmatrix} \mu C & DE \\ AB & -\varepsilon_2 F \end{vmatrix} + AD \begin{vmatrix} AB & \mu F \\ -\varepsilon_1 C & DE \end{vmatrix} + D^2 \begin{vmatrix} AB & \mu C \\ -\varepsilon_1 C & AB \end{vmatrix} = 0$$

Evaluating the above, one has

$$A^2 \left(D^2 E^2 + \mu \varepsilon_2 F^2 \right) - AD \left(-\mu \varepsilon_2 CF - ABED \right) + AD \left(ABDE + \mu \varepsilon_1 CF \right) + D^2 \left(A^2 B^2 + \mu \varepsilon_1 C^2 \right) = 0$$

Rearranging terms

$$\mu \varepsilon_2 \left(A^2 F^2 + ADCF \right) + \mu \varepsilon_1 CD \left(AF + CD \right) + 2A^2 D^2 BE + A^2 D^2 \left(E^2 + B^2 \right) = 0$$

and finally

$$\left(\frac{F}{D} + \frac{C}{A} \right) \left(\mu \varepsilon_2 \frac{F}{D} + \mu \varepsilon_1 \frac{C}{A} \right) + (E + B)^2 = 0$$

Table 5.2 List of MATLAB functions for Chapter 5.

Listing	Function name	Description
5C.1.1	bessel_J.m	Plots Bessel function J_m
5C.1.2	bessel_Y.m	Plots Bessel function Y_m
5C.1.3	bessel_K.m	Plots Bessel function K_m
5C.1.4	bessel_I.m	Plots Bessel function I_m
5C.2.1	HE11.m	Driver function for universal relations
5C.2.2	func_HE11.m	Function used by $driver_H E11$
5C.3	sellmeier.m	Plots refractive index using Sellmeier equation
5C.4	disp_mat.m	Determines and plots material dispersion

Substitute original definitions and extracting common factors

$$\left[\frac{J'_m(\kappa a)}{\kappa J_m(\kappa a)} + \frac{K'_m(\gamma a)}{\gamma K_m(\gamma a)}\right] \left[\frac{k_0^2 \bar{n}_1^2 J'_m(\kappa a)}{\kappa J_m(\kappa a)} + \frac{k_0^2 \bar{n}_2^2 K'_m(\gamma a)}{\gamma K_m(\gamma a)}\right] = \frac{\beta^2 m^2}{a^2}\left(\frac{1}{\kappa^2} + \frac{1}{\gamma^2}\right)$$

(5.88)

Appendix 5C: MATLAB listings

In Table 5.2 we provide a list of MATLAB files created for Chapter 5 and a short description of each function.

Listing 5C.1.1   Program bessel_J.m. Plots Bessel function J_m.

```
% File name: bessel_J.m
% Plot of Bessel functions of the first kind J_m(z)
clear all
N_max = 101;                    % number of points for plot
z = linspace(0,10,N_max);       % creation of z arguments
%
hold on
for m = [0 1 2]                 % m - order of Bessel function
    J = BESSELJ(m,z);
    h = plot(z,J);
% Redefine figure properties
xlabel('z','FontSize',22);
ylabel('J_m(z)','FontSize',22);
text(1.1, 0.85, 'm = 0','Fontsize',18);
text(1.8, 0.65, 'm = 1','Fontsize',18);
text(3,   0.55, 'm = 2','Fontsize',18);
```

```
grid on
box on
%
set(h,'LineWidth',1.5);            % new thickness of plotting lines
set(gca,'FontSize',22);            % new size of tick marks on both axes
end
pause
close all
```

Listing 5C.1.2   Program bessel_Y.m. Plots Bessel function Y_m.

```
% File name: bessel_Y.m
% Plot of Bessel functions of the second kind Y_m(z)
clear all
N_max = 101;                       % number of points for plot
z = linspace(0.01,10,N_max);       % creation of z arguments
%
hold on
for m = [0 1 2]                    % m - order of Bessel function
    Y = BESSELY(m,z);
    h = plot(z,Y);
% Redefine figure properties
xlabel('z','FontSize',22);
ylabel('Y_m(z)','FontSize',22);
text(2.0, 0.7, 'm = 0','Fontsize',18);
text(3.5, 0.7, 'm = 1','Fontsize',18);
text(5.0, 0.7, 'm = 2','Fontsize',18);
axis([0 10 -5 1.5]);
grid on
box on
%
set(h,'LineWidth',1.5);            % new thickness of plotting lines
set(gca,'FontSize',22);            % new size of tick marks on both axes
end
pause
close all
```

Listing 5C.1.3   Program bessel_K.m. Plots Bessel function K_m.

```
% File name: bessel_K.m
% Plot of modified Bessel functions of the second kind K_m(z)
clear all
N_max = 101;                       % number of points for plot
z = linspace(0.01,3,N_max);        % creation of z arguments
```

```
%
hold on
for m = [0 1 2]                    % m - order of Bessel function
    K = BESSELK(m,z);
    h = plot(z,K);
% Redefine figure properties
xlabel('z','FontSize',22);
ylabel('K_m(z)','FontSize',22);
text(0.2, 0.8, 'm = 0','Fontsize',18);
text(1.0, 0.8, 'm = 1','Fontsize',18);
text(1.5, 0.8, 'm = 2','Fontsize',18);
axis([0 3 0 5]);
grid on
box on
%
set(h,'LineWidth',1.5);            % new thickness of plotting lines
set(gca,'FontSize',22);            % new size of tick marks on both axes
end
pause
close all
```

Listing 5C.1.4 Program bessel_I.m. Plots Bessel function I_m.

```
% File name: bessel_I.m
% Plot of modified Bessel functions of the first kind I_m(z)
clear all
N_max = 101;                       % number of points for plot
z = linspace(0,3,N_max);           % creation of z arguments
%
hold on
for m = [0 1 2]                    % m - order of Bessel function
    Y = BESSELI(m,z);
    h = plot(z,Y);
% Redefine figure properties
xlabel('z','FontSize',22);
ylabel('I_m(z)','FontSize',22);
text(1.1,  1.7, 'm = 0','Fontsize',18);
text(1.75, 1.7, 'm = 1','Fontsize',18);
text(2.4,  1.7, 'm = 2','Fontsize',18);
grid on
box on
%
set(h,'LineWidth',1.5);            % new thickness of plotting lines
set(gca,'FontSize',22);            % new size of tick marks on both axes
```

```
end
pause
close all
```

Listing 5C.2.1 Program HE11.m. MATLAB code for finding a universal relation. The program determines and plots the following relations: $b = b(V)$, $\frac{d(bV)}{dV}$, $V\frac{d^2(bV)}{dV^2}$.

```
% File name: HE11.m
% Function determines and plots universal relation for mode HE_11
clear all
format long
% It works in the range of V = [0.50 2.4]
N_max = 190;
for n = 1:N_max
    VV(n) = 0.50 + n*0.01;
    V = VV(n);
    b_c(n) = fzero(@(b) func_HE11(b,V),[0.0000001 0.8]);
end
%
hold on
h1 = plot(VV,b_c);              % plots b =b(V)
%
temp1 = VV.*b_c;
dy1 = diff(temp1)./diff(VV);
Vnew1 = VV(1:length(VV)-1);
h2 = plot(Vnew1,dy1);           % plots first derivative
%
temp2 = dy1;
dy2 = diff(temp2)./diff(Vnew1);
Vnew2 = Vnew1(1:length(Vnew1)-1);
h3 = plot(Vnew2,Vnew2.*dy2);    % plots second derivative
%
text(2.3, 0.6, 'b','Fontsize',16);
text(1.8, 1.2, '{d(bV)}/{dV}','Fontsize',16);
text(0.25, 1.3, 'V{d^{2}(bV)}/{dV^{2}}','Fontsize',16);
axis([0 2.5 0 1.5]);
% Redefine figure properties
xlabel('V','FontSize',22);
grid on
box on
set(h1,'LineWidth',1.5);        % new thickness of plotting lines
set(h2,'LineWidth',1.5);
set(h3,'LineWidth',1.5);
set(gca,'FontSize',22);         % new size of tick marks on both axes
```

```
pause
close all
```

Listing 5C.2.2 Program func_HE11.m. MATLAB function for finding a universal relation.

```
function result = func_HE11(b,V)
% Function name: func_HE11.m
% Function which defines basic relation
u= V*sqrt(1-b);
J1 = BESSELJ(1,u);
J0 = BESSELJ(0,u);
%
w = V*sqrt(b);
K1 = BESSELK(1,w);
K0 = BESSELK(0,w);

lhs = J1./J0;
rhs = (w./u).*(K1./K0);
result = lhs - rhs;
```

Listing 5C.3 Program sellmeier.m. Plots refractive index for SiO_2 using the Sellmeier equation.

```
% File name: sellmeier.m
% Plot of refractive index based on Sellmeier equation
% for Si O_2
clear all
N_max = 101;                         % number of points for plot
lambda = linspace(0.5,1.8,N_max);    % creation of lambda arguments
                                     % between 0.5 and 1.8 microns
%
% Data for SiO_2
G_1 = 0.696749; G_2 = 0.408218; G_3 = 0.890815;
lambda_1=0.0690606; lambda_2=0.115662; lambda_3=9.900559; % in microns
%
term1 = (G_1*lambda.^2)./(lambda.^2 - lambda_1^2);
term2 = (G_2*lambda.^2)./(lambda.^2 - lambda_2^2);
term3 = (G_3*lambda.^2)./(lambda.^2 - lambda_3^2);
ref_index_sq = 1.0 + term1 + term2 + term3;
ref_index = sqrt(ref_index_sq);
%
h = plot(lambda,ref_index,'LineWidth',1.5);
% Redefine figure properties
```

```
xlabel('wavelength (\mum)','FontSize',14);
ylabel('refractive index','FontSize',14);
set(gca,'FontSize',14);              % size of tick marks on both axes
pause
close all
```

Listing 5C.4 Program disp_mat.m. Determines coefficient D_{mat} material dispersion. Refractive index is determined using the Sellmeier equation.

```
% File name: disp_mat.m
% Plots D_mat
% Material dispersion is determined using Sellmeier equation
% describing refractive index for pure and doped silica
% Data from Kasap2001, p.45 for SiO_2 - GeO_2
clear all
c_light = 3d5;                       % velocity of light, km/s
disp_min = -20; disp_max =20;
N_max = 101;                         % number of points for plot
lambda_min = 1.200; lambda_max = 1.600;  % in microns
% creation of lambda arguments between 1.2 and 1.6 microns
lambda = linspace(lambda_min,lambda_max,N_max);
G_1 = 0.711040; G_2 = 0.451885; G_3 = 0.704048;
lambda_1 = 0.0642700; lambda_2 = 0.129408; lambda_3 = 9.425478;
%
term1 = (G_1*lambda.^2)./(lambda.^2 - lambda_1^2);
term2 = (G_2*lambda.^2)./(lambda.^2 - lambda_2^2);
term3 = (G_3*lambda.^2)./(lambda.^2 - lambda_3^2);
%
ref_index_sq = 1.0 + term1 + term2 + term3;
ref_index = sqrt(ref_index_sq);
%
%---------- Determination of material dispersion D_mat -------------
ttt1 = ref_index;
dy1_lam = diff(ttt1)./diff(lambda);  % first derivative
lambda1 = lambda(1:length(lambda)-1);
ttt2 = dy1_lam;
dy2_lam = diff(ttt2)./diff(lambda1);
lambda2 = lambda1(1:length(lambda1)-1);
D_mat = (-lambda2.*dy2_lam/c_light)*1d9;
plot(lambda2,D_mat,'LineWidth',1.5);
xlabel('wavelength (\mum)','FontSize',14);
ylabel('Material dispersion (ps/km nm)','FontSize',14);
set(gca,'FontSize',14);              % size of tick marks on both axes
axis([lambda_min, lambda_max, disp_min, disp_max])
```

```
grid on
pause
close all
```

References

[1] G. Keiser. *Optical Fiber Communications. Third Edition.* McGraw-Hill, Boston, 2000.

[2] F. L. Pedrotti and L. S. Pedrotti. *Introduction to Optics. Third Edition.* Prentice Hall, Upper Saddle River, NJ, 2007.

[3] T. Miya, Y. Terunuma, T. Hosaka, and T. Miyashita. Ultimate low-loss single-mode fibre at 1.55 μm. *Electronics Letters*, **15**:106–8, 1979.

[4] J. A. Buck. *Fundamentals of Optical Fibers. Second Edition.* Wiley-Interscience, Hoboken, NJ, 2004.

[5] J. C. Palais. *Fiber Optic Communications. Fifth Edition.* Prentice Hall, Upper Saddle River, NJ, 2005.

[6] D. K. Cheng. *Fundamentals of Engineering Electromagnetics.* Addison-Wesley, Reading, MA, 1993.

[7] G. Arfken. *Mathematical Methods for Physicists.* Academic Press, Orlando, FL, 1985.

[8] T. Okoshi. *Optical Fibers.* Academic Press, New York, 1982.

[9] W. van Etten and J. van der Plaats. *Fundamentals of Optical Fiber Communications.* Prentice Hall, New York, 1991.

[10] C. R. Pollock and M. Lipson. *Integrated Photonics.* Kluwer Academic Publishers, Boston, 2003.

[11] A. W. Snyder and J. D. Love. *Optical Waveguide Theory.* Kluwer Academic Publishers, 1983.

[12] D. Marcuse. *Theory of Dielectric Optical Waveguides. Second Edition.* Academic Press, New York, 1991.

[13] D. Gloge. Weakly guiding fibers. *Appl. Opt.*, **10**:2252–8, 1971.

[14] G. Ghatak and K. Thyagarajan. *Introduction to Fiber Optics.* Cambridge University Press, Cambridge, 1998.

[15] A. H. Cherin. *An Introduction to Optical Fibers.* McGraw-Hill, New York, 1983.

6 Propagation of linear pulses

In this chapter we outline basic issues associated with the description of linear pulses. Basic principles of two fundamental modulation formats, namely return-to-zero (RZ) and non-return-to-zero (NRZ) modulations, will also be explained.

Formats of data transmission are one of the critical factors in optical communication systems. We therefore have implemented several waveforms corresponding to basic modulation formats and various profiles of individual bits. Later on, some of this information will be used to analyse performance of communication systems. We start with a summary of the basic pulses.

6.1 Basic pulses

In this section we will review properties of the basic pulses. Basic approximations and properties will be described. We start with rectangular pulses.

6.1.1 Rectangular pulses

The simplest pulse, namely the rectangular one, is shown in Fig. 6.1. It is described as

$$s_{rect}(t) = \begin{Bmatrix} s_0, & 0 < t < T \\ 0, & \text{otherwise} \end{Bmatrix} \quad (6.1)$$

where T is the pulse duration and s_0 its magnitude. It can be represented as a Fourier series

$$s_{rect}(t) = \sum_{n=1}^{\infty} s_n \sin\left(\frac{n\pi}{T}t\right) \quad (6.2)$$

where

$$s_n = \begin{cases} 0, & n \text{ even} \\ \frac{4s_0}{\pi n}, & n \text{ odd} \end{cases} \quad (6.3)$$

In Fig. 6.1 we plotted three possible approximations to a rectangular pulse for $n = 1$, $n = 3$, $n = 5$ corresponding to different number of terms in Eq. (6.2). One can observe that with the increasing number of terms in Fourier expansion more accurate approximation to a rectangular pulse is possible. MATLAB code used to create Fig. 6.1 is shown in the Appendix, Listing 6A.1.

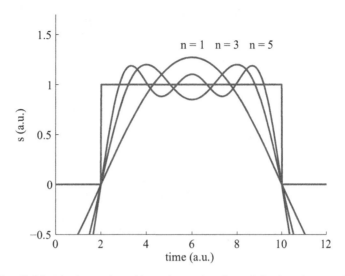

Fig. 6.1 Rectangular pulse of height 1 (a.u.) approximated by varying number of terms in Fourier series expansion.

In practice, one often considers a symmetrical rectangular pulse shown in Fig. 6.3. It is defined as

$$s_{rect,symm}(t) = \begin{cases} 1, & |t| < \tau \\ 0, & |t| > \tau \end{cases} \quad (6.4)$$

Its Fourier transform can be evaluated analytically as follows:

$$\begin{aligned} S_{rect,symm}(\omega) &= \int_{-\infty}^{+\infty} x(t)\, e^{-i\omega t}\, dt = \int_{-\tau}^{+\tau} e^{-i\omega t}\, dt = \frac{1}{-i\omega} e^{-i\omega t} \Big|_{-\tau}^{+\tau} \\ &= \frac{1}{-i\omega}\left(e^{-i\omega\tau} - e^{i\omega\tau}\right) = \frac{2\sin\omega\tau}{\omega} \end{aligned} \quad (6.5)$$

where we have used Euler identity. A plot of Eq. (6.5) is shown in Fig. 6.2.

To sum up the discussion of rectangular pulses, we have evaluated Fourier transform of a rectangular pulse using MATLAB built-in functions. A plot of a symmetric rectangular pulse and its MATLAB generated absolute value of Fourier transform are shown in Fig. 6.3. Plots were created by MATLAB code shown in Appendix, Listing 6A.2.

6.1.2 Gaussian pulses

A Gaussian pulse is considered an accurate model of the data waveforms generated in practical optical communication systems. The Gaussian profile is described as

$$s_G(t) = \frac{A}{\sigma\sqrt{2\pi}} e^{-\frac{1}{2}\left(\frac{t}{\sigma}\right)^2} \quad (6.6)$$

The pulse has an area $A = P_0 \tau_0$, where $\tau = \sigma\sqrt{2\pi}$ and P_0 is its maximum value (at $t = 0$). Its spectral function $S_G(\omega)$ is obtained by Fourier transform and it is also a Gaussian

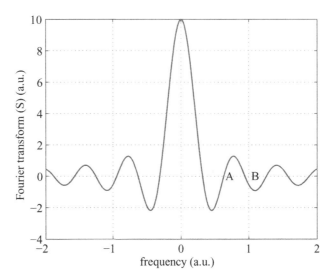

Fig. 6.2 Fourier transform of rectangular pulse of duration τ and area A. Point A corresponds to the frequency $\nu = 2/\tau$ and point B to frequency $\nu = 3/\tau$.

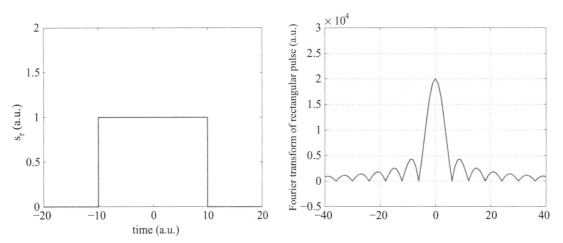

Fig. 6.3 Rectangular pulse (left) and the absolute value of its Fourier transform determined using MATLAB functions.

function, see Fig. 6.4:

$$S_G(\omega) = A e^{-\frac{1}{2}(\omega\sigma)^2} \tag{6.7}$$

6.1.3 Super-Gaussian pulse

A super-Gaussian pulse is a generalization of the usual Gaussian and is

$$s_{SG}(t) = \frac{A}{\sigma\sqrt{2\pi}} e^{-\frac{1}{2}\left(\frac{t}{\sigma}\right)^{2m}} \tag{6.8}$$

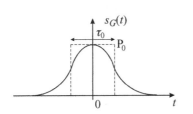
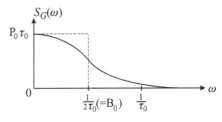

Fig. 6.4 Schematic illustration of Gaussian pulse in time and its Fourier transform.

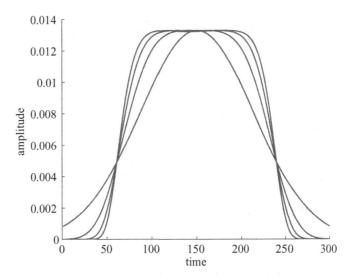

Fig. 6.5 Super-Gaussian pulses for several values of $m = 1, 2, 3, 4$.

The parameter m controls the degree of edge sharpness. For $m = 1$ one recovers the ordinary Gaussian pulse. For larger values of m the pulse becomes closer to a rectangular pulse, see Fig. 6.5. (Figure generated by MATLAB code, in Appendix, Listing 6A.3.)

6.1.4 Chirped Gaussian pulse

Chirp refers to a process which changes the frequency of a pulse with time. Chirp pulses are most commonly created during direct modulation of a semiconductor laser diode (semiconductor lasers are discussed in Chapter 7). Mathematical expression for a chirped Gaussian pulse at $z = 0$ is [1], [2]

$$\begin{aligned} s(z=0, t) &= \mathbf{Re}\left\{\exp\left[-\frac{1+iC}{2}\left(\frac{t}{T_0}\right)^2\right]\exp\left(-i\omega_0 t\right)\right\} \\ &= e^{-\frac{1}{2}\left(\frac{t}{T_0}\right)^2}\cos\left[\omega_0 t + \frac{1}{2}C\left(\frac{t}{T_0}\right)^2\right] \end{aligned} \qquad (6.9)$$

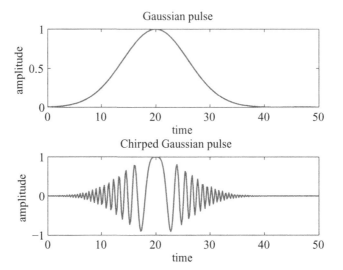

Fig. 6.6 Comparison of Gaussian pulse (upper plot) and chirped Gaussian pulse (lower plot).

where T_0 determines the width of the pulse, C is the chirp factor which determines the degree of chirp of the pulse and ω_0 is the carrier frequency of the pulse. $s(0, t)$ is the dimensionless function describing pulse.

Phase of such pulses is

$$\phi(\omega) = \omega_0 t + \frac{1}{2} C \left(\frac{t}{T_0} \right)^2 \qquad (6.10)$$

Change in frequency $\delta\omega(t)$ is therefore

$$\delta\omega(t) = -\frac{\partial \phi(\omega)}{\partial t} = \omega_0 + C \frac{t}{T_0^2} \qquad (6.11)$$

Fourier spectrum of a chirped pulse is broader compared to unchirped one. Evaluation of the Fourier transform of (6.9) gives

$$s(0, \omega) = \left(\frac{2\pi T_0^2}{1 + iC} \right)^{1/2} \exp \left[-\frac{\omega^2 T_0^2}{2(1 + iC)} \right] \qquad (6.12)$$

The spectral half-width is defined as

$$\frac{|s(0, \omega_e)|^2}{|s(0, 0)|^2} = e^{-1} \qquad (6.13)$$

For a symmetric pulse centered at $t = 0$, one has $\Delta\omega = \omega_e$. Detailed calculations for spectral half-width (point where the intensity drops $1/e$) give [3] (see Problem 2)

$$\Delta\omega = \left(1 + C^2 \right) T_0^{-1} \qquad (6.14)$$

Chirped Gaussian pulse is plotted in Fig. 6.6 (lower plot). For comparison, regular Gaussian pulse for the same parameters is also shown (upper plot). The plots were obtained using MATLAB code provided in Appendix, Listing 6A.4.

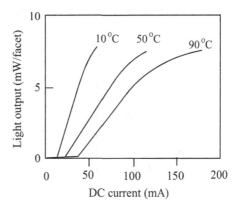

Fig. 6.7 Schematic illustration of light-current laser characteristics at various temperatures.

6.2 Modulation of a semiconductor laser

In practical applications optical pulses are created by semiconductor laser. Typical light-current (L-I) characteristics of the semiconductor laser are shown in Fig. 6.7. We show here an output of power-current characteristics (L-I) for several temperatures. One can observe a flat region to the left until the so-called threshold and then sudden increase in power. Threshold current I_{th} is the forward injection current at which optical gain in the laser cavity is equal to losses. Above the threshold, stimulated emission is the dominant process in converting current to light; below the threshold is the light emission which is caused by spontaneous processes with characteristics (speed, spectrum and efficiency) similar to those from light emitting diodes (LEDs).

In Fig. 6.8 we illustrate principles of direct modulation of a semiconductor laser. It is a simple and inexpensive process since no other components are required. The disadvantage of direct modulation is that the resulting pulses are considerably chirped.

6.2.1 Modulation formats

Various modulation formats are often discussed in literature. For detailed analysis consult books by Keiser [4], Liu [5], Binh [6], Ramaswami or Sivarajan [7]. Here, we discuss briefly two important cases only as illustrated in Figs. 6.9 and 6.10.

At a digital format light pulse intensity (say) larger than zero, represents logical ONE and its absence, logical ZERO. In the return-to-zero (RZ) format, signal drops to zero between the pulses, see Fig. 6.10. Each bit has allocated time T. In RZ, pulses occupy half of the time slot reserved for each bit, see Fig. 6.10. Also, power spectrum is shown. Since most of the signal power is at the lower frequency, the RZ signal is transmitted by a system having a bandwidth of $\frac{2}{T}$ Hz.

In the non-return-to-zero (NRZ) format, the pulses occupy entire time slot reserved for each bit. NRZ pulses are twice as long as the RZ pulses. In NRZ modulation format the

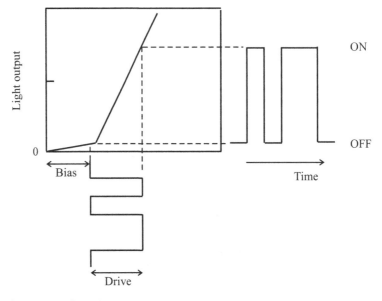

Fig. 6.8 Illustration of the principle of modulation in a semiconductor laser.

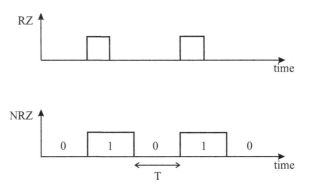

Fig. 6.9 Comparison of RZ and NRZ modulation formats. Digital bit stream is: 01010.

intensity of light remains at its high value whenever a number of consecutive ONEs are transmitted. The bandwidth is $1/T$, only half of that required for RZ system.

For encoding, practical current pulse driving the semiconductor laser is assumed to be of the form [8]

$$I(t) = \begin{cases} 0 & t < 0 \\ I_m \left[1 - \exp\left(-t/t_r\right)\right] & 0 \leq t \leq T' \\ I_m \exp\left(-(t - T)/t_r\right) & T' < t \end{cases} \qquad (6.15)$$

where I_m is the peak modulation, t_r determines the pulse rise time, and $T' = T$ for NRZ encoding and $T' = T/2$ for RZ coding. The bias current I_{bias} is typically 1.1 times the laser threshold current.

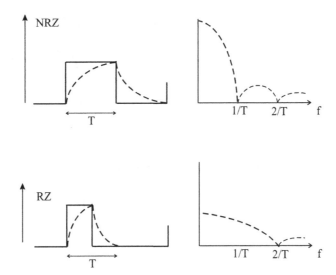

Fig. 6.10 Comparison of modulation formats (RZ and NRZ) (schematically) and their corresponding bandwidths.

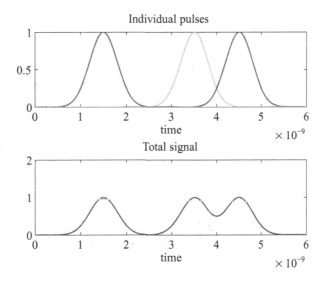

Fig. 6.11 Train of Gaussian pulses for logical combination of bits 010110. Individual pulses are shown (top) and also complete signal (bottom).

6.2.2 Creation of waveforms

In practice, to evaluate photonic systems for some modulation formats one needs to generate waveforms consisting of pulses of various shapes.

The generated sequence of Gaussian pulses (waveform) for some particular logical combination is illustrated in Fig. 6.11. MATLAB code is provided in Appendix, Listings 6A.5 and 6A.5.1.

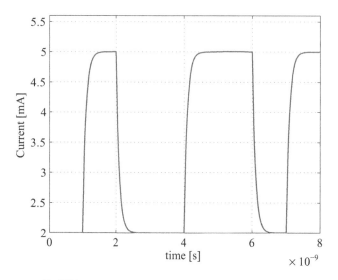

Fig. 6.12 Eight-bit pattern generated in NRZ format.

Another wavefront which is used to represent driving current of semiconductor lasers in NRZ (non-return-to-zero) format can be expressed as [9]

$$J_{mod}(t) = \begin{cases} I_{bias} + I_m \left(1 - e^{-2.2t/\tau_r}\right) & \text{current bit 1, previous 0} \\ I_{bias} + I_m e^{-2.2t/\tau_r} & \text{current bit 0, previous 1} \\ I_{bias} & \text{current bit 0, previous 0} \\ I_{bias} + I_m & \text{current bit 1, previous 1} \end{cases} \quad (6.16)$$

Implementation of the above modulation format is shown in Appendix, Listing 6A.6 and a typical result in Fig. 6.12.

6.3 Simple derivation of the pulse propagation equation in the presence of dispersion

Assume the existence of a propagating electric field $\mathbf{E}(\mathbf{r}, t)$ at point $\mathbf{r}$ and time t. It can be Fourier decomposed as

$$\mathbf{E}(\mathbf{r}, t) = \frac{1}{2\pi} \int_{-\infty}^{\infty} \mathbf{E}(\mathbf{r}, \omega) e^{-j\omega t} d\omega \quad (6.17)$$

Therefore electric field can be interpreted as an infinite sum of monochromatic waves, each having an angular frequency ω. The inverse Fourier transform is

$$\mathbf{E}(\mathbf{r}, \omega) = \int_{-\infty}^{\infty} \mathbf{E}(\mathbf{r}, t) e^{j\omega t} dt \quad (6.18)$$

We further assume that Fourier component $\mathbf{E}(\mathbf{r}, \omega)$ can be expressed as

$$\mathbf{E}(\mathbf{r}, \omega) = \widehat{x} B(0, \omega) F(x, y) e^{j\beta z} \tag{6.19}$$

where unit vector $\widehat{x}$ represents the state of polarization of propagating field, $B(0, \omega)$ is the initial amplitude of the field propagating along z-direction, $F(x, y)$ describes the transverse distribution of the field and β is the propagation constant. We made a similar assumption earlier when we discussed propagation of the EM field in planar waveguides.

For the field propagating in an optical fibre, the transverse distribution $F(x, y)$ is often assumed to be Gaussian, namely

$$F(x, y) = \exp\left(-\pi \frac{x^2 + y^2}{A_{eff}}\right) \tag{6.20}$$

where A_{eff} is the effective area of the fibre.

The effect of dispersion results in the frequency dependence of the propagation constant β; i.e. $\beta = \beta(\omega)$. As a result, each frequency component in Eq. (6.17) will propagate with a different propagation constant.

To derive propagation equation for each frequency component, write Eq. (6.19) as

$$\mathbf{E}(\mathbf{r}, \omega) = \mathbf{E}(x, y, z = 0, \omega) e^{j\beta z} \tag{6.21}$$

From above, by direct differentiation with respect to z, one finds

$$\frac{\partial \mathbf{E}(\mathbf{r}, \omega)}{\partial z} = j\beta(\omega) \mathbf{E}(\mathbf{r}, \omega) \tag{6.22}$$

The above equation describes propagation of EM field with the dispersion induced, broadening included. More rigorous derivation of the above equation, starting from Maxwell's equations, will be performed in the next section.

Propagation constant $\beta(\omega)$ in terms of the frequency dependent mode index $n(\omega)$ is [3]

$$\beta(\omega) = n(\omega) \frac{\omega}{c} \tag{6.23}$$

It is customary to expand $\beta(\omega)$ around the centre frequency of the pulse ω_0 in a Taylor series

$$\beta(\omega) = \beta_0 + \beta_1 (\omega - \omega_0) + \frac{1}{2}\beta_2 (\omega - \omega_0)^2 + \frac{1}{6}\beta_3 (\omega - \omega_0)^3 \tag{6.24}$$

where $\beta_i = \left.\frac{\partial^i \beta}{\partial \omega^i}\right|_{\omega = \omega_0}$. Expansion up to the third order is sufficient to describe typical effects existing in optical fibre. More practical expressions for expansion coefficients are [3]

$$\beta_1 = \frac{1}{c}\left[n(\omega) + \omega \frac{dn(\omega)}{d\omega}\right] = \frac{n_g}{c} = \frac{1}{v_g} \tag{6.25}$$

where n_g is the group index and v_g group velocity. For coefficient β_2 one finds

$$\beta_2 = \frac{\partial^2 \beta(\omega)}{\partial \omega^2} = \frac{\partial \beta_1}{\partial \omega} = \frac{1}{c}\left[2\frac{dn(\omega)}{d\omega} + \omega \frac{d^2 n(\omega)}{d\omega^2}\right] \tag{6.26}$$

Coefficient β_2 is related to dispersion parameter D as

$$D = \frac{1}{L}\frac{d\tau}{d\lambda} = -\frac{2\pi c}{\lambda^2}\beta_2 \qquad (6.27)$$

Here τ is the group delay through a fibre of length L, and λ is the wavelength. Units of parameter D are $[\frac{ps}{nm \cdot km}]$.

To account for parameter β_3 in practice, the so-called dispersion slope S is introduced:

$$S = \frac{dD}{d\lambda} \qquad (6.28)$$

Parameter S describe the variation of D with the wavelength. Units of it are $[\frac{ps}{nm^2 \cdot km}]$.

6.4 Mathematical theory of linear pulses

Here, we will outline a full derivation of the equation describing linear pulses based on Maxwell equations. This more formal derivation illustrates needed approximations.

Maxwell's equations were introduced in Chapter 3; for optical fibre where there is no flow of current and no free charges these are

$$\nabla \times \mathbf{E} = -\frac{\partial \mathbf{B}}{\partial t}, \quad \nabla \times \mathbf{H} = \frac{\partial \mathbf{D}}{\partial t} \qquad (6.29)$$

and

$$\nabla \cdot \mathbf{D} = 0, \quad \nabla \cdot \mathbf{B} = 0 \qquad (6.30)$$

Constitutive relations are expressed as

$$\mathbf{D} = \varepsilon_0 \mathbf{E} + \mathbf{P}, \quad \mathbf{B} = \mu_0 \mathbf{H} + \mathbf{M} \qquad (6.31)$$

Here $\mathbf{P}$ is polarization which is divided into linear and nonlinear parts:

$$\mathbf{P}(\mathbf{r}, t) = \mathbf{P}_L(\mathbf{r}, t) + \mathbf{P}_{NL}(\mathbf{r}, t) \qquad (6.32)$$

For a nonmagnetic medium such as fibre, the magnetization $\mathbf{M} = 0$. In this chapter we discuss only linear effects.

Linear polarization is related to electric field as

$$\mathbf{P}_L(\mathbf{r}, t) = \varepsilon_0 \int_{-\infty}^{\infty} \chi^{(1)}(t - t') \, \mathbf{E}(\mathbf{r}, t') \, dt' \qquad (6.33)$$

Fourier transform and its inverse are introduced as

$$\mathbf{E}(\mathbf{r}, \omega) = \int_{-\infty}^{\infty} \mathbf{E}(\mathbf{r}, t) \, e^{i\omega t} dt \qquad (6.34)$$

$$\mathbf{E}(\mathbf{r}, t) = \frac{1}{2\pi} \int_{-\infty}^{\infty} \mathbf{E}(\mathbf{r}, \omega) \, e^{i\omega t} \, d\omega \qquad (6.35)$$

Fourier transformation of linear polarization, Eq. (6.33) gives

$$\mathbf{P}_L(\mathbf{r}, \omega) = \int_{-\infty}^{\infty} \mathbf{P}_L(\mathbf{r}, t) \, e^{i\omega t} dt$$

$$= \varepsilon_0 \int_{-\infty}^{\infty} \int_{-\infty}^{\infty} \chi^{(1)}(t - t') \, \mathbf{E}(\mathbf{r}, t') \, e^{i\omega t} \, dt' dt$$

$$= \varepsilon_0 \int_{-\infty}^{\infty} \int_{-\infty}^{\infty} \chi^{(1)}(t'') \, \mathbf{E}(\mathbf{r}, t') \, e^{i\omega t''} e^{i\omega t'} \, dt' dt''$$

$$= \varepsilon_0 \int_{-\infty}^{\infty} dt' \mathbf{E}(\mathbf{r}, t') \, e^{i\omega t'} \int_{-\infty}^{\infty} dt'' \chi^{(1)}(t'') \, e^{i\omega t''}$$

$$= \varepsilon_0 \, \chi^{(1)}(\omega) \, \mathbf{E}(\mathbf{r}, \omega) \qquad (6.36)$$

where we have made change of variables $t - t' = t''$. $\chi^{(1)}(\omega)$ is the Fourier transform of the susceptibility $\chi^{(1)}(t)$

$$\chi^{(1)}(\omega) = \int_{-\infty}^{\infty} dt \chi^{(1)}(t) \, e^{i\omega t} \qquad (6.37)$$

To derive wave equation take $\nabla \times \ldots$ operation of the first Maxwell's equation and use the second Maxwell's equation and constitutive relations. In a few steps, one obtains

$$\nabla \times \nabla \times \mathbf{E}(\mathbf{r}, t) = -\frac{\partial}{\partial t} \nabla \times \mathbf{B}(\mathbf{r}, t)$$

$$= -\frac{\partial}{\partial t} \mu_0 \nabla \times \mathbf{H}(\mathbf{r}, t)$$

$$= -\mu_0 \frac{\partial^2 \mathbf{D}(\mathbf{r}, t)}{\partial t^2}$$

$$= -\mu_0 \varepsilon_0 \frac{\partial^2 \mathbf{E}(\mathbf{r}, t)}{\partial t^2} - \mu_0 \frac{\partial^2 \mathbf{P}(\mathbf{r}, t)}{\partial t^2}$$

In the linear regime

$$\nabla \times \nabla \times \mathbf{E}(\mathbf{r}, t) = -\mu_0 \varepsilon_0 \frac{\partial^2 \mathbf{E}(\mathbf{r}, t)}{\partial t^2} - \mu_0 \frac{\partial^2 \mathbf{P}_L(\mathbf{r}, t)}{\partial t^2}$$

Using mathematical identity

$$\nabla \times \nabla \times \mathbf{E} = \nabla (\nabla \cdot \mathbf{E}) - \nabla^2 \mathbf{E}$$

and third Maxwell's equation ($\nabla \cdot \mathbf{D} = \varepsilon \nabla \cdot \mathbf{E} = 0$), one finally obtains

$$\nabla^2 \mathbf{E}(\mathbf{r}, t) + \frac{1}{c^2} \frac{\partial^2 \mathbf{E}(\mathbf{r}, t)}{\partial t^2} + \mu_0 \frac{\partial^2 \mathbf{P}_L(\mathbf{r}, t)}{\partial t^2} = 0 \qquad (6.38)$$

The above is a linear equation in field $\mathbf{E}(\mathbf{r}, t)$. Taking Fourier transform gives

$$\nabla^2 \mathbf{E}(\mathbf{r}, \omega) + \frac{\omega^2}{c^2} \mathbf{E}(\mathbf{r}, \omega) + \mu_0 \omega^2 \mathbf{P}_L(\mathbf{r}, \omega) = 0 \qquad (6.39)$$

Finally, substituting the result in (6.36) for linear polarization gives the wave equation in the frequency domain

$$\nabla^2 \mathbf{E}(\mathbf{r}, \omega) + \varepsilon(\omega) \frac{\omega^2}{c^2} \mathbf{E}(\mathbf{r}, \omega) = 0 \qquad (6.40)$$

The above three-dimensional wave equation will be developed in one dimension next.

6.4.1 One-dimensional approach

A one-dimensional wave equation (in time domain) for a linear medium with zero dispersion is

$$\frac{\partial^2 E(z,t)}{\partial z^2} - \mu_0 \varepsilon_0 \bar{n}_0^2 \frac{\partial^2 E(z,t)}{\partial t^2} = 0 \tag{6.41}$$

where $\bar{n}_0$ is refractive index which in the absence of dispersion is frequency independent. It is related to propagation constant β_0 as

$$\beta_0 = \bar{n}_0 k_0 \tag{6.42}$$

The wave equation describes pulses with central frequency ω_0. The following relation holds:

$$k_0 = \omega_0 \sqrt{\mu_0 \varepsilon_0} \tag{6.43}$$

Solution of the wave equation which describes linear pulse is

$$E(z,t) = A(z,t) \, e^{i(\omega_0 t - \beta_0 z)} \tag{6.44}$$

where $A(z,t)$ is the pulse envelope. In the following we will work within slowly varying envelope approximation (SVEA), which is based on the following approximations:

$$\left| \frac{\partial^2 A(z,t)}{\partial z^2} \right| \ll \left| \beta_0 \frac{\partial A(z,t)}{\partial z} \right|$$

and

$$\left| \frac{\partial^2 A(z,t)}{\partial t^2} \right| \ll \left| \omega_0 \frac{\partial A(z,t)}{\partial t} \right|$$

One needs to evaluate derivatives appearing in Eq. (6.41) using solution given by Eq. (6.44) and then apply SVEA. The resulting linear wave equation describes pulses propagating without dispersion and it is

$$\frac{\partial A(z,t)}{\partial z} + \frac{\beta_0}{\omega_0} \frac{\partial A(z,t)}{\partial t} = 0 \tag{6.45}$$

In the presence of dispersion it is convenient to represent pulse in terms of Fourier components. Each such component travels with different velocity.

Taking Fourier transformation of the original pulse described by Eq. (6.44) gives

$$E(z,t) = \frac{1}{2\pi} \int_{-\infty}^{\infty} E(z,\omega) \, e^{i\omega t} d\omega \tag{6.46}$$

$E(z,\omega)$ describes propagation of spectral component with angular frequency ω inside the fibre as follows [1]:

$$E(z,\omega) = E(0,\omega) \, e^{-i\beta z} \tag{6.47}$$

Combining Eqs. (6.44), (6.34) and (6.47) gives

$$A(z,t) \, e^{i(\omega_0 t - \beta_0 z)} = \frac{1}{2\pi} \int_{-\infty}^{\infty} E(0,\omega) \, e^{-i\beta z} \, e^{i\omega t} d\omega$$

From the above, the expression for pulse envelope $A(z,t)$ is

$$A(z,t) = \frac{1}{2\pi}\int_{-\infty}^{\infty} E(0,\omega)\, e^{i(\omega-\omega_0)t}\, e^{-i(\beta-\beta_0)z} d\omega \qquad (6.48)$$

Next, let us introduce $\Delta\omega = \omega - \omega_0$ and assume $\Delta\omega \ll \omega_0$.

Pulse broadening results from frequency dependence of the propagation constant β; i.e. $\beta = \beta(\omega)$. Because of $\Delta\omega \ll \omega_0$, it is customary to expand β around frequency ω_0 as follows [the same equation as (6.24)]:

$$\beta(\omega) = \beta_0 + \beta_1 \Delta\omega + \frac{1}{2}\beta_2(\Delta\omega)^2 + \frac{1}{6}\beta_3(\Delta\omega)^3 + \ldots \qquad (6.49)$$

where higher order terms are neglected. The following notation has been introduced: $\beta_0 = \beta(\omega_0)$, $\beta_1 = \frac{d\beta}{d\omega}\big|_{\omega=\omega_0}$, $\beta_2 = \frac{d^2\beta}{d\omega^2}\big|_{\omega=\omega_0}$, $\beta_3 = \frac{d^3\beta}{d\omega^3}\big|_{\omega=\omega_0}$. Substituting the above expansion into Eq. (6.48) gives

$$A(z,t) = \frac{1}{2\pi}\int_{-\infty}^{\infty} \tilde{E}(0,\omega)\, e^{i\Delta\omega t}\, e^{-i\Delta\omega\beta_1 z}\, e^{-i(\Delta\omega)^2\frac{1}{2}\beta_2 z}\, e^{-i(\Delta\omega)^3\frac{1}{6}\beta_3 z} d\omega$$

The above expression will be used to evaluate derivatives $\frac{\partial A(z,t)}{\partial z}$ and $\frac{\partial A(z,t)}{\partial t}$. One finds

$$\frac{\partial A(z,t)}{\partial t} = \frac{1}{2\pi}\int_{-\infty}^{\infty} \tilde{E}(0,\omega)\, i\Delta\omega\, e^{i\Delta\omega t}\, e^{-i\Delta\omega\beta_1 z}\, e^{-i(\Delta\omega)^2\frac{1}{2}\beta_2 z}\, e^{-i(\Delta\omega)^3\frac{1}{6}\beta_3 z} d\omega \qquad (6.50)$$

and

$$\begin{aligned}\frac{\partial A(z,t)}{\partial z} &= \frac{1}{2\pi}\int_{-\infty}^{\infty} \tilde{E}(0,\omega)(-i\Delta\omega\beta_1)\, e^{i\Delta\omega t}\, e^{-i\Delta\omega\beta_1 z}\, e^{-i(\Delta\omega)^2\frac{1}{2}\beta_2 z}\, e^{-i(\Delta\omega)^3\frac{1}{6}\beta_3 z} d\omega \\ &+ \frac{1}{2\pi}\int_{-\infty}^{\infty} \tilde{E}(0,\omega)\left[-i\frac{1}{2}(\Delta\omega)^2\beta_2\right] e^{i\Delta\omega t}\, e^{-i\Delta\omega\beta_1 z}\, e^{-i(\Delta\omega)^2\frac{1}{2}\beta_2 z}\, e^{-i(\Delta\omega)^3\frac{1}{6}\beta_3 z} d\omega \\ &+ \frac{1}{2\pi}\int_{-\infty}^{\infty} \tilde{E}(0,\omega)\left[-i\frac{1}{6}(\Delta\omega)^3\beta_3\right] \\ &\times e^{i\Delta\omega t}\, e^{-i\Delta\omega\beta_1 z}\, e^{-i(\Delta\omega)^2\frac{1}{2}\beta_2 z}\, e^{-i(\Delta\omega)^3\frac{1}{6}\beta_3 z} d\omega \end{aligned} \qquad (6.51)$$

Terms on the right hand side of the above equation can be expressed in terms of the time derivatives as follows:

$$\frac{1}{2\pi}\int_{-\infty}^{\infty} \tilde{E}(0,\omega)(-i\Delta\omega\beta_1)\, e^{i\Delta\omega t}\, e^{-i\Delta\omega\beta_1 z}\, e^{-i(\Delta\omega)^2\frac{1}{2}\beta_2 z}\, e^{-i(\Delta\omega)^3\frac{1}{6}\beta_3 z} d\omega = -\beta_1 \frac{\partial A(z,t)}{\partial t}$$

$$\frac{1}{2\pi}\int_{-\infty}^{\infty} \tilde{E}(0,\omega)\left[-i\frac{1}{2}(\Delta\omega)^2\beta_2\right] e^{i\Delta\omega t}\, e^{-i\Delta\omega\beta_1 z}\, e^{-i(\Delta\omega)^2\frac{1}{2}\beta_2 z}\, e^{-i(\Delta\omega)^3\frac{1}{6}\beta_3 z} d\omega$$
$$= i\frac{1}{2}\beta_2 \frac{\partial^2 A(z,t)}{\partial t^2}$$

$$\frac{1}{2\pi}\int_{-\infty}^{\infty} \tilde{E}(0,\omega)\left[-i\frac{1}{6}(\Delta\omega)^3\beta_3\right] e^{i\Delta\omega t}\, e^{-i\Delta\omega\beta_1 z}\, e^{-i(\Delta\omega)^2\frac{1}{2}\beta_2 z}\, e^{-i(\Delta\omega)^3\frac{1}{6}\beta_3 z} d\omega$$
$$+ \frac{1}{6}\beta_3 \frac{\partial^3 A(z,t)}{\partial t^3}$$

Using the above replacements in Eq. (6.51) finally gives

$$\frac{\partial A(z,t)}{\partial z} = -\beta_1 \frac{\partial A(z,t)}{\partial t} - i\frac{1}{2}\beta_2 \frac{\partial^2 A(z,t)}{\partial t^2} + \frac{1}{6}\beta_3 \frac{\partial^3 A(z,t)}{\partial t^3} \qquad (6.52)$$

This is the basic equation which describes pulse evolution in the linear regime in the presence of dispersion. Its generalization to include the nonlinear effect (Kerr effect) which describes optical solitons will be discussed in Chapter 15.

In the absence of dispersion, i.e. when $\beta_2 = \beta_3 = 0$, the optical pulse propagates without changing its shape. In such case pulse envelope changes as $A(z,t) = A(0, t - \beta_1 z)$.

The general equation describing pulse propagation can be simplified by going to a reference frame moving with the pulse. The transformation to the new coordinates is

$$t' = t - \beta_1 z \quad \text{and} \quad z' = z \qquad (6.53)$$

In the reference frame moving with the pulse, Eq. (6.52) takes the form

$$\frac{\partial A(z',t')}{\partial z'} + \frac{i}{2}\beta_2 \frac{\partial^2 A(z',t')}{\partial t'^2} - \frac{1}{6}\beta_3 \frac{\partial^3 A(z',t')}{\partial t'^3} \qquad (6.54)$$

The above equation found many applications which have been discussed extensively, see [3], [1], [9].

6.5 Propagation of pulses

6.5.1 Analytical description of the propagation of a chirp Gaussian pulse

Pulse behaviour is described by the Eq. (6.54) derived previously. For $\beta_3 = 0$ the equation is

$$\frac{\partial A}{\partial z'} + \frac{i}{2}\beta_2 \frac{\partial^2 A}{\partial t'^2} = 0$$

and pulse evolution can be determined analytically. The solution is

$$A(z,t) = \frac{1}{2\pi} \int_{-\infty}^{+\infty} A(0,\omega) \, e^{\frac{i}{2}\beta_2 \omega^2 z - i\omega t} d\omega \qquad (6.55)$$

In the solution, the initial Gaussian pulse $A(0, \omega)$ is given by Eq. (6.12). Substituting Eq. (6.12) into (6.55) and evaluating the Gaussian integral gives

$$A(z,t) = \frac{T_0}{\left[T_0^2 - i\beta_2 z (1 + iC)\right]^{1/2}} \exp\left\{-\frac{(1+iC)\,t^2}{T_0^2 - i\beta_2 z (1+iC)}\right\} \qquad (6.56)$$

The above result defines new pulse width T_1 for a propagating pulse at position z. Its value is

$$\frac{T_1}{T_0} = \left[\left(1 + \frac{C\beta_2 z}{T_0^2}\right)^2 + \left(\frac{\beta_2 z}{T_0^2}\right)^2\right]^{1/2} \qquad (6.57)$$

Evolution of chirped and unchirped pulses is shown in Fig. 6.13. Consult listing 6A.7 for MATLAB code.

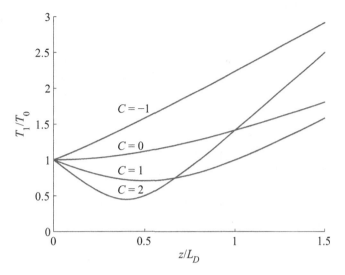

Fig. 6.13 Evolution of pulse width as a function of normalized distance z/L_D for chirped and unchirped pulses. Dispersian length is defined by Eq. (15.14).

6.5.2 Numerical method using Fourier transform

We finish this chapter by outlining the numerical method of using Eq. (6.54) to describe propagation of a Gaussian pulse (and also a super-Gaussian pulse) in the presence of dispersion. The method consists of the following steps.

1. The single pulse $p = exp(-t^2/T_0)$ is used with sampling period of $T = 0.08s$ in the time domain. (If needed it can be plotted using $plot(t, p)$ MATLAB function.)

2. The Fourier transform of this single pulse is found using $P = fftshift(fft(p))$ code where p is in time domain and P is in the frequency domain.

3. The frequency domain pulse is sampled using sampling frequency $=1/64$. The solution in frequency domain is found by multiplying Fourier transform of single pulse and transfer function $H(\omega)$ where spectral function is

$$H(\omega) = exp\left[\left(-\frac{\alpha}{2} - \frac{i}{2} \cdot \beta_2 \cdot \omega^2 - \frac{i}{6} \cdot \beta_3 \cdot \omega^3\right) \cdot L\right] \quad (6.58)$$

where α are losses, L is the propagation length and β_2 and β_3 are previously defined dispersion parameters.

4. The solution in time domain is obtained by using inverse Fourier transform $ifft(fftshift(P\_prop))$ and it is plotted. Consult Listings 6A.8 and 6A.8.1 for MATLAB code.

The propagation of Gaussian pulse over a distance of 4000 km is shown in Fig. 6.14 for the following parameters: $\alpha = 0$, $\beta_2 = 60\,ps^2/km$ and $\beta_3 = 0.01\,ps^3/km$. MATLAB code is provided in Appendix, Listings 6A.8.1 and 6A.8.2.

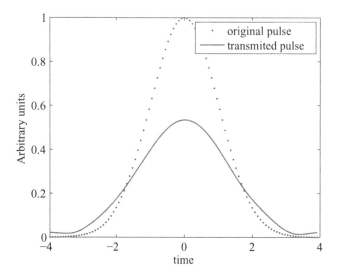

Fig. 6.14 Dispersion induced broadening of a Gaussian pulse. $\beta_2 = 60$ ps^2/km and $\beta_3 = 0.01$ ps^3/km.

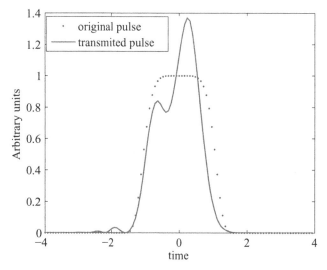

Fig. 6.15 Evolution of a super-Gaussian pulse with $m = 3$. $\beta_2 = 0$ and $\beta_3 = 0.01$ ps^3/km.

In Fig. 6.15 we show propagation of a super-Gaussian pulse with $m = 3$ over the same distance with $\beta_2 = 0$. The remaining parameters remain unchanged.

6.5.3 Fourier transform split-step method

This method will be extensively discussed in later Chapter 12 describing beam propagation method (BPM). Here we use it to illustrate propagation of Gaussian pulse. First, we briefly describe the method.

Table 6.1 Values of parameters.

Name	Symbol	Value
Second-order dispersion	β_2	60 ps/nm^{-1} km^{-1}
Third-order dispersion	β_3	0.01 ps/nm^{-1} km^{-1}
Losses	α	0
Width of Gaussian pulse	w_0	20 ps

We use the linear Schroedinger equation derived previously, Eq. (6.54). Taking Fourier transform results in

$$\frac{\partial \widetilde{A}(z,\omega)}{\partial z} = -\frac{i\beta_2}{2}(-i\omega)^2 \widetilde{A}(z,\omega) + \frac{\beta_3}{6}(-i\omega)^3 \widetilde{A}(z,\omega) - \frac{\alpha}{2}\widetilde{A}(z,\omega) \quad (6.59)$$

where α describes losses. Fourier transform is defined in a usual way as

$$\widetilde{A}(z,\omega) = \int_{-\infty}^{+\infty} A(z,t)\, e^{i\omega t}\, dt$$

Eq. (6.59) can be expressed as

$$\frac{\partial \widetilde{A}(z,\omega)}{\partial z} = P \cdot \widetilde{A}(z,\omega) \quad (6.60)$$

where the propagator operator P is defined as

$$P = \frac{i\beta_2}{2}\omega^2 + \frac{i\beta_3}{6}\omega^3 - \frac{\alpha}{2} \quad (6.61)$$

Eq. (6.60) can be integrated over small distance h along the z-axis with the result

$$\widetilde{A}(z+h,\omega) = e^{P \cdot h}\, \widetilde{A}(z,\omega) \quad (6.62)$$

The above equation is directly implemented in MATLAB. In Table 6.1 we summarized values used in simulations.

The results are shown in Figs. 6.16 and 6.17. We show evolution of the Gaussian pulse and also a three-dimensional view of the evolving pulse. MATLAB code is provided in Appendix, Listing 6A.9.

6.6 Problems

1. Evaluate Fourier series for rectangular pulse represented by Eq. (6.1). Determine coefficients s_n.
2. Evaluate Fourier transform of chirped Gaussian pulse and its spectral half-width $\Delta\omega$ as given by Eq. (6.14).

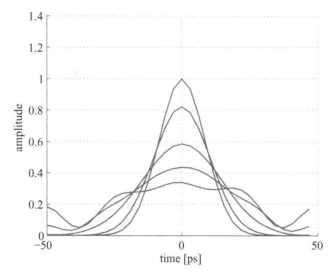

Fig. 6.16 Evolution of Gaussian pulse by Fourier transform split-step method.

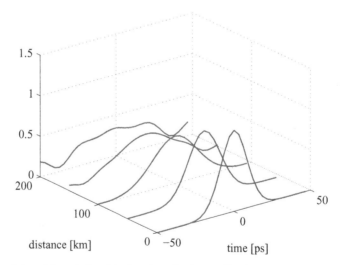

Fig. 6.17 Three-dimensional view of the evolution of the Gaussian pulse shown in Fig. 6.16.

Appendix 6A: MATLAB listings

In Table 6.2 we provide a list of MATLAB files created for Chapter 6 and a short description of each function.

Table 6.2 List of MATLAB functions for Chapter 6.

Listing	Function name	Description
6A.1	*rp.m*	Creates approximations to rectangular pulse
6A.2	*FT_rp.m*	Evaluates FT of rectangular pulse using analytical expression and also MATLAB functions
6A.3	*super_gauss.m*	Plots super Gaussian pulses
6A.4	*gauss_chirp.m*	Plots chirp Gaussian pulses
6A.5.1	*ptrain.m*	Generates sequence of Gaussian pulses
6A.5.2	*fgauss.m*	Function used by *ptrain.m*
6A.6	*bit_gen.m*	Illustrates NRZ modulation format
6A.7	*pulse_evol.m*	Evolution of pulses (chirped and unchirped)
6A.8.1	*pdisp.m*	Width of Gaussian pulse with dispersion
6A.8.2	*gauss_m.m*	Function used by *pdisp.m*. Defines Gaussian pulses
6A.9	*sslg.m*	Evolution of Gaussian pulse by Fourier transform split-step method

Listing 6A.1 Function rp.m. Function which draws rectangular pulse and calculates approximations based on Fourier series for three values of n.

```
% File name: rp.m (rectangular pulse)
clear all
N_max = 100;                            % number of points for plot
% Pulse characteristics
s_0 = 1.0;                              % pulse amplitude
t_min = 2;                              % start of pulse
t_max = 10.0;                           % end of pulse
A = 0.5;
dt = 1/N_max;
t = 0:dt:2*t_max;
x = A*(sign(t-t_min)-sign(t-t_max));    % creates rectangular pulse
hold on
h = plot(t,x);
set(h,'LineWidth',2);
axis([0 1.2*t_max -0.5 1.7]);           % plots rectangular pulse
%
s = 0.0;
for n = [1 3 5]
    s = s + 4*s_0/(pi*n)*sin(n*pi*(t-t_min)./(t_max-t_min));
    h = plot(t,s);
    set(h,'LineWidth',2);
end
axis([0 1.2*t_max -0.5 1.7]);
xlabel('time (a.u.)','FontSize',14);    % size of x label
ylabel('s (a.u)','FontSize',14);        % size of y label
set(gca,'FontSize',14);                 % size of tick marks on both axes
```

```
% Add description
text(5.5, 1.4, 'n = 1','Fontsize',13)
text(7.0, 1.4, 'n = 3','Fontsize',13)
text(8.5, 1.4, 'n = 5','Fontsize',13)
pause
close all
```

Listing 6A.2 Function FT_rp.m. Function plots rectangular pulse, its Fourier transform obtained analytically, and also evaluates its Fourier transform using MATLAB built-in functions.

```
% File name: FT_rp.m
% Plots symmetric rectangular pulse of duration tau and its
% Fourier transform
clear all
N_max = 1000;                    % number of points for plot
% Pulse characteristics
tau = 10.0;                      % 1/2 pulse width
A = 0.5;                         % pulse height
% Plotting Fourier transform of rectangular pulse using analytical formula
omega = linspace(-20/tau, 20/tau, N_max);  % creation of frequency argument
S = 2*A*sin(tau.*omega)./omega;
h = plot(omega,S);
set(h,'LineWidth',1.5);          % thickness of plotting lines
grid
xlabel('frequency (a.u.)','FontSize',14);
ylabel('Fourier transform (S) (a.u.)','FontSize',14);
set(gca,'FontSize',14);          % size of tick marks on both axes
text(0.65, 0, 'A','Fontsize',13);
text(1, 0, 'B','Fontsize',13);
pause
close all
%---- Plotting rectangular pulse ----------------------------------------
dt = 1/N_max;
t = -6*tau:dt:6*tau;             % Creation of points for pulse plotting
s = A*(sign(t+tau)-sign(t-tau));
h = plot(t,s);
set(h,'LineWidth',1.5);          % thickness of plotting lines
axis([-2*tau 2*tau 0 2]);
xlabel('time (a.u.)','FontSize',14); ylabel('s_r (a.u.)','FontSize',14);
set(gca,'FontSize',14);          % size of tick marks on both axes
pause
close all
%----- Plotting its Fourier transform using Matlab functions ----------
y = fft(s);        % discrete Fourier transform
SM = fftshift(y);  % shift zero-frequency component to center of spectrum
N = length(SM);
```

```
k = -(N-1)/2:(N-1)/2;
f = abs(SM);
h = plot(k,f);
set(h,'LineWidth',1.5);              % thickness of plotting lines
axis([-4*tau 4*tau -5000 30000]);
ylabel('Abs FT of rectangular pulse (a.u)','FontSize',14);
grid
set(gca,'FontSize',14);              % size of tick marks on both axes
pause
close all
```

Listing 6A.3 Function super_gauss.m. Plots super-Gaussian pulses for several values of parameter *m*.

```
% File name: super_gauss.m
% Plots super-Gaussian pulses
clear all
A = 1;
sigma = 30;
N = 300.0;
t = linspace(-50,50,N);
hold on
for m = [1 2 3 4]
    p = A/(sigma*sqrt(2*pi))*exp(-t.^(2*m)/(sigma^(2*m)));
    h = plot(p);
    set(h,'LineWidth',1.5);          % thickness of plotting lines
end
xlabel('time','FontSize',14); ylabel('amplitude','FontSize',14);
set(gca,'FontSize',14);              % size of tick marks on both axes
pause
close all
```

Listing 6A.4 Function gauss_chirp.m. Generates Gaussian and chirp Gaussian pulses.

```
% File name: gauss_chirp.m
% Creation of Gaussian pulse and also chirped Gaussian pulse
clear all;
%
t_zero = 20.0;         % center of incident pulse
width = 6.0;           % width of the incident pulse
C = 15;                % chirped parameter
%
N     = 300.0;
time = linspace(0,50,N);
pulse = exp(-0.5*((t_zero - time)/width).^2);           % Gaussian pulse
```

```
pulse_ch = exp(-(0.5+1i*C)*((t_zero - time)/width).^2); % chirped pulse
%
subplot (2,1,1), h = plot(time,pulse);
set(h,'LineWidth',1.5);                    % thickness of plotting lines
title('Gaussian pulse','FontSize',14)
xlabel('time','FontSize',14); ylabel('amplitude','FontSize',14);
set(gca,'FontSize',14);                    % size of tick marks on both axes
%
subplot (2,1,2), h = plot(time,pulse_ch);
set(h,'LineWidth',1.5);                    % thickness of plotting lines
title('Chirped Gaussian pulse','FontSize',14)
xlabel('time','FontSize',14); ylabel('amplitude','FontSize',14);
set(gca,'FontSize',14);                    % size of tick marks on both axes
pause
close all
```

Listing 6A.5.1 Function ptrain.m. Generates a train of pulses.

```
% File name: ptrain.m
% Generates train of pulses
% Allows for overlap of pulses
clear all
bits = [0 1 0 1 1 0];                      % definition of logical pattern
T = 1d-9;                                  % pulse period [s]
num_pulses = length(bits);                 % number of pulses
N = 1000;                                  % number of time points
width = 0.3*T;                             % width of pulse
time=linspace(0,num_pulses*T,N);
signal = zeros(1,N);
pulses = zeros(num_pulses,N);
%
t_0= T/2;                                  %position of peak of first impulse in signal
for i=1:num_pulses
    if (bits(i)==1)
        pulses(i,:)= fgauss(time,width,t_0);
        signal = signal + fgauss(time,width,t_0);
    end
    t_0 = t_0 + T;
end
%
subplot (2,1,1); h = plot(time,pulses);
set(h,'LineWidth',1.5);                    % thickness of plotting lines
xlabel('time','FontSize',14);              % size of x label
title('Individual pulses','FontSize',14)
set(gca,'FontSize',14);                    % size of tick marks on both axes
subplot (2,1,2); h = plot(time,signal);
set(h,'LineWidth',1.5);                    % thickness of plotting lines
```

```
xlabel('time','FontSize',14);              % size of x label
set(gca,'FontSize',14);                    % size of tick marks on both axes
title('Total signal','FontSize',14)
pause
close all
```

Listing 6A.5.2 Function fgauss.m. Function used by program *ptrain.m*. Function creates individual Gaussian pulses.

```
function pulse = fgauss(time,width,t_0)
% function pulse = wave_gauss(t,T_period)
% Generates individual Gaussian pulse used in the creation of
% train of pulses
pulse = exp(-0.5*((t_0 - time)/width).^2);
```

Listing 6A.6 Function bits_gen.m. Function generates an 8-bits long bit pattern.

```
% File name: bits_gen.m
%------------------------------------------------------------------
% Purpose:
% Generates 8-bits long pattern
% generates single bit which is then repeated 8 times
% Source
% R.Sabella and P.Lugli
% "High Speed Optical Communications"
% Kluwer Academic Publishers 1999
% p.32
%********************************
%
clear all
bits = [0 1 0 0 1 1 0 1];              % Definition of logical pattern
%
T_period = 1d-9;   % pulse period [s]
I_bias = 2;       % mA
I_m = 3;          % mA
tau_r = 0.2*T_period;   % rise time
%
% Generation of current pattern corresponding to bit pattern
I_p = 0;
N_div = 50;   % number of divisions within each bit interval
t = linspace(0, T_period, N_div);    % the same time interval is
                                     % generated for each bit
%******** first bit *************************************
    if bits(1)==0,      I_p_1 = I_bias + 0*t;
    elseif bits(1)==1,  I_p_1 = I_bias + I_m + 0*t;
    end
%-------------------------------------------
```

```
% Generates single, arbitrary bit
%t = linspace(0, T_period, N_div);
temp_I = I_p_1;
%
number_of_bits = length(bits);
%
for k = 2:number_of_bits
    if bits(k)==1 && bits(k-1)==0
            I_p = I_bias + I_m*(1 - exp(-2.2*t./tau_r));
    elseif bits(k)==0 && bits(k-1)==1
                            I_p = I_bias + I_m*exp(-2.2*t./tau_r);
    elseif bits(k)==0 && bits(k-1)==0
                            I_p = I_bias + 0*t;
    else bits(k)==1 && bits(k-1)==1
                            I_p = I_bias + I_m + 0*t;
    end
    I_p = [temp_I,I_p];
    temp_I = I_p;
end
temp_t = t;
for k = 2:number_of_bits
    t = linspace(0, T_period, N_div);
    t = [temp_t,(k-1)*T_period+t];
    temp_t = t;
end
%
x_min = 0;
x_max = max(t);
y_min = I_bias;
y_max = I_bias + 1.2*I_m;
%
h = plot(t,I_p);
grid on
set(h,'LineWidth',1.5);                  % thickness of plotting lines
xlabel('time [s]','FontSize',14);        % size of x label
ylabel('Current [mA]','FontSize',14);    % size of y label
set(gca,'FontSize',14);                  % size of tick marks on both axes
axis([x_min x_max y_min y_max])
pause
close all
```

Listing 6A.7 Function pulse_evol.m. Function describes evolution width of a chirped pulse.

```
% File name: pulse_evol.m
% Describes evolution of pulse width as a function of normalized
```

```
% distance for chirped and unchirped pulses
%
clear all
% Data
beta_2 = -20;                       % [-20ps^2/km] at 1.55 microns
T_0 = 200;                          % [ps] = 0.2 ns, bit rate 10Gb/s
L_D = (T_0^2)/abs(beta_2);
N_max = 401;                        % number of points for plot
x = linspace(0,1.5,N_max);          % normalized distance; x = z/L_D
%
hold on
for C = [-1 0 1 2]                  % chirped coefficient
    T = sqrt((1+sign(beta_2)*C*x).^2 + (x).^2);
    h = plot(x,T);
    set(h,'LineWidth',1.5);         % thickness of plotting lines
end
% Redefine figure properties
ylabel('T_1/T_0','FontSize',14)
xlabel('z/L_D','FontSize',14)
text(0.35, 1.65, 'C = -1','Fontsize',14)
text(0.35, 1.2, 'C = 0','Fontsize',14)
text(0.35, 0.8, 'C = 1','Fontsize',14)
text(0.35, 0.55, 'C = 2','Fontsize',14)
set(gca,'FontSize',14);             % size of tick marks on both axes
pause
close all
```

Listing 6A.8.1 Function pdisp.m. Function describes the evolution of a Gaussian pulse with dispersion.

```
% File name: pdisp.m
% Propagation of Gaussian pulse in the presence of dispersion
% using Fourier transform
clear all;
% Creation of input pulse
T_0 = 1.0;                          % pulse width
T_s = 0.08;                         % sampling period
t = -4:T_s:4-T_s;                   % creation of time interval
p = gauss_m(t,T_0);
P=fftshift(fft(p));                 % Fourier transform of the original pulse
Fs=1/64;                            % sampling frequency
N=length(t);                        % length of time interval
f = -N/2*Fs:Fs:N/2*Fs-Fs;           % frequency range
omega = 2*pi*f;
% Parameters of optical fiber
```

```
alpha = 0.0;                            % losses
beta_2 = 60.0;                          % coefficient beta_2 [ps^2/km]
%beta_2 = 0.0;                          % coefficient beta_2 [ps^2/km]
beta_3 = 0.01;                          % coefficient beta_3 [ps^3/km]
% Transfer function of optical fiber
distance = 4000;
H=exp((alpha/2+1i/2*beta_2*omega.^2+1i/6*beta_3*omega.^3)*distance);
%
P_prop = P.*H;                          % Fourier transform of final pulse
p_prop = ifft(fftshift(P_prop));        % time dependence of final pulse
p_plot = abs(p_prop).^2;
%
h = plot(t, p, '.', t, p_plot);
set(h,'LineWidth',1.5);                 % thickness of plotting lines
xlabel('time','FontSize',14);           % size of x label
ylabel('Arbitrary units','FontSize',14);% size of y label
set(gca,'FontSize',14);                 % size of tick marks on both axes
legend('original pulse', 'transmited pulse')
pause
close all
```

Listing 6A.8.2 Function gauss_m.m. Function used by *pdisp.m*.

```
function p = gauss_m(t,T_0)
% Definition of Gaussian and super-Gaussian pulses
m = 1;                                  % m=1, usual Gauss; m=3, super-Gauss
%m = 3;
p = exp(-t.^(2*m)/(2*T_0^(2*m)));
```

Listing 6A.9 Function sslg.m. Function creates a Gaussian pulse in the time domain which is subsequently Fourier transformed to the frequency domain and then propagated over a linear system with dispersion characterized by β_2 and β_3.

```
% File name: sslg.m
% Calculates and plots the evolution of a Gaussian pulse in optical fiber
% using Fast Fourier Transform split-step method with linear terms only
clear all
% Input parameters
N= 32;                                  % number of points along time axis
T_domain = 100;                         % total time domain kept [in ps]
beta_2 = 2;                             % dispersion coefficient [in ps/nm-km]
beta_3 = 1.01;
%
Delta_t = T_domain/N;                   % node spacing in time
Delta_om = 2*pi/T_domain;               % node spacing in radial frequency
t = Delta_t*(-N/2:1:(N/2)-1);           % array of time points
omega = Delta_om*(-N/2:1:(N/2)-1);      % array of radial frequency points
```

```
t_FWHM_0 = 20;                         % initial pulse FWHM [in ps]
P_0 = 1;                               % initial peak power [in mW]
A_0 = sqrt(P_0);                       % initial amplitude of the Gaussian
T_0 = t_FWHM_0/(2*sqrt(log(2)));       % initial pulse standard deviation
gauss = A_0*exp(-t.^2/(2*T_0^2));      % initial Gauss pulse in time
%
L_D = 200;
z_plot = [0 0.25 0.5 0.75 1.0]*L_D;    % z-values to plot [in km]
gauss_F = fftshift(abs(fft(gauss)));   % initial Gauss in frequency
z = 0;                                 % starting distance
n = 0;                                 % controls stepping
hold on
for z_val = z_plot                     % for selected z-values
    n = n + 1;                         % creates new step
    % P -propagator function
    P = exp(((1j/2)*beta_2*(omega.^2)+(1j/6)*beta_3*(omega.^3))*z_val);
    u_F_z = gauss_F.*P;                % propagation at point z
    u_z = ifft(u_F_z,N);               % takes inverse Fourier transform
    u_abs_z = abs(u_z).^2;
    u = fftshift(u_abs_z);             % shifts frequency components
    u_3D(:,n) = u';                    % create array for 3D plot
    plot(t,u,'LineWidth',1.5)
end
grid on
xlabel('time [ps]','FontSize',14); ylabel('amplitude','FontSize',14);
set(gca,'FontSize',14);                % size of tick marks on both axes
pause
close all
%
% Make 3D plot
for k = 1:1:length(z_plot)             % choosing 3D plots every step
    y = z_plot(k)*ones(size(t));       % spread out along y-axis
    plot3(t,y,u_3D(:,k),'LineWidth',1.5)
    hold on
end
xlabel('time [ps]','FontSize',14); ylabel('distance [km]','FontSize',14);
set(gca,'FontSize',14);                % size of tick marks on both axes
grid on
pause
close all
```

References

[1] G. P. Agrawal. *Fiber-Optic Communication Systems. Second Edition.* Wiley, New York, 1997.

[2] E. Iannone, F. Matera, A. Mocozzi, and M. Settembre. *Nonlinear Optical Communication Networks*. Wiley, New York, 1998.

[3] G. P. Agrawal. *Nonlinear Fiber Optics*. Academic Press, Boston, 1989.

[4] G. Keiser. *Optical Fiber Communications. Fourth Edition*. McGraw-Hill, Boston, 2011.

[5] M. M.-K. Liu. *Principles and Applications of Optical Communications*. Irwin, Chicago, 1996.

[6] L. N. Binh. *Optical Fiber Communications Systems*. CRC Press, Boca Raton, 2010.

[7] R. Ramaswami and K. N. Sivarajan. *Optical Networks: A Practical Perspective. Second Edition*. Morgan Kaufmann Publishers, San Francisco, 1998.

[8] J. C. Cartledge. Improved transmission performance resulting from the reduced chirp of a semiconductor laser coupled to an external high-Q resonator. *J. Lightwave Technol.*, **8**:716–21, 1990.

[9] R. Sabella and P. Lugli. *High Speed Optical Communications*. Kluwer Academic Publishers, Dordrecht, 1999.

7 Optical sources

In this chapter, we will give a basic introduction to optical sources with the main emphasis on semiconductor lasers. The bulk of our description is based on the rate equations approach. We start with a general overview of lasers.

7.1 Overview of lasers

Generic laser structure is shown in Fig. 7.1 [1], [2]. It consists of a resonator (cavity), here formed by two mirrors, and a gain medium where the amplification of electromagnetic radiation (light) takes place. A laser is an oscillator analogous to an oscillator in electronics. To form an oscillator, an amplifier (where gain is created) and feedback are needed. Feedback is provided by two mirrors which also confine light. One of the mirrors is partially transmitting which allows the light to escape from the device. There must be an external energy provided into the gain medium (a process known as pumping). Most popular (practical) pumping mechanisms are by optical or electrical means.

Gain medium can be created in several ways. Conceptually, the simplest one is the collection of gas molecules. Such systems are known as gas lasers. In a gas laser one can regard the active medium effectively as an ensemble of absorption or amplification centers (e.g. like atoms or molecules) with only some electronic energy levels which couple to the resonant optical field. Other electronic states are used to excite or pump the system. The pumping process excites these molecules into a higher energy level.

A schematic of a pumping cycle of a typical laser is shown in Fig. 7.2. The system consists of a ground state level (here labelled as '0'), two levels with energies E_1 and E_2 and a band labelled as '3'. The system is pumped by an external source with the frequency ω_{30}. Fast decay takes place between band '3' and energy level '2'. In this way the population inversion is achieved (occupancy of level '2' is larger than level '1'), and at some point laser action can take place and light with frequency ω_{21} can be emitted.

The system consisting of level '2' and '1' can be further isolated. Popular visualization of such systems is known as a two-level system (TLS), see Fig. 7.3. Only two energies (out of many in the case of a molecule) are selected and the transitions are considered within those energies. As illustrated, three basic processes are possible: absorption, stimulated emission and spontaneous emission.

Such TLS are met often in Nature. Generally, for an atomic system, in the case under consideration, we can always separate just two energy levels, upper level and ground state, thereby forming TLS.

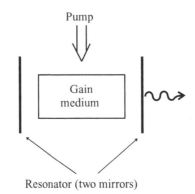

Fig. 7.1 Generic laser structure: two mirrors with a gain medium in between. The two mirrors form a cavity, which confines the light and provides the optical feedback. One of the mirrors is partially transmitted and thus allows light to escape. The resulting laser light is directional, with a small spectral bandwidth.

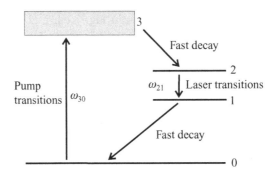

Fig. 7.2 Pumping cycle of a typical laser.

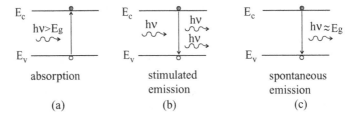

Fig. 7.3 Illustration of possible transitions in two-level system.

As mentioned above, an electron can be excited into the upper level due to external interactions (for lasers, process known as pumping). Electrons can lose their energies radiatively (emitting photons) or non-radiatively, say by collisions with phonons.

For laser action to occur, the pumping process must produce population inversion, meaning that there are more molecules in the excited state (here, upper level with energy E_2) than in the ground state. If the population inversion is present in the cavity, incoming light

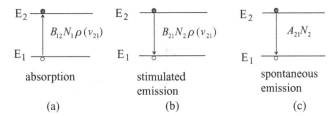

Fig. 7.4 Illustration of the notation of electron transitions in a TLS.

can be amplified by the system, see Fig. 7.3b where one incoming photon generates two photons at the output.

The way how TLS is practically utilized results in various types of lasers; like gaseous, solid state, semiconductor. Also, different types of resonators are possible which will be discussed in subsequent sections.

7.1.1 Transitions in a TLS

Assume the existence of two energy levels E_1 and E_2 (forming TLS) which are occupied with probabilities N_1 and N_2. Also introduce:

A_{21} the probability of spontaneous emission
B_{21} the probability of stimulated emission
B_{12} the probability of absorption.

The notation associated with those processes is illustrated in Fig. 7.4.

We introduce $\nu_{21} = (E_2 - E_1)/h$ and $\rho(\nu_{21})$ is the density of photons with frequency ν_{21}. Coefficients A_{21}, B_{21}, B_{12} are known as Einstein coefficients. The density of photons $\rho(\nu_{21})$ of frequency ν_{21} can be determined from the Planck's distribution of the energy density in the black body radiation as [2]

$$\rho(\nu_{21}) = \frac{8\pi h \nu_{21}^3}{c^3} \frac{1}{\exp \frac{h\nu_{21}}{kT} - 1} \tag{7.1}$$

For the laser we require that amplification be greater than absorption. Therefore, the number of stimulated transitions must be greater than the absorption transitions. Thus the net amplification can be created. In the following we will determine the condition for the net amplification in such systems [2].

The change in time of the occupancy of the upper level is

$$\frac{dN_2}{dt} = -A_{21}N_2 - B_{21}N_2\rho(\nu_{21}) + B_{12}N_1\rho(\nu_{21}) \tag{7.2}$$

First term on the right hand side describes spontaneous emission; second term is responsible for stimulated emission and the last one for absorption. In thermal equilibrium which requires that $\frac{dN_2}{dt} = 0$, using Eq. (7.1) one obtains

$$N_2 \left\{ B_{21} \frac{8\pi h \nu_{21}^3}{c^3 \left(\exp \frac{h\nu_{21}}{kT} - 1\right)} + A_{21} \right\} = N_1 B_{12} \frac{8\pi h \nu_{21}^3}{c^3 \left(\exp \frac{h\nu_{21}}{kT} - 1\right)} \tag{7.3}$$

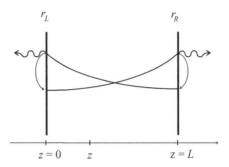

Fig. 7.5 Schematic illustration of the amplification in a Fabry-Perot (FP) semiconductor laser with homogeneously distributed gain.

Next, assume that N_1 and N_2 are given by the Maxwell-Boltzmann statistics; i.e.

$$\frac{N_1}{N_2} = \exp\left(-\frac{h\nu_{21}}{kT}\right) \tag{7.4}$$

Substitute (7.4) into Eq. (7.3) and have

$$\exp\left(-\frac{h\nu_{21}}{kT}\right)\left\{B_{21}\frac{8\pi h\nu_{21}^3}{c^3\left(\exp\frac{h\nu_{21}}{kT}-1\right)}+A_{21}\right\} = B_{12}\frac{8\pi h\nu_{21}^3}{c^3\left(\exp\frac{h\nu_{21}}{kT}-1\right)}$$

From the above equation, the density of photons $\rho(\nu_{21})$ can be expressed in terms of Einstein coefficients as

$$\frac{8\pi h\nu_{21}^3}{c^3\left(\exp\frac{h\nu_{21}}{kT}-1\right)} = \frac{A_{21}}{B_{12}\exp\left(\frac{h\nu_{21}}{kT}\right)-B_{21}} \tag{7.5}$$

In the above formula, both sides are equal if

$$B_{21} = B_{12} \tag{7.6}$$

In such case, one finds

$$\frac{A_{21}}{B_{21}} = \frac{8\pi h\nu_{21}^3}{c^3} \tag{7.7}$$

The relations (7.6) and (7.7) are known as Einstein relations [2].

7.1.2 Laser oscillations and resonant modes

Light propagation with amplification is illustrated in Fig. 7.5. Mathematically it is described by assuming that there is no phase change on reflection at either end (left and right). Left end is defined as $z = 0$ and right end as $z = L$. At the right facet, the forward optical wave has a fraction r_R reflected (amplitude reflection) and after reflection that fraction travels back (from right to left).

In order to form a stable resonance, the amplitude and phase of the wave after a single round trip must match the amplitude and phase of the starting wave. At arbitrary point z inside the cavity, see Fig. 7.5, the forward wave is

$$E_0 e^{gz} e^{-j\beta z} \qquad (7.8)$$

where we have dropped $e^{i\omega t}$ term which is common and defined $g = g_m - \alpha_m$, where g_m describes gain (amplification) of the wave and α_m its losses. Also r_R and r_L are, respectively, right and left reflectivities, L length of the cavity and β propagation constant.

The wave travelling one full round will be

$$\{E_0 e^{gz} e^{-j\beta z}\} \{e^{g(L-z)} e^{-j\beta(L-z)}\} \{r_R e^{gL} e^{-j\beta L}\} \{r_L e^{gz} e^{-j\beta z}\} \qquad (7.9)$$

The above terms are interpreted as follows. In the first bracket there is an original forward propagating wave which started at z, in the second bracket there is a wave travelling from z to L, the third bracket describes a wave propagating from $z = L$ to $z = 0$, and the last one contains a wave travelling from $z = 0$ to the starting point z. At that point the wave must match the original wave as given by Eq. (7.8). From the above, one obtains a condition for stable oscillations

$$r_R r_L e^{2gL} e^{-2j\beta L} = 1 \qquad (7.10)$$

That condition can be split into an amplitude condition

$$r_R r_L e^{2(g_m - \alpha_m)L} = 1 \qquad (7.11)$$

and phase condition

$$e^{-2j\beta L} = 1 \qquad (7.12)$$

From the amplitude condition one obtains

$$g_m = \alpha_m + \frac{1}{2L} \ln \frac{1}{r_R r_L} \qquad (7.13)$$

From the phase condition, it follows

$$2\beta L = 2\pi n \qquad (7.14)$$

where n is an integer. The last equation determines wavelengths of oscillations since

$$\beta = \frac{2\pi}{\lambda_n} = \frac{\omega_n}{c} \qquad (7.15)$$

with λ_n being the wavelength. Typical gain spectrum and location of resonator modes are shown in Fig. 7.6a. Longitudinal modes with angular frequencies ω_{n-1}, ω_n and ω_{n+1} are shown. In time, the mode which has the largest gain will survive; the other modes will diminish, see Fig. 7.6b.

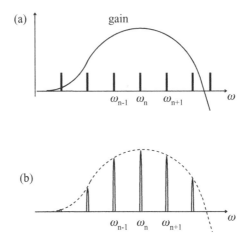

Fig. 7.6 Gain spectrum of semiconductor laser and location of longitudinal modes. ω_n are FP resonances determined from phase condition.

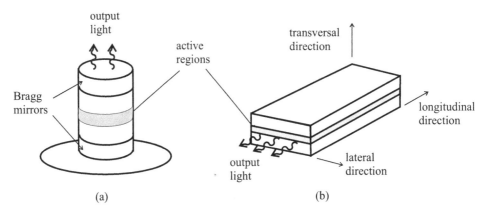

Fig. 7.7 VCSEL (left) and in-plane laser (right).

7.2 Semiconductor lasers

A significant percentage of today's lasers are fabricated using semiconductor technology. Those devices are known as semiconductor lasers. Over the last 15 years or so, several excellent books describing different aspects and different types of semiconductor lasers have been published [3], [4], [5], [6], [7].

The operation of semiconductor lasers as sources of electromagnetic radiation is based on the interaction between EM radiation and the electrons and holes in semiconductors. Typical semiconductor laser structures are shown in Fig. 7.7. Those are: vertical cavity surface emitting laser (VCSEL) [8] where light propagates perpendicularly to the main plane and in-plane laser where light propagates in the main plane [9]. The largest dimension

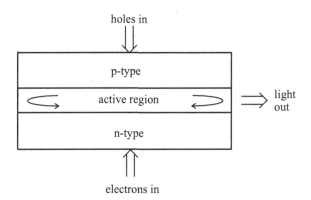

Fig. 7.8   The basic p-n junction laser.

of in-plane structures is typically in the range of 250 μm (longitudinal direction) whereas the typical diameter of VCSEL cylinder is about 10 μm.

The basic semiconductor laser is just a p-n junction, see Fig. 7.8 where cross-section along lateral-transversal directions is shown. Current flows (holes on p-side and electrons on n-side) along the transversal direction, whereas light travels along the longitudinal direction and leaves device at one or both sides.

In VCSEL, the cavity is formed by the so-called Bragg mirrors and an active region typically consists of several quantum well layers separated by barrier layers, see Fig. 7.7. (For a detailed discussion of quantum wells see [1] and [3]. This discussion is beyond the scope of this book.) Bragg mirrors consist of several layers of different semiconductors which have different values of refractive index. Due to the Bragg reflection such structure shows a very large reflectivity (around 99.9%). Such large values are needed because a very short distance of propagation of light does not allow to build enough amplification when propagating between distributed mirrors.

A three-dimensional perspective view of some generic semiconductor lasers is shown in Fig. 7.7. The structures consists of many layers of various materials, each engaged in a different role. Those layers are responsible for the efficient transport of electrons and holes from electrodes into an active region and for confinement of carriers and photons so they can strongly interact. Modern structures contain so-called quantum wells which form an active region and where conduction-valance band transitions are taking place. It is possible to have different types of mirrors as they also provide mode selectivity. Two basic types are illustrated in Fig. 7.9 and Fig. 7.10 [9]. They are known as distributed feedback (DFB) and distributed Bragg (DBR) structures.

In DFB lasers, grating (corrugation) is produced in one of the cladding layers, thus creating Bragg reflections at such periodic structure. The structure causes a wavelength sensitive feedback. It should be emphasized that the grating extends over the entire laser structure. When one restricts corrugation to the mirror regions only and leaves a flat active region in the middle, then the so-called DBR structure is created, see Fig. 7.10, which also provides wavelength sensitive feedback.

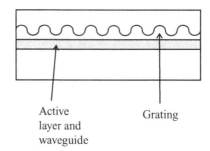

Fig. 7.9 Basic DFB laser structure.

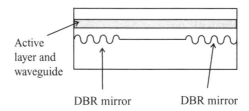

Fig. 7.10 Basic DBR laser structure.

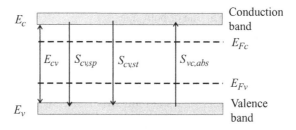

Fig. 7.11 Electron transitions between conduction and valence bands.

7.2.1 Electron transitions in semiconductors

Previous discussion of TLS will now be extended to describe transitions between bands in semiconductors, see Fig. 7.11. For that purpose, we introduce:

- f_v the probability of the state of energy E_v in the valence band being filled.
- f_c the probability of the state of energy E_c in the conduction band being filled.

Rates of transitions can now be determined, similarly to the TLS description. Here we use index notation appropriate to semiconductors, i.e. c, v instead of 1, 2. The rates are

$$S_{spon} = S_{cv,sp} = A_{21} f_c (1 - f_v) \tag{7.16}$$

$$S_{stim} = S_{cv,st} = B_{21} f_c (1 - f_v) \rho(E_{cv}) \tag{7.17}$$

$$S_{abs} = S_{cv,abs} = B_{12} f_v (1 - f_c) \rho(E_{cv}) \tag{7.18}$$

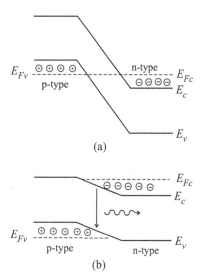

Fig. 7.12 Energy-band diagram of a p-n junction. (a) thermal equilibrium, (b) forward bias.

Fermi-Dirac statistics are assumed for the appropriate probabilities as

$$f_v = \frac{1}{\exp\left(\frac{E_v - E_{Fv}}{kT} + 1\right)} \quad (7.19)$$

$$f_c = \frac{1}{\exp\left(\frac{E_c - E_{Fc}}{kT} + 1\right)} \quad (7.20)$$

Here E_{Fc} and E_{Fv} are the quasi-Fermi levels for electrons and holes.

The condition for stimulated emission to exceed absorption is

$$S_{stim} > S_{abs}$$

which gives

$$B_{21} f_c (1 - f_v) > B_{12} f_v (1 - f_c)$$

Since the coefficients B_{21} and B_{12} are equal, after substituting expressions (7.19) and (7.20) one obtains

$$\exp\left(\frac{E_v - E_{Fv}}{kT}\right) > \exp\left(\frac{E_c - E_{Fc}}{kT}\right)$$

which can be written in the form

$$E_{Fc} - E_{Fv} > E_c - E_v = h\nu \quad (7.21)$$

The above inequality is known as Bernard-Duraffourg condition [10].

7.2.2 Homogeneous p-n junction

Energy-band diagrams of a p-n junction in thermal equilibrium and under forward bias conditions are shown in Fig. 7.12.

Electrons at the n-type side do not have enough energy to climb over the potential barrier to the left. Similar situation exists for holes in the p-type region, see Fig. 7.12a. One needs to apply forward voltage to lower the potential barriers for both electrons and holes, see Fig. 7.12b.

An application of the forward voltage also separates Fermi levels. Thus two different so-called quasi-Fermi levels E_{Fc} and E_{Fv} are created which are connected with an external bias voltage as

$$E_{Fc} - E_{Fv} = eV_{bias} \qquad (7.22)$$

With the lowered potential barriers, electrons and holes can penetrate central region where they can recombine and produce photons. However, the confinement of both electrons and holes into the central region is very poor (there is no mechanism to confine those carriers). Also, there is no confinement of photons (light) into the region where electrons and holes recombine (the region is known as an active region). Therefore, the interaction between carriers and photons is weak, which makes homojunctions a very poor light source. One must, therefore, provide some mechanism which will confine both carriers and photons into the same physical region where they will strongly interact. Such a concept is possible with the invention of heterostructures, which will be discussed next.

7.2.3 Heterostructures

A heterojunction is formed by joining dissimilar semiconductors. The basic type is formed by two heterojunctions and it is known as a double-heterojunction (more popular name is double-heterostructure).

The materials forming double heterostructures have different bandgap energies and different refractive indices. Therefore, in a natural way potential wells for both electrons and holes are created. Schematic energy-band diagrams of a double-heterostructure p-n junction in thermal equilibrium and under forward bias conditions are shown in Fig. 7.13. The bandgap and refractive index for *InGaAsP* material will now be summarized.

Bandgap

For the $In_{1-x}Ga_xAs_yP_{1-y}$ system, the relation between compositions x and y which results in the lattice-match to *InP* is

$$y = \frac{0.1894y}{0.4184 - 0.013y} \qquad (7.23)$$

In such case the bandgap is [11], [12]

$$E_{gap}[eV] = 1.35 - 0.72y + 0.12y^2 \qquad (7.24)$$

with the extra relations $y \approx 2.20x$ and $0 \leq x \leq 0.47$. The material for such compositions thus covers bandgaps in the range of wavelengths 0.92 μm ⇔ 1.65 μm. For example, the material $In_{0.74}Ga_{0.26}As_{0.57}P_{0.43}$ (i.e. $x = 0.26, y = 0.57$) has bandgap $E_g = 0.97$eV, which corresponds to the wavelength $\lambda = 1.27$ μm.

Table 7.1 Typical values of refractive indices for various compositions.

Wavelength	Range of y compositions	Refractive index
$\lambda = 1.3$ μm	$0 \leq y \leq 0.6$	$\bar{n}(y) = 3.205 + 0.34y + 0.21y^2$
$\lambda = 1.55$ μm	$0 \leq y \leq 0.9$	$\bar{n}(y) = 3.166 + 0.26y + 0.09y^2$

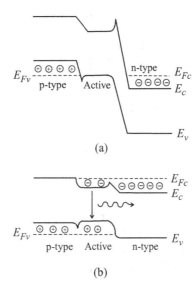

Fig. 7.13 Energy-band diagram of a double-heterostructure p-n junction. (a) Thermal equilibrium, (b) forward bias.

Refractive index

As was explained before, the dependence of refractive index as a function of wavelength is described by Sellmeier equation. For important telecommunication wavelengths, namely 1.3 μm and 1.55 μm, the y dependence of refractive index of $In_{1-x}Ga_xAs_yP_{1-y}$ that is lattice matched to InP is shown in Table 7.1 [13].

In the wells formed by heterostructures, electrons and holes recombine thus generating light. Also, differences in the values of refractive indices such that central region has larger value of refractive index create planar optical waveguide where light propagates and efficiently interacts with carriers.

Some illustrative values of bandgaps and refractive indices for $InGaAsP/InP$ heterostructure are shown in Fig. 7.14.

7.2.4 Optical gain

For efficient and reliable numerical simulations, an exact mathematical expression of the material gain is critical. Such an expression should also be subject to an experimental verification. This problem has been investigated since the early developments of semiconductor lasers [14], [15], [16]. The first step is usually the determination of gain spectra

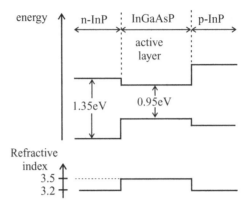

Fig. 7.14 Band structure and refractive index for InGaAsP/InP heterostructure.

and its comparison with experiment. The simplest approach will be provided in the next section. Typical curves of gain spectra for a four quantum well system and comparison with experimental measurements are shown in Fig. 7.15, from [17]. Mathematics involved in determining optical gain of semiconductor quantum-well structures is complex and has been discussed extensively, for example [18], [19], [20]. As a result, analytic approximations of the optical gain which can be used in fast calculations were determined, see [21], [22], [23]. From an extensive discussion it was determined that the peak material gain for bulk materials varies linearly with carrier density

$$g(N, \lambda) = a(\lambda) \min(N, P) - b(\lambda) \qquad (7.25)$$

The parameter $a(\lambda)$ is commonly called the differential gain.

On the basis of experimental observations, Westbrook [24], [25] extended the linear gain peak model to allow for wavelength dependence. In its simplest form it can be written as [24], [26], [27]

$$g(N, \lambda) = a(\lambda_p)N - b(\lambda_p) - b_a (\lambda - \lambda_p)^2 \qquad (7.26)$$

where λ_p is the wavelength of the peak gain and b_a governs the base width of the gain spectrum. Wavelength peak λ_p can be also carrier density dependent.

Gain (absorption) in a semiconductor is a function of carrier density n [cm^{-3}] (see Fig. 7.16, which shows gain spectrum for several values of carrier concentration) [28]. When n is below *transparency density* n_{tr}, the medium absorbs optical signal. For $n > n_{tr}$, optical gain in the material exceeds loss. The dependence of the so-called *gain peak* on the carrier density for quantum well systems is logarithmic, see Fig. 7.17. It is described by the expression [4]

$$g = g_0 \ln \frac{N}{N_{tr}} \qquad (7.27)$$

For modelling purposes, a linear approximation is often employed. In the above formulas g_0 is the differential gain, N_{tr} transparency carrier density and N_{th} is the carrier density at threshold.

Semiconductor lasers

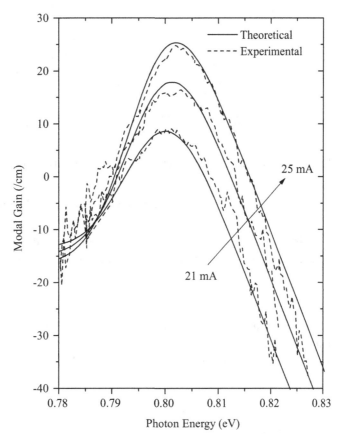

Fig. 7.15 Fitted theoretical and experimental net modal gain versus photon energy including leakage terms for three values of injected current, from [17].

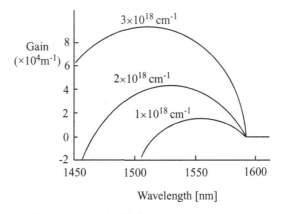

Fig. 7.16 Schematic optical gain spectra for various carrier densities.

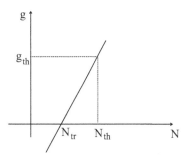
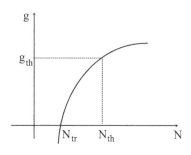

Fig. 7.17 Illustration of threshold and transparency densities for linear (left) and logarithmic (right) gain models.

7.2.5 Determination of optical gain

The simplest approach to determine optical gain is based on Fermi golden rule. Here we derive it for $T=0$. Transition rate between two levels a and b is [1]

$$W_{ab} = \frac{2\pi}{\hbar} \left|H'_{ab}\right|^2 \delta\left(E_b - E_a - \hbar\omega\right) \quad (7.28)$$

where H'_{ab} is the matrix element describing transitions between levels a and b. Assume parabolic bands for both electrons and holes with the effective mass m^* taking the value of m_c^* for conduction band and m_v^* for a valence band as follows:

$$E(k) = \frac{\hbar^2 k^2}{2m^*} \quad (7.29)$$

From the above

$$E_b - E_a = \frac{\hbar^2 k^2}{2}\left(\frac{1}{m_c} + \frac{1}{m_v}\right) + E_g \quad (7.30)$$

Due to conservation of the crystal momentum in the transition, one has $\vec{k}_{in} = \vec{k}_{fi}$ [29]. For a single transition at specific value of k, one therefore has

$$W(k) = \frac{2\pi}{\hbar}\left|H'_{vc}(k)^2\right|\delta\left(\frac{\hbar^2 k^2}{2m_r} + E_g - \hbar\omega\right) \quad (7.31)$$

where $m_r = \frac{m_v m_c}{m_v + m_c}$ is reduced effective mass. If N is a total number of transitions per second in crystal volume V and $g(k) = \frac{k^2 V}{\pi^2}$ is number of states per unit k in volume V, the total number of transitions per second is

$$N = \int_0^\infty W(k) \cdot g(k)\, dk$$

$$= \frac{2V}{\pi\hbar}\int_0^\infty \left|H'_{vc}(k)\right|^2 \delta\left(\frac{\hbar^2 k^2}{2m_r} + E_g - \hbar\omega\right) k^2\, dk \quad (7.32)$$

To evaluate integral, let us introduce new variable $X \equiv \frac{\hbar^2 k^2}{2m_r} + E_g - \hbar\omega$. Integral takes the form

$$\begin{aligned} N &= \frac{2v}{\pi \hbar} \int |H'_{vc}(k)|^2 \frac{m_r}{\hbar} \delta(X) \sqrt{\frac{2m_v}{\hbar^2}(X + \hbar\omega - E_g)} dX \\ &= \frac{V}{\pi} |H'_{vc}(k)|^2 \frac{(2m_r)^{3/2}}{\hbar^4} (\hbar\omega - E_g)^{1/2} \end{aligned} \quad (7.33)$$

where $\frac{\hbar^2 k^2}{2m_r} + E_g = \hbar\omega$. We are in a position now to determine absorption coefficient α. It is defined as

$$\alpha(\omega) \equiv \frac{\text{power absorbed per unit volume}}{\text{power crossing a unit area}} = \frac{(N) \cdot (\hbar\omega)/V}{\varepsilon_0 \bar{n} E_0^2 c/2} \quad (7.34)$$

where $\bar{n}$ is refractive index, c is the velocity of light in a vacuum and E_0 is the amplitude of electric field. Use

$$H'_{vc}(k) = \frac{eE_0 \chi_{vc}}{2}, \quad \chi_{vc} = \langle u_{vk'} | x | u_{ck} \rangle \quad (7.35)$$

Using the above and expression for N, one finds

$$\alpha_0(\omega) = \frac{\omega e^2 \chi_{vc}^2 (2m_r)^{3/2}}{2\pi \varepsilon_0 nc\hbar^3} (\hbar\omega - E_g)^{1/2} = K(\hbar\omega - E_g)^{1/2} \quad (7.36)$$

Example Estimates of K.

Data for GaAs. $\hbar\omega = 1.5$ eV, $m_v = 0.46 m_0$, $m_c = 0.067 m_0$, $\chi_{vc} = 3.2$ Å, $n = 3.64$. Those data give $K \approx 11\,700$ cm^{-1} $(eV)^{-1/2}$, and $\alpha_0(\omega) = 1170$ cm^{-1} [1].

At T = 0 depending on the value of $\hbar\omega$, one has the following possibilities summarized next:

$$\begin{array}{lll} \hbar\omega < E_g & \alpha(\omega) = 0 & \\ E_g < \hbar\omega < E_{Fc} - E_{Fv} & \alpha(\omega) = -\alpha_0(\omega) = -K(\hbar\omega - E_g)^{1/2} & \text{amplification} \\ E_{Fc} - E_{Fv} < \hbar\omega & \alpha(\omega) = \alpha_0(\omega) = K(\hbar\omega - E_g)^{1/2} & \text{absorption} \end{array} \quad (7.37)$$

7.3 Rate equations

With all basic elements in place we are now ready to provide the simplest (phenomenological) description of semiconductor lasers. It is based on rate equations [4]. The main role in those devices is played by two subsystems: carriers (electrons and holes) and photons. They interact in the so-called **active region**, defined as part of the structure where recombining carriers contribute to useful gain and photon emission. We describe both subsystems separately, starting with carriers.

In Fig. 7.18 (adapted from [4]), we schematically showed model of a laser operating below threshold. It resembles a tank partially filled with water continuously flowing in and at the same time water leaves the tank. The water models carriers which are continuously

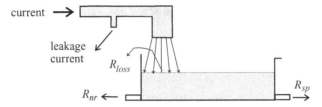

Fig. 7.18 Model of a laser operating below threshold. From L. A. Coldren and S. W. Corzine, *Diode Lasers and Photonic Integrated Circuits*, Wiley (1995). Reprinted with permission of John Wiley & Sons, Inc.

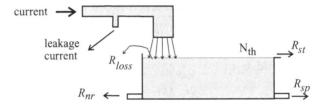

Fig. 7.19 Model of a laser operating above threshold. From L. A. Coldren and S. W. Corzine, *Diode Lasers and Photonic Integrated Circuits*, Wiley (1995). Reprinted with permission of John Wiley & Sons, Inc.

provided by current flowing in. Not all the current at electrodes actually reaches the device (tank). Some of it is lost as so-called leakage current. Below threshold, carriers disappear through losses in the device (R_{loss}), non-radiative recombination (R_{nr}) and spontaneous emission (R_{sp}). Above threshold (shown schematically in Fig. 7.19), the tank is completely filled with water (in laser the situation corresponds to the so-called threshold with density N_{th}). Its operation is mostly dominated by yet another process, namely stimulated emission (here characterized by coefficient R_{st}).

Next, based on the above picture we will formulate equations describing dynamics of carriers and photons.

7.3.1 Carriers

Inside the laser there exist two types of carriers: electrons and holes. Both are described in a similar way; however, parameters used will be different for both types.

The rate of change of carrier density is governed by

$$\frac{dN}{dt} = G_{gener} - R_{recom} \tag{7.38}$$

where the terms on right are responsible for generation and recombination of carriers and are given by

$$G_{gener} = \eta_i \frac{I}{q \cdot V}$$

with V being the volume of the active region, q is the electron's charge and I is the electric current. The internal efficiency η_i describes the fraction of terminal current that generates carriers in the active region. The recombination term consists of several contributions

$$R_{recom} = R_{sp} + R_{nr} + R_l + R_{st}$$

The meaning of terms is as follows:

R_{sp} — spontaneous recombination rate
R_{nr} — nonradiative recombination rate
R_l — carrier leakage rate
R_{st} — net stimulated recombination rate

Recombination processes are described phenomenologically as [4]

$$R_{recom} = \frac{N}{\tau} + v_g g(N) S \qquad (7.39)$$

7.3.2 Photons

Let S be the photon density. We postulate the following rate of change of photon density:

$$\frac{dS}{dt} = \Gamma R_{st} - \frac{S}{\tau_p} + \Gamma \beta_{sp} R_{sp}$$

with the following definitions:

τ_p — photon life-time
β_{sp} — spontaneous emission factor (reciprocal of the number of optical modes)
R_{st} — stimulated recombination.
Γ — confinement factor which is the ratio of the active layer volume to the volume of the optical mode.

Consider growth of a photon's density over the active region, (assume $\Gamma = 1$)

$$S + \Delta S = S e^{g \cdot \Delta z}$$

where g is gain.

$$\text{if } \Delta z \ll 1 \text{ then } \exp(g \cdot \Delta z) \approx 1 + g \Delta z$$

Using the relation $\Delta z = v_g \Delta t$ (v_g—group velocity), one finds $S = S g \cdot v_g \cdot \Delta t$.

Thus, the generation term can be written as

$$\left(\frac{dS}{dt}\right)_{gen} = R_{st} = \frac{\Delta S}{\Delta t} = v_g g S$$

Finally, the rate equations used in this section are

$$\frac{dN}{dt} = \eta_i \frac{I}{qV} - \frac{N}{\tau} - v_g g(N) S \qquad (7.40)$$

$$\frac{dS}{dt} = \Gamma v_g g(N) S - \frac{S}{\tau_p} + \Gamma \beta_{sp} R_{sp} \qquad (7.41)$$

We have explicitely indicated that gain g depends on a carrier's concentration.

7.3.3 Rate equation parameters

Other important parameters which in the simple model can be taken as constants, in fact have complicated dependencies. Those parameters are:

1. the carrier's lifetime τ which strongly depends on carrier's density. Typical dependence is shown below:

$$\frac{1}{\tau} = A + BN + CN^2$$

where coefficient A describes non-radiative processes, B is responsible for spontaneous recombination and C describes non-radiative Auger recombinations.

2. photon lifetime τ_p which is

$$\frac{1}{v_g \tau_p} = \alpha_i + \alpha_m = \Gamma g_{th}$$

Here α_m describes mirror reflectivity

$$\alpha_m = \frac{1}{L} \ln \frac{1}{R}$$

and α_i account for all losses.

The meaning of all parameters appearing in rate equations and gain models is summarized in Table 7.2 along with typical values for those symbols. The parameters were collected from [30], [31] and [32].

In the following, rate equations will provide a starting point for analysis of dynamical properties of semiconductor lasers. Before we start detailed analysis based on rate equations, we will establish a rate equation for electric field.

7.3.4 Derivation of rate equation for electric field

The relevant equation will be derived starting from the wave equation. We concentrate on Fabry-Perot lasers. The starting wave equation is

$$\nabla^2 \mathbf{E}(\mathbf{r}, t) - \frac{1}{c^2} \frac{\partial^2}{\partial t^2} \mathbf{E}(\mathbf{r}, t) = \mu_0 \frac{\partial^2}{\partial t^2} \mathbf{P}(\mathbf{r}, t) \tag{7.42}$$

Assume the following decomposition:

$$\mathbf{E}(\mathbf{r}, t) = \mathbf{a} E_t(x, y) \sin(\beta_z z) E(t) e^{j\omega t} \tag{7.43}$$

where $\mathbf{a}$ is a unit vector describing polarization, β_z propagation constant in the z-direction. The transversal field $E_t(x, y)$ obeys the following equation:

$$\left(\frac{\partial^2}{\partial x^2} + \frac{\partial^2}{\partial y^2} \right) E_t(x, y) = -\kappa_t^2 E_t(x, y) \tag{7.44}$$

It is further assumed that $E(t)$ is slowly varying compared to $e^{j\omega t}$. Substituting solution (7.43) into (7.42), neglecting fast-varying terms in time and using (7.44), one obtains

$$\left\{ \kappa_t^2 - \beta_z^2 - \frac{1}{c^2} \left[2j\omega \frac{\partial E(t)}{\partial t} - \omega^2 E(t) \right] \right\} \mathbf{a} E_t(x, y) \sin(\beta_z z) e^{j\omega t} = \mu_0 \frac{\partial^2}{\partial t^2} \mathbf{P}(\mathbf{r}, t)$$

Table 7.2 Basic parameters appearing in rate equation approach and their typical values.

Symbol	Description	Value and unit
N	carrier density	cm^{-3}
S	photon density	cm^{-3}
I	current	mA
q	elementary charge	1.602×10^{-19} C
L	cavity length	250 μm
w	width of active region	2 μm
d	thickness of active region	80Å
η_i	fraction of the injected current I into active region	0.8
V_{active}	volume of active region	$L \cdot w \cdot d$
τ	carrier lifetime	2.71 ns
v_g	group velocity	c/n_{ref}
n_{ref}	refractive index	3.4
τ_p	photon lifetime	2.77 ps
Γ	confinement factor	0.01
β_{sp}	spontaneous emission factor	10^{-4}
a	differential gain (linear model)	5.34×10^{-16} cm^2
N_{tr}	carrier density at transparency	3.77×10^{18} cm^{-3}
I_{th}	current at threshold	1.11 mA
α_m	facet loss	45 cm^{-1}
λ_{ph}	laser wavelength	1.3 μm

To evaluate terms on the right hand side, we need to account for a nonlocal relation between polarization and susceptibility:

$$\mathbf{P}(\mathbf{r},t) = \varepsilon_0 \int \chi(\mathbf{r},t') E(\mathbf{r}, t-t') dt' \qquad (7.45)$$

In the above $E(\mathbf{r}, t-t')$ is given by (7.43). As $E(t)$ is slowly varying in time, we can expand it into Taylor series around t. The general formula for Taylor expansion of the function $f(x)$ around a is $f(x) = f(a) + \frac{df}{dx}(x-a)$. In our case, substituting $x = t-t'$ and $a = t$, we obtain

$$E(t-t') \approx E(t) + \frac{dE(t)}{dt}(-t')$$

Full electric field is therefore

$$E(\mathbf{r}, t-t') = E_t(x,y) \sin(\beta_z z) E(t-t') e^{j\omega(t-t')}$$

Substituting the last result and Taylor expansion for $E(t-t')$ into expression for polarization (7.45), one finally obtains the wave equation for $E(t)$:

$$\left[-\beta_0^2 + \frac{\omega^2}{c^2} \varepsilon_r(\omega) \right] E(t) - \frac{2j\omega}{c^2} \left[\varepsilon_r(\omega) + \frac{1}{2}\omega \frac{d\varepsilon_r(\omega)}{d\omega} \right] \frac{dE(t)}{dt} = 0 \qquad (7.46)$$

Here $\varepsilon_r(\mathbf{r}, \omega) = 1 + \chi(\mathbf{r}, \omega)$ and Fourier transform of susceptibility is (we have dropped $\mathbf{r}$ dependence)

$$\chi(\omega) = \int dt' \chi(t') e^{-j\omega t'}$$

Susceptibility is further separated into terms which account for various physical effects as follows:

$$\chi(\omega) = 1 + \chi_b + \chi_p - j\chi_{loss}$$

where $\chi_b = \chi_b' + j\chi_b''$ is due to background, $\chi_p = \chi_p' + j\chi_p''$ is induced by pump and χ_{loss} accounts for losses. The relative dielectric constant is written as (dropping argument dependencies) $\varepsilon_r = \varepsilon_r' + j\varepsilon_r''$. The real part of ε_r is approximated in terms of refractive index $\varepsilon_r' = (\bar{n}_0 + \Delta \bar{n}_p)^2 \approx \bar{n}_0^2 + 2\bar{n}_0 \Delta \bar{n}_p$, where $\bar{n}_0$ is the background refractive index and $\Delta \bar{n}_p$ is the change induced by pump. Using the above results, ε_r takes the form

$$\begin{aligned}\varepsilon_r &= \varepsilon_r' + j\varepsilon_r'' \approx \bar{n}_0^2 + 2\bar{n}_0 \Delta \bar{n}_p + j\varepsilon_r'' \\ &= 1 + \chi_b' + \chi_p' + j\left(\chi_b'' + \chi_p'' - \chi_{loss}\right)\end{aligned} \quad (7.47)$$

Explicitly, one has the following identifications:

$$\begin{aligned}1 + \chi_b' &= \bar{n}_0^2 \\ \chi_p' &= 2\bar{n}_0 \Delta \bar{n}_p \\ \chi_b'' + \chi_p'' - \chi_{loss} &= \varepsilon_r''\end{aligned}$$

Using the above relations, the term in the equation for slowly varying amplitude $E(t)$, Eq. (7.46) is

$$\frac{\omega^2}{c^2}\varepsilon_r(\omega) - \beta_0^2 = \frac{\omega^2 - \omega_0^2}{c^2}\bar{n}_0^2 + \frac{\omega^2}{c^2}\chi_p' + j\frac{\omega^2}{c^2}\left(\chi_b'' + \chi_p'' - \chi_{loss}\right) \quad (7.48)$$

where we have used the approximate relation

$$\beta_0 \approx \frac{\omega_0}{c}\bar{n}_0$$

Imaginary parts of background and pump susceptibilities are related to gain via an experimental relation

$$\frac{\omega}{c\bar{n}_0}\left(\chi_b'' + \chi_p''\right) = \Gamma g(N)$$

where $g(N)$ is gain. Assume also

$$\chi_{loss} = \frac{\omega}{c\bar{n}_0}\alpha_{loss}$$

With the last two relations, term given by Eq. (7.48) can be expressed as

$$\frac{\omega^2}{c^2}\varepsilon_r(\omega) - \beta_0^2 = \frac{\omega^2 - \omega_0^2}{c^2}\bar{n}_0^2 + \frac{\omega^2}{c^2}2\bar{n}_0 \Delta \bar{n}_p + j\frac{\omega \bar{n}_0}{c}[\Gamma g(N) - \alpha_{loss}]$$

Using $\bar{n}^2(\omega) = \varepsilon_r(\omega)$, the dispersion term in (7.46) is evaluated as

$$\varepsilon_r(\omega) + \frac{1}{2}\omega\frac{d\varepsilon_r(\omega)}{d\omega} = \bar{n}^2(\omega) + \frac{1}{2}\omega\bar{n}(\omega)\frac{d\bar{n}(\omega)}{d\omega}$$
$$= \bar{n}(\omega) + \omega\bar{n}_g(\omega) \qquad (7.49)$$

where we have defined group index $\bar{n}_g(\omega)$ as

$$\bar{n}_g(\omega) = \bar{n}(\omega) + \omega\frac{d\bar{n}(\omega)}{d\omega} \qquad (7.50)$$

Using the results (7.48) and (7.49), Eq. (7.46) is

$$\frac{dE(t)}{dt} = \left\{-j(\omega - \omega_0)\frac{\bar{n}_0}{\bar{n}_g} - j\frac{\omega}{\bar{n}_g}\Delta\bar{n}_p + \frac{1}{2}v_g[\Gamma g(N) - \alpha_{loss}]\right\}E(t) \qquad (7.51)$$

where we approximated $\omega^2 - \omega_0^2 = (\omega - \omega_0)(\omega + \omega_0) \simeq 2\omega(\omega - \omega_0)$ and also $\frac{\bar{n}_0(\omega)}{\bar{n}(\omega)} \simeq 1$ and used the relation $\frac{c}{\bar{n}_g} = v_g$. Our Eq. (7.51) is very similar to Eq. (6.2.9) of Agrawal and Dutta [9].

7.4 Analysis based on rate equations

Rate equations just introduced will now be used to study some dynamical properties of semiconductor lasers. We start with steady-state.

7.4.1 Steady-state analysis

In this situation one has no change in time; i.e.

$$\frac{dN}{dt} = \frac{dS}{dt} = 0$$

Rate equations take the form

$$\eta_i\frac{I_0}{qV} - \frac{N_0}{\tau} - v_g g(N_0)S_0 = 0 \qquad (7.52)$$

and

$$\Gamma v_g g(N_0)S_0 - \frac{S_0}{\tau_p} + \Gamma\beta_{sp}R_{sp} = 0 \qquad (7.53)$$

If in the last equation we neglect the small spontaneous emission term, we obtain an expression for a photon life-time

$$\frac{1}{\tau_p} = \Gamma v_g g(N_0) \qquad (7.54)$$

7.4.2 Small-signal analysis with the linear gain model

When neglecting spontaneous emission and using linear gain model where gain is given by $g = a(N - N_{tr})$, one obtains

$$\frac{dN}{dt} = \eta_i \frac{I}{qV} - \frac{N}{\tau} - v_g a(N - N_{tr})S \qquad (7.55)$$

and

$$\frac{dS}{dt} = \Gamma v_g a(N - N_{tr})S - \frac{S}{\tau_p} \qquad (7.56)$$

Assume that all time-dependent quantities oscillate as

$$I = I_0 + i(\omega)e^{j\omega t}$$
$$N = N_0 + n(\omega)e^{j\omega t}$$
$$S = S_0 + s(\omega)e^{j\omega t}$$

where ω is the angular frequency of an external (small) perturbation and I_0 is the bias value of current. Here N_0 and S_0 are the solutions in the steady-state. Substitute into the first rate equation and after multiplication and neglecting second-order term ($n(\omega)s(\omega)$) one obtains

$$j\omega n(\omega)e^{j\omega t} = \frac{\eta_i}{qV}\left[I_0 + i(\omega)e^{j\omega t}\right] - \frac{N_0}{\tau} - \frac{1}{\tau}n(\omega)e^{j\omega t}$$
$$- v_g a\left((N_0 - N_{tr})S_0 + (N_0 - N_{tr})s(\omega)e^{j\omega t} + S_0 n(\omega)e^{j\omega t}\right) + O(n^2)$$

Using steady-state results, expression (7.54) for photon life-time and dropping $e^{j\omega t}$ dependence, one finally obtains

$$j\omega n(\omega) = \frac{\eta_i}{qV}i(\omega) - \frac{n(\omega)}{\tau} - \frac{s(\omega)}{\Gamma \tau_p} - v_g a S_0 n(\omega) \qquad (7.57)$$

The second equation for photons is obtained in a similar way. Substituting small-signal expressions, neglecting the second-order term, using the photon life-time expression and finally dropping $e^{j\omega t}$ dependence, one obtains

$$j\omega s(\omega) = \Gamma v_g a S_0 n(\omega) \qquad (7.58)$$

From the above equations, we want to determine *modulation response* which is defined as [33], [34]

$$M(\omega) = \frac{s(\omega)}{i(\omega)}$$

Expressing $n(\omega)$ from Eq. (7.58) and substituting into (7.57), one obtains

$$\frac{s(\omega)}{i(\omega)} = \frac{\frac{\eta_i}{eV}\Gamma v_g a S_0}{D(\omega)}$$

where $D(\omega)$ is given by

$$D(\omega) = -\omega^2 + j\omega\left(\frac{1}{\tau} + v_g a S_0\right) + \frac{v_g a S_0}{\tau_p}$$

Define response function as

$$r(\omega) = \left|\frac{s(\omega)}{i(\omega)}\right| = \frac{\frac{\eta_i}{eV}\Gamma v_g a S_0}{|D(\omega)|}$$

If we write $D(\omega) = a + jb$, we have $|D(\omega)|^2 = a^2 + b^2$. Therefore

$$|D(\omega)|^2 = \left(\omega^2 - \frac{v_g a S_0}{\tau_p}\right)^2 + \omega^2 \left(\frac{1}{\tau} + v_g a S_0\right)^2$$

We will see soon that the response function has a peak at frequency f_R. To find that frequency, we evaluate first derivative $\frac{\partial |D(\omega)|^2}{\partial \omega^2}$ and set it to zero. Explicitly

$$\left.\frac{\partial |D(\omega)|^2}{\partial \omega^2}\right|_{\omega=\omega_R} = -2\left(\frac{v_g a S_0}{\tau_p} - \omega_R^2\right) - \frac{1}{2}\left(\frac{1}{\tau} + v_g a S_0\right) = 0$$

From the above

$$\omega_R^2 = \frac{v_g a S_0}{\tau_p} - \frac{1}{2}\left(\frac{1}{\tau} + v_g a S_0\right) \quad (7.59)$$

which is known as a relaxation-oscillation frequency and it describes the rate at which energy is exchanged between photon system and carriers. The second term in (7.59) is usually very small and can be neglected. Relaxation-oscillation frequency is thus approximated as

$$\omega_R = \sqrt{\frac{1}{\tau_p}\frac{S_0 v_g a}{1+\varepsilon S_0}} \quad (7.60)$$

To estimate the typical values of the relaxation-oscillation frequency, let us assume [1]: $L = 300$ μm, $\tau_{ph} \approx 10^{-12}s$ and $\tau \sim 4 \times 10^{-9}s$. One finds $AP_0 \sim 10^9 \, s^{-1}$. So, the zeroth-order expression is a very good one.

Using MATLAB code and parameters shown in Listings 7A.1.1 and 7A.1.2, one obtains the results shown in Fig. 7.20.

7.4.3 Small-signal analysis with gain saturation

Let us remind ourselves of the rate equations which were established in the previous section:

$$\frac{dN}{dt} = \eta_i \frac{I}{qV} - \frac{N}{\tau} - v_g g(N) S \quad (7.61)$$

$$\frac{dS}{dt} = \Gamma v_g g(N) S - \frac{S}{\tau_p} + \beta_{sp} R_{sp} \quad (7.62)$$

Use small-signal assumptions

$$I(t) = I_0 + i(t)$$
$$N(t) = N_0 + n(t)$$
$$S(t) = S_0 + s(t)$$

Also, generalize the gain model to include saturation as follows:

$$g(n) = \frac{g_0 + g'(N - N_0)}{1 + \epsilon S}$$

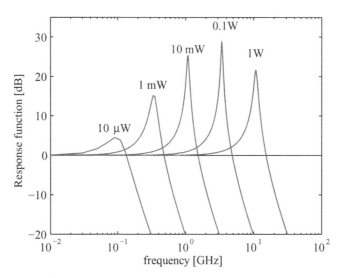

Fig. 7.20 Modulation characteristics of semiconductor lasers from small-signal analysis.

where $g' = \frac{\partial g}{\partial N}\big|_{N=N_0}$ is differential gain evaluated at $N = N_0$ and $g_0 = g(N_0)$. The factor $1 + \epsilon S$ introduces nonlinear gain saturation. The gain compression parameter ϵ has a small value and the term ϵS is small compared with the one even at very high optical power. The effect of this term on dc properties is small and can be neglected. However, it significantly affects dynamics of a semiconductor laser. The reason is that the laser's dynamics depends on the small difference between gain and cavity loss. This difference is only about a few percent. Therefore, even small gain compression due to ϵ produces significant effect.

Evaluate the gain compression term as follows:

$$1 + \epsilon S = 1 + \epsilon(S_0 + s(t)) = (1 + \epsilon S_0)\left(1 + \frac{\epsilon s(t)}{1 + \epsilon S_0}\right)$$

With the above result, gain can be approximated as

$$g(n) = \frac{g_0}{1 + \epsilon S_0} + \frac{g' n(t)}{1 + \epsilon S_0} - \frac{g_0}{1 + \epsilon S_0}\frac{\epsilon s(t)}{1 + \epsilon S_0} \tag{7.63}$$

Using the above results, stimulated emission term takes the form

$$g(n)S = \frac{g_0 S_0}{1 + \epsilon S_0} + \frac{g' S_0}{1 + \epsilon S_0} n(t) - \frac{g_0 S_0}{(1 + \epsilon S_0)^2}\epsilon s(t) + \frac{g_0}{1 + \epsilon S_0} s(t) + O(s^2) \tag{7.64}$$

As usual, the following analysis is separated into two parts: dc and ac analysis.

dc analysis

With the assumption that $\frac{d}{dt} = 0$ from rate equations, one obtains

$$\eta_i \frac{I_0}{qV} - \frac{N_0}{\tau} - v_g \frac{g_0 S_0}{1 + \epsilon S_0} = 0$$

and
$$\Gamma v_g \frac{g_0 S_0}{1 + \epsilon S_0} - \frac{S_0}{\tau_p} + \beta_{sp} R_{sp} = 0$$

Neglecting spontaneous emission ($\beta_{sp} = 0$), from the second equation one obtains an expression for the photon life-time τ_p:

$$\frac{1}{\tau_p} = \Gamma v_g \frac{g_0}{1 + \epsilon S_0} \qquad (7.65)$$

ac analysis

Substitute small-signal assumptions into rate equations, use an approximation for stimulated emission (7.64) and eliminate dc terms. The results are

$$\frac{dn(t)}{dt} = \eta_i \frac{i(t)}{qV} - \frac{n(t)}{\tau} - \frac{v_g g' S_0}{1 + \epsilon S_0} n(t) + \frac{1}{\tau_p \Gamma} \frac{S_0}{1 + \epsilon S_0} \epsilon s(t) - \frac{1}{\tau_p \Gamma} s(t)$$

$$\frac{ds(t)}{dt} = \frac{\Gamma v_g g' S_0}{1 + \epsilon S_0} n(t) - \frac{1}{\tau_p} \frac{S_0}{1 + \epsilon S_0} \epsilon s(t)$$

In the above equations we used expression (7.65). Those equations can be written in matrix form:

$$\frac{d}{dt} \begin{bmatrix} n(t) \\ s(t) \end{bmatrix} + \begin{bmatrix} A & B \\ -C & D \end{bmatrix} \begin{bmatrix} n(t) \\ s(t) \end{bmatrix} = \begin{bmatrix} \eta_i \frac{i(t)}{qV} \\ 0 \end{bmatrix}$$

where

$$A = \frac{1}{\tau} + \frac{v_g S_0 g'}{1 + \epsilon S_0}, \quad B = \frac{1}{\tau_p \Gamma} - \frac{1}{\tau_p \Gamma} \frac{S_0}{1 + \epsilon S_0} \epsilon, \quad C = \frac{\Gamma v_g S_0 g'}{1 + \epsilon S_0}, \quad D = \frac{1}{\tau_p} \frac{S_0}{1 + \epsilon S_0} \epsilon$$

Assuming harmonic time dependence of the form $exp(j\omega t)$, the matrix equation is

$$\begin{bmatrix} j\omega + A & B \\ -C & j\omega + D \end{bmatrix} \begin{bmatrix} n(t) \\ s(t) \end{bmatrix} = \begin{bmatrix} \eta_i \frac{i(t)}{qV} \\ 0 \end{bmatrix}$$

Solving the above system, one obtains modulation response which is expressed as

$$\frac{s(t)}{i(t)} = \eta_i \frac{1}{qV} \frac{C}{H(\omega)}$$

where

$$H(\omega) = (j\omega + A)(j\omega + D) + CB$$

The function $H(\omega)$ can be written as

$$H(\omega) = -\omega^2 + j\omega\gamma + \omega_R^2$$

Here ω_R is the relaxation-oscillation frequency and γ is the damping factor. One defines modulation response $r(\omega)$ as [30]

$$r(\omega) = \frac{H(\omega)}{H(0)}$$

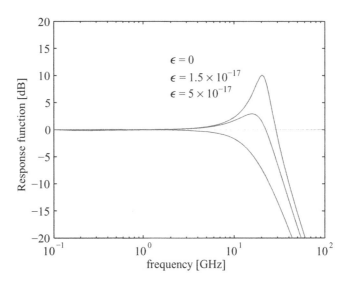

Fig. 7.21 The effect of gain compression on modulation response.

The plot of response $r(\omega)$ is shown in Fig. 7.21 for three values of ϵ. The effect of gain compression parameter ϵ is clearly visible. MATLAB code used to obtain those results is provided in Listings 7A.2.1 and 7A.2.2.

7.4.4 Large-signal analysis for QW lasers

Large-signal analysis based on rate equations for semiconductor lasers has been reported extensively [35], [36], [37]. Main equations without spontaneous emission are

$$\frac{dN}{dt} = \eta_i \frac{I}{eV} - \frac{N}{\tau} - v_g g(N) \cdot S \qquad (7.66)$$

$$\frac{dS}{dt} = \Gamma \cdot v_g g(N) \cdot S - \frac{S}{\tau_p} \qquad (7.67)$$

All symbols have been explained previously and their values are summarized in Table 7.2.

Typical results for step current are shown in Fig. 7.22. MATLAB code used to obtain those results is provided in Appendix, Listings 7A.3.1 and 7A.3.2.

One can observe that when we neglect spontaneous emission term in the rate equation for photons, photon density tends to zero (steady-state value) as $t \longrightarrow \infty$, the result which is also evident from the steady-state analysis.

7.4.5 Frequency chirping

Frequency chirping arises during direct modulation of semiconductor laser when carrier density undergoes abrupt change. As a result, material gain also changes. This change results in a variation of the refractive index, which in turn affects the phase of the electric field.

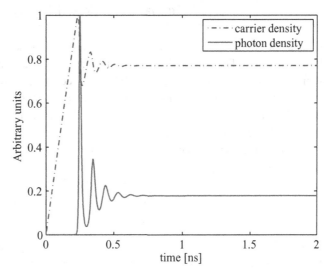

Fig. 7.22 Numerical solutions of large-signal rate equations after applying rectangular current pulse at $t = 0$. Normalized values of electron density N(t) and photon density S(t) are shown.

The rate of change of the phase is

$$\frac{d\phi}{dt} = \frac{1}{2}v\alpha_H \left(\Gamma g - \alpha_{loss} \right)$$

An associated change in frequency is created which is described by

$$\Delta v(t) = \frac{1}{2\pi}\frac{d\phi}{dt} = \frac{1}{4\pi}v\alpha_{enh} \left(\Gamma g - \alpha_{loss} \right)$$

This change in frequency is called 'chirp'. During direct modulation it results in a shift in frequency of the order of 10–20 GHz.

To eliminate chirp effect, an external modulator should be used. In such cases a laser produces constant output power and a separate modulator provides modulation.

Another way to reduce chirp is to fabricate devices with small values of line-width enhancement factors α_H.

$$\alpha_H \equiv -\frac{\frac{d\bar{n}_r}{dN}}{\frac{d\bar{n}_i}{dv}}$$

where $\bar{n} = \bar{n}_r + j\bar{n}_i$ (complex refractive index).

7.4.6 Equivalent circuit models

Often in engineering practice there is a need to operate with circuit models, instead of more complicated physical-based models. Here we outline the method of constructing such a model for a semiconductor laser.

7.4.7 Equivalent circuit model for a bulk laser

Equivalent circuit models of semiconductor lasers provide further understanding of the laser properties. They are derived from rate equations. Development of those models was initialized a long time ago [38], [39], [40]. More recent works which also include additional effects like carrier transport or proper representation of optical gain, are [41], [42], [43]. In the following, we outline the creation of equivalent circuit model for a bulk semiconductor laser following Kibar *et al.* [41].

For bulk lasers, rate equations are [41], [9]

$$\frac{dS}{dt} = (G - \gamma)S + R_{sp} \tag{7.68}$$

$$\frac{dN}{dt} = \frac{I}{q} - \gamma_e N - G \cdot S \tag{7.69}$$

Here N, S are the total number of electrons inside the cavity and total number of photons in the lasing mode. Those are dimensionless quantities. We introduce small deviations from equilibrium as

$$S(t) = S_0 + \delta S(t) \tag{7.70}$$

$$N(t) = N_0 + \delta N(t) \tag{7.71}$$

Optical gain is approximated as

$$G = G_0 + \frac{\partial G}{\partial N}\delta N + \frac{\partial G}{\partial S}\delta S = G_0 + G_N \delta N + G_S \delta S \tag{7.72}$$

In the steady-state one obtains

$$0 = (G_0 - \gamma)S_0 + R_{sp,0}$$

$$0 = \frac{I_0}{q} - \gamma_e N_0 - G_0 \cdot S_0$$

Observing that spontaneous emission $R_{sp} = R_{sp}(N)$ depends on N, and performing an expansion

$$R_{sp} = R_{sp,0} + \frac{\partial R_{sp}}{\partial N}\delta N \tag{7.73}$$

one can derive small-signal equations

$$\frac{d\delta S}{dt} = -\Gamma_S \delta S + \sigma_N \delta N \tag{7.74}$$

$$\frac{d\delta N}{dt} = -\Gamma_N \delta N - \sigma_S \delta S \tag{7.75}$$

where

$$\Gamma_S = \frac{R_{sp}}{S_0} - S_0 G_S, \quad \Gamma_N = \gamma_e + N_0 \frac{\partial \gamma_e}{\partial N} + S_0 G_N \tag{7.76}$$

and

$$\sigma_N = S_0 G_N + \frac{\partial R_{sp}}{\partial N}, \quad \sigma_S = G_0 + S_0 G_S \tag{7.77}$$

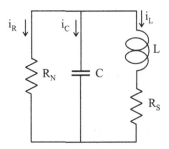

Fig. 7.23 Equivalent circuit model for a bulk laser.

Using these equations, one creates equivalent circuit model. It is based on an observation that the small-signal rate equations are similar to the voltage and current equations of the RLC circuit. We therefore assume (following Kibar et al., [41]) that electrons in a laser cavity can be represented by the charge across a capacitor and the photons are represented by the magnetic flux of an inductor. Formally one introduces the following equivalence:

$$\delta S(t) = \frac{\phi(t)}{q \cdot unit}, \quad \delta N(t) = \frac{Q(t)}{q} \tag{7.78}$$

where q is an electron's charge, unit is a parameter in $[\frac{H}{s}]$ to ensure proper units for the components of the electrical circuit and $\phi(t)$ is the magnetic flux in $[Wb]$. We also recall standard relations from electromagnetism:

$$Q = C \cdot v_C \tag{7.79}$$

$$i_C = \frac{dQ}{dt} \tag{7.80}$$

$$\phi(t) = L \cdot i_L \tag{7.81}$$

$$v_L = \frac{d\phi}{dt} \tag{7.82}$$

Now, we use (7.78) and relations (7.79)–(7.82) to convert small-signal equations (7.74) and (7.75) into voltage and current relations. After simple algebra, from (7.74) one finds

$$v_L(t) = -\Gamma_S \cdot L \cdot i_L(t) + \sigma_N \cdot C \cdot unit \cdot v_C \tag{7.83}$$

This equation can be identified with loop equation for the RLC circuit shown in Fig. 7.23. A loop equation gives

$$v_C = v_L + v_{Rp} \tag{7.84}$$

In order for the above equations to be identical, one makes the following identifications:

$$R_S = \Gamma_S \cdot L = \Gamma_S \cdot \frac{unit}{\sigma_S}, \quad C = \frac{1}{unit \cdot \sigma_N} \tag{7.85}$$

Similarly, from (7.75) one obtains

$$i_C(t) = -\Gamma_N \cdot C \cdot v_C(t) - \frac{\sigma_S}{unit} \cdot L \cdot i_L(t) \qquad (7.86)$$

That equation can be compared with the node equation from a circuit in Fig. 7.23:

$$i_C = -i_R - i_L \qquad (7.87)$$

For those equations to be identical, one sets

$$R_N = \frac{1}{\Gamma_N \cdot C}, \quad \text{and} \quad L = \frac{unit}{\sigma_S} \qquad (7.88)$$

One also has the following equivalence:

$$\delta N = \frac{C \cdot v_C}{q}, \quad \text{and} \quad \delta S = \frac{L \cdot i_L}{q \cdot unit} \qquad (7.89)$$

7.5 Problems

1. Perform large-signal analysis for the log gain model described by Eq. (7.27).
2. Analyse large-signal response to a Gaussian pulse.
3. Based on rate equations for multimode operation, create the small-signal equivalent circuit model of a semiconductor laser. Draw the corresponding electrical circuit and derive equations which establish links between circuit parameters and rate equations. *Hint:* Consult the paper by Kibar *et al.* [41].
4. Construct equivalent circuit model for single quantum well laser with carrier transport (follow Lau's work in [42]). Introduce carrier density in an SCH region and allow for its interaction with densities in the quantum well.

7.6 Project

1. Conduct multimode analysis of a semiconductor laser. Formulate multimode rate equations. Gain spectrum approximate as described by Coldren and Corzine [4].

Appendix 7A: MATLAB listings

In Table 7.3 we provide a list of MATLAB files created for Chapter 7 and a short description of each function.

Table 7.3 List of MATLAB functions for Chapter 7.

Listing	Function name	Description
7A.1.1	*small_signal.m*	Small-signal analysis
7A.1.2	*param_rate_eq_bulk.m*	Parameters for bulk active region in rate eqs.
7A.2.1	*small_epsilon.m*	Small-signal response with gain compression
7A.2.2	*param_rate_eq_QW.m*	Parameters for QW active region in rate eq.
7A.3.1	*large_signal.m*	Large-signal analysis
7A.3.2	*eqs_large.m*	Defines rate equations for large-signal analysis

Listing 7A.1.1 Program small_signal.m. MATLAB program to perform small-signal analysis for a quantum well semiconductor laser.

```
% File name: small_signal.m
% Purpose:
% Determines response function
clear all
param_rate_eq_bulk                      % input data
%
% loop over frequency in GHz
N_max = 5000;                           % number of points for plot
f_min = 0.01; f_max = 100;
freq_GHz = linspace(f_min,f_max,N_max); % From 0.1 to 2 GHz
semilogx(freq_GHz,0);
%
freq = freq_GHz*1d9;                    % convert frequency to 1/s
omega = 2*pi*freq;
%
hold on
for power_out = [1d-4 1d-3 0.01 0.1 1.0]   % values of output power [W]
% Determine steady-state photon density from a given output power
S_zero = 2*power_out/(v_g*alpha_m*h_Planck*freq_ph*V_p);
%
% Construct denominator
D = - omega.^2 + 1j*omega*(1/tau +v_g*a*S_zero) + v_g*a*S_zero/tau_p;
%
response = (v_g*a*S_zero/tau_p)./abs(D);
response_dB = 10*log(response);
%
h = semilogx(freq_GHz,response_dB);
end
xlabel('frequency [GHz]','FontSize',14);      % size of x label
ylabel('Response function [dB]','FontSize',14); % size of y label
```

```
set(gca,'FontSize',14);   % size of tick marks on both axes
axis([f_min f_max -20 35]);
text(0.05, 7, '10 \muW','Fontsize',14)
text(0.2, 18, '1 mW','Fontsize',14)
text(0.6, 27, '10 mW','Fontsize',14)
text(2.5, 33, '0.1W','Fontsize',14)
text(8, 26, '1W','Fontsize',14)
pause
close all
```

Listing 7A.1.2 Function param_rate_eq_bulk.m. Function contains parameters for the rate equation model for the bulk active region.

```
% File name: param_rate_eq_bulk.m
% Purpose:
% Contains parameters for rate equation model for bulk
% active region
% Source
% G.P. Agrawal and N.K. Dutta
% Long-wavelength semiconductor lasers
% Van Nostrand 1986
% Table 6.1, p. 227
%
% General constants
c = 3d10;                 % velocity of light [cm/s]
q = 1.6021892d-19;        % elementary charge [C]
h_Planck = 6.626176d-34;  % Planck constant [J s]
hbar = h_Planck/(2.0*pi); % Dirac constant [J s]
% Geometrical dimensions bulk active region
length = 250d-3;          % cavity length [cm]; 250 microns
width = 2d-3;             % active region width [cm]; 2 microns
thickness = 0.2d-3;       % thickness of an active region [cm]; 0.2 microns
volume_active = length*width*thickness;  % volume of active region
%
conf = 0.3;               % confinement factor [dimensionless]
V_p = volume_active/conf; % cavity volume [cm^3]
ref_index = 3.4;          % effective mode index
%
v_g = c/ref_index;        % group velocity [cm/s]
tau_p = 1.6d-12;          % photon life-time [s]
tau = 2.2d-9;             % carrier life-time [2.71 ns]
%
a = 2.5d-16;              % differential gain (linear model) [cm^2]
% Parameters needed to determine output power
alpha_m = 45;             % mirror reflectivity [cm^-1]
lambda_ph = 1.3d-3;       % laser wavelength [microns]; 1.3 microns
freq_ph = v_g/lambda_ph;  % phonon frequency
```

Listing 7A.2.1 Function small_epsilon.m. Driver program which determines small-signal response with gain compression for a single quantum well. For large-signal analysis based on rate equations.

```
% File name: small_epsilon.m
% Purpose:
% Determines response function for quantum well with epsilon
clear all
param_rate_eq_QW       % input data
%
% loop over frequency in GHz
N_max = 100;                               % number of points for plot
f_min = 0.1; f_max = 100;
freq_GHz = linspace(f_min,f_max,N_max);    % From 0.1 to 2 GHz
semilogx(freq_GHz,0);                      % make axis and force log scale
%
freq = freq_GHz*1d9;                       % convert frequency to 1/s
omega = 2*pi*freq;
power_out = 0.001;                         % output power (10 mW/facet)
%
hold on
for epsilon = [0 1.5d-17 5d-17]            % values of epsilon
% Determine steady-state photon density from a given output power
S_zero = 2*power_out/(v_g*alpha_m*h_Planck*freq_ph*V_p);
%
% Construct denominator
A = 1/tau + v_g*S_zero*a/(1+epsilon*S_zero);
B = 1/(conf*tau_p) - (1/(conf*tau_p))*S_zero*epsilon/(1+epsilon*S_zero);
C = conf*v_g*S_zero*a/(1+epsilon*S_zero);
D = (1/tau_p)*S_zero*epsilon/(1+epsilon*S_zero);
damping = A + D;
omega_R2 = A*D + C*B;

H = - omega.^2 + 1j*omega.*damping + omega_R2;
%
response = (A*D+C*B)./abs(H);
response_dB = 10*log(response);
h = plot(freq_GHz,response_dB);
end
xlabel('frequency [GHz]','FontSize',14);          % size of x label
ylabel('Response function [dB]','FontSize',14);   % size of y label
set(gca,'FontSize',14);                    % size of tick marks on both axes
axis([f_min f_max -20 20]);
text(2, 13, '\epsilon = 0','Fontsize',14)
text(2, 10, '\epsilon = 1.5 \times 10^{-17}','Fontsize',14)
text(2, 7, '\epsilon = 5 \times 10^{-17}','Fontsize',14)
pause
close all
```

Listing 7A.2.2 Function param_rate_eq_QW.m. Function which contains rate equation parameters for a single quantum well.

```
% File name: param_rate_eq_QW.m
% Purpose:
% Contains parameters for rate equation model for QW active region
% Source
% L.A. Coldren and S.W. Corzine,
% "Diode Lasers and Photonic Integrated Circuits", Wiley 1995.
% General constants
c = 3d10;                 % velocity of light [cm/s]
q = 1.6021892d-19;        % elementary charge [C]
h_Planck = 6.626176d-34;  % Planck constant [J s]
hbar = h_Planck/(2.0*pi); % Dirac constant [J s]
% Geometrical dimensions
% QW active region
L = 250d-4;               % cavity length [cm];         250 microns
w = 2d-4;                 % active region width [cm];   2 microns
d = 80d-8;                % thickness of an active region [cm]; 80 Angstroms
%
%V = L*w*d;
V = 4d-12;                % volume of active region
%
conf = 0.03;              % confinement factor [dimensionless]
V_p = V/conf;             % cavity volume [cm^3]
ref_index = 4.2;          % effective mode index
%
v_g = c/ref_index;        % group velocity [cm/s]
tau_p = 2.77d-12;         % photon life-time [s]
tau = 2.71d-9;            % carrier life-time [s]
beta_sp = 0.8d-4;         % spontaneous emission factor
%
a = 5.34d-16;             % differential gain (linear model) [cm^2]
N_tr = 1.8d18;            % carrier density at transparency [cm^-3]
eta_i = 0.8;
epsilon = 1d-17;
current_th = 1.11d-3;     % current at threshold [A]; 1.11 mA
%
% Parameters needed to determine output power
alpha_m = 45;             % mirror reflectivity [cm^-1]
lambda_ph = 1.3d-3;       % laser wavelength [microns]; 1.3 microns
freq_ph = v_g/lambda_ph;  % phonon frequency
```

Listing 7A.3.1 Function large_signal.m. Driver program for large-signal analysis based on rate equations.

```
% File name: large_signal.m
% Purpose:
% Driver which controls computations for large signal rate equations
% function z = large_signal
%
clear all
tspan = [0 2d-9];                    % time interval, up to 2 ns
y0 = [0, 0];                         % initial values of N and S
%
[t,y] = ode45('eqs_large',tspan,y0);
%
size(t);
t=t*1d9;
y_max = max(y);
y1 = y_max(1);
y2 = y_max(2);
h = plot(t,y(:,1)/y1,'-.', t, y(:,2)/y2);   % divided to normalize
set(h,'LineWidth',1.5);                     % thickness of plotting lines
xlabel('time [ns]','FontSize',14);   % size of x label
ylabel('Arbitrary units','FontSize',14);    % size of y label
set(gca,'FontSize',14);                     % size of tick marks on both axes
legend('carrier density', 'photon density')  % legend inside the plot
pause
close all
```

Listing 7A.3.2 Function eqs_large.m. Function defines rate equations equations for large-signal analysis.

```
% File name: eqs_large.m
% Purpose:
% Large signal rate equations are established
% N == y(1)
% S == y(2)
% assumed linear gain model
function ydot = eqs_large(t,y)
%
param_rate_eq_QW                 % input of needed parameters
%
current =  3d-2;                 % bias current (step function) [A]; 3 mA
A = v_g*a*(y(1) - N_tr)/(1+epsilon*y(2));
ydot(1) = eta_i*current/(q*V) - y(1)/tau - A*y(2);
ydot(2) = conf*A*y(2) - y(2)/tau_p + conf*beta_sp*y(1)/tau;
ydot = ydot';                    % must return column vector
```

References

[1] A. Yariv. *Quantum Electronics. Third Edition*. Wiley, New York, 1989.

[2] C. C. Davis. *Lasers and Electro-Optics. Fundamentals and Engineeing*. Cambridge University Press, Cambridge, 2002.

[3] B. Mroziewicz, M. Bugajski, and W. Nakwaski. *Physics of Semiconductor Lasers*. Polish Scientific Publishers/North-Holland, Warszawa, Amsterdam, 1991.

[4] L. A. Coldren and S. W. Corzine. *Diode Lasers and Photonic Integrated Circuits*. Wiley, New York, 1995.

[5] J. Carroll, J. Whiteaway, and D. Plumb. *Distributed Feedback Semiconductor Lasers*. The Institution of Electrical Engineers, London, 1998.

[6] E. Kapon, ed. *Semiconductor Lasers*. Academic Press, San Diego, 1999.

[7] D. Sands. *Diode Lasers*. Institute of Physics Publishing, Bristol and Philadelphia, 2005.

[8] C. W. Wilmsen, H. Temkin, and L. A. Coldren, eds. *Vertical-Cavity Surface-Emitting Lasers*. Cambridge University Press, Cambridge, 1999.

[9] G. P. Agrawal and N. K. Dutta. *Semiconductor Lasers. Second Edition*. Kluwer Academic Publishers, Boston/Dordrecht/London, 2000.

[10] M. G. A. Bernard and G. Duraffourg. Laser conditions in semiconductors. *Physica Status Solidi*, **1**:699–703, 1961.

[11] G. Keiser. *Optical Fiber Communications. Third Edition*. McGraw-Hill, Boston, 2000.

[12] S.-L. Chuang. *Physics of Optoelectronic Devices*. Wiley, New York, 1995.

[13] J.-M. Liu. *Photonic Devices*. Cambridge University Press, Cambridge, 2005.

[14] G. Lasher and F. Stern. Spontaneous and stimulated recombination radiation in semiconductors. *Phys. Rev.*, **133**:A553–63, 1964.

[15] C. J. Hwang. Properties of spontaneous and stimulated emission in GaAs junction lasers. I. Densities of states in the active regions. *Phys. Rev. B*, **2**:4117–25, 1970.

[16] F. Stern. Calculated spectral dependence of gain in excited GaAs. *J. Appl. Phys.*, **47**:5382–6, 1976.

[17] M. S. Wartak, P. Weetman, T. Alajoki, J. Aikio, V. Heikkinen, N. Pikhin, and P. Rusek. Optical modal gain in multiple quantum well semiconductor lasers. *Canadian Journal of Physics*, **84**:53–66, 2006.

[18] S.-L. Chuang. *Physics of Photonic Devices*. Wiley, New York, 2009.

[19] J. P. Loehr. *Physics of Strained Quantum Well Lasers*. Kluwer Academic Publishers, Boston, 1998.

[20] P. S. Zory, Jr., ed. *Quantum Well Lasers*. Academic Press, Boston, 1993.

[21] T.-A. Ma, Z.-M. Li, T. Makino, and M. S. Wartak. Approximate optical gain formulas for 1.55 μm strained quaternary quantum well lasers. *IEEE J. Quantum Electron.*, **31**:2934, 1995.

[22] T. Makino. Analytical formulas for the optical gain of quantum wells. *IEEE J. Quantum Electron.*, **32**:493–501, 1996.

[23] S. Balle. Simple analytical approximations for the gain and refractive index spectra in quantum-well lasers. *Phys. Rev. A*, **57**:1304–12, 1998.

[24] L. D. Westbrook. Measurements of dg/dN and dn/dN and their dependence on photon energy in $\lambda = 1.5$ µm InGaAsP laser diodes. *IEE Proceedings J*, **133**:135–42, 1986.

[25] L. D. Westbrook. Measurements of dg/dN and dn/dN and their dependence on photon energy in $\lambda = 1.5$ µm InGaAsP laser diodes. Addendum. *IEE Proceedings J.*, **134**:122, 1987.

[26] H. Ghafouri-Shiraz and B. S. L. Lo. *Distributed Feedback Laser Diodes*. Wiley, Chichester, 1996.

[27] M.-C. Amann and J. Buus. *Tunable Laser Diodes*. Artech House, Boston, 1998.

[28] M. J. Connelly. *Semiconductor Optical Amplifiers*. Kluwer Academic Publishers, Boston, 2002.

[29] K. J. Ebeling, ed. *Integrated Optoelectronics*. Springer-Verlag, Berlin, 1989.

[30] R. S. Tucker. High-speed modulation of semiconductor lasers. *J. Lightwave Technol.*, **3**:1180–92, 1985.

[31] M. M.-K. Liu. *Principles and Applications of Optical Communications*. Irwin, Chicago, 1996.

[32] M. Ahmed and A. El-Lafi. Analysis of small-signal intensity modulation of semiconductor lasers taking account of gain suppression. *PRAMANA*, **71**:99–115, 2008.

[33] R. S. Tucker. High-speed modulation of semiconductor lasers. *IEEE Trans. Electron Devices*, **32**:2572–84, 1985.

[34] A. Yariv and P. Yeh. *Photonics. Optical Electronics in Modern Communications. Sixth Edition*. Oxford University Press, New York, Oxford, 2007.

[35] K. Otsuka and S. Tarucha. Theoretical studies on injection locking and injection-induced modulation of lase diodes. *IEEE J. Quantum Electron.*, **17**:1515, 1981.

[36] D. Marcuse and T.-P. Lee. On approximate analytical solutions of rate equations for studying transient spectra of injection lasers. *IEEE J. Quantum Electron.*, **19**:1397, 1983.

[37] M. S. Demokan and A. Nacaroglu. An analysis of gain-switched semiconductor lasers generating pulse-code-modulated light with a high bit rate. *IEEE J. Quantum Electron.*, **20**:1016, 1984.

[38] J. Katz, S. Margalit, C. Harder, D. Wilt, and A. Yariv. The intrinsic electrical equivalent circuit of a laser diode. *IEEE J. Quantum Electron.*, **17**:4–7, 1981.

[39] C. Harder, J. Katz, S. Margalit, J. Shacham, and A. Yariv. Noise equivalent circuit of a semiconductor laser diode. *IEEE J. Quantum Electron.*, **18**:333–7, 1982.

[40] R. S. Tucker and D. J. Pope. Circuit modeling of the effect of diffusion on damping in a narrow-stripe semiconductor laser. *IEEE J. Quantum Electron.*, **19**:1179–83, 1983.

[41] O. Kibar, D. Van Blerkom, C. Fan, P. J. Marchand, and S. C. Esener. Small-signal-equivalent circuits for a semiconductor laser. *Applied Optics*, **37**:6136–9, 1998.

[42] K. Y. Lau. Dynamics of quantum well lasers. In P. S. Zory, Jr, ed., *Quantum Well Lasers*, pages 217–75. Academic Press, Boston, 1993.

[43] E. J. Flynn. A note on the semiconductor laser equivalent circuit. *J. Appl. Phys.*, **85**:2041–5, 1999.

8. Optical amplifiers and EDFA

As signals travel in optical communication systems, they are attenuated by optical fibre. Eventually, after some distance they can become too weak to be detected. One possible way to avoid this situation is to use optical amplifiers to increase the amplitude of the signal. An optical amplifier is an optical device for amplifying signals propagating over several channels in optical fibre to compensate for their loss during propagation. Signals also get distorted, say by presence of noise, and the amplification process does not clean them nor reshape them. The cleaning process is commonly known as regeneration and will not be discussed here.

In this chapter we will concentrate on optical amplifiers. We review their general properties and will also discuss erbium doped fibre amplifiers (EDFA). Semiconductor optical amplifiers (SOA), which are essentially semiconductor lasers without mirrors so they do not lase, will be discussed in the next chapter.

Possible applications of optical amplifiers are illustrated in Fig. 8.1. Some of the basic applications are listed below:

a) as in-line amplifiers for power boosting,
b) as pre-amplifiers to increase the received power at the receiver,
c) as power amplifiers to increase transmitted power,
d) as a power booster in a local area network.

The main types of optical amplifiers are:

- doped fibre amplifiers
- Raman and Brillouin amplifiers, and
- semiconductor optical amplifiers.

Issues which will be discussed in relation to those amplifiers are: gain and gain bandwidth, gain saturation, and noise and noise figure.

In short, the operation and characteristics of all types of amplifiers are:

1. population inversion is created, which means that more systems (atoms, molecules) are in a high energy state than in a lower one,
2. the incoming pulses of signal induce stimulated emission,
3. amplifiers saturate above a certain signal power,
4. amplifiers add noise to the signal.

In this book we will only describe two types of optical amplifiers: semiconductor optical amplifiers (SOA) and erbium doped fibre amplifiers (EDFA). Their general characteristics are compared in Table 8.1 where we summarize essential properties of EDFA and

Table 8.1 Some properties of EDFA and SOA.

Properties of amplifiers	EDFA	SOA
Active medium	Er^{3+} ion in silica	Electron-hole in semiconductors
Typical length	Few metres	500 μm
Pumping	Optical	Electrical
Gain spectrum	1.5–1.6 μm	1.3–1.5 μm
Gain bandwidth	24–35 nm	100 nm
Relaxation time	0.1–1 ms	<10–100 ps
Maximum gain	3–50 dB	25–30 dB
Saturation power	>10 dBm	0–10 dBm
Crosstalk	–	For bit rate 10 GHz
Polarization	Insensitive	Sensitive
Noise figure	3–4 dB	6–8 dB
Insertion loss	<1 dB	4–6 dB
Optics	Pump laser diode couplers, fibre splice	Antireflection coatings, fibre-waveguide coupling
Integration	No	Yes

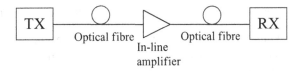

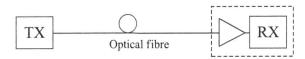

Fig. 8.1 Possible applications of optical amplifiers: as an in-line amplifier (top) or as a preamplifier (bottom). TX stands for transmitter, RX for receiver.

SOA, after Yariv and Yeh [1] (reproduced with permission). Next, we will discuss general characteristics of optical amplifiers.

Some books on the subject are: Desurvire [2], Becker *et al.* [3] and Bjarklev [4].

8.1 General properties

Amplification in optical amplifiers is through stimulated emission (the same as in lasers). One can therefore consider similar characteristic parameters, like gain and its spectrum, bandwidth, etc. We will discuss them in some detail.

8.1.1 Gain spectrum and bandwidth

To model gain, one typically starts with a homogeneously broadened two-level system. Local gain coefficient for such system is [5]

$$g(\nu, P) = \frac{g_0}{1 + \frac{(\nu-\nu_0)^2}{\Delta\nu_0^2} + \frac{P}{P_{sat}}} \tag{8.1}$$

where g_0 is the peak value of the unsaturated gain, ν_0 atomic transition frequency, $\Delta\nu_0$ 3-dB local gain bandwidth, P_{sat} saturation power and P and ν are optical power and frequency of the amplified signal.

Local gain can also be written as

$$g(\omega, P) = \frac{g_0}{1 + (\omega - \omega_0)^2 T_2^2 + P/P_{sat}} \tag{8.2}$$

where $\omega = 2\pi\nu$ is the angular frequency and T_2 is known as dipole relaxation time [6].

For small signal power one can consider a unsaturated regime defined as $P \ll P_{sat}$. In this limit local gain is

$$g(\omega) = \frac{g_0}{1 + (\omega - \omega_0)^2 T_2^2} \tag{8.3}$$

The following conclusions can be derived from the above equation:

1. maximum gain corresponds to transitions with angular frequency $\omega = \omega_0$,
2. for $\omega \neq \omega_0$ gain spectrum is described by Lorentzian profile,
3. local gain bandwidth, which is defined as the full width at half maximum (FWHM), is

$$\Delta\omega_0 = \frac{2}{T_2} \tag{8.4}$$

Local gain bandwidth $\Delta\omega_0$ is defined by points in frequency where local gain takes half the value at the maximum. In terms of frequency it can be written as

$$\Delta\nu_0 = \frac{\Delta\omega_0}{2\pi} = \frac{1}{\pi T_2} \tag{8.5}$$

Let $P(z)$ be the optical power at a distance z from the input end. Its change is described as

$$\frac{dP(z)}{dz} = g(\nu, P) \cdot P(z) \tag{8.6}$$

Assume a linear device (here for power levels $P \ll P_{sat}$) where local gain is independent of the signal power. Integration of the above equation gives

$$P(z) = P(0) e^{g \cdot z} \tag{8.7}$$

where $P(0) = P_{in}$ is the signal input power. Linear amplifier gain is defined as

$$G = \frac{P(L)}{P(0)} \tag{8.8}$$

where $P(L) = P_{out}$ is the output power. From the above solution one obtains

$$G = \frac{P(L)}{P(0)} = e^{gL} = \exp\left[\frac{g_0 \cdot L}{1 + \frac{(\nu-\nu_0)^2}{\Delta\nu_0^2}}\right] \qquad (8.9)$$

Amplifier bandwidth B_0 is evaluated using the above solution. It is defined by two frequency points where power drops by 50%, i.e. $P_{3dB} = \frac{1}{2}P_{max}(L)$, which translates into $G_{3dB} = \frac{1}{2}G_{max}$. Here G_{max} is the maximum value of gain evaluated at $\nu = \nu_0$. In detail

$$\exp\left[\frac{g_0 \cdot L}{1 + \frac{B_0^2}{\Delta\nu_0^2}}\right] = \frac{1}{2}\exp\left[\frac{g_0 \cdot L}{1 + \frac{0}{\Delta\nu_0^2}}\right] = \frac{1}{2}\exp(g_0 \cdot L)$$

where we have introduced 3 − dB bandwidth B_0 as $B_0 = \nu - \nu_0$. By the straightforward algebra, from the above relation one finds

$$B_0 = \Delta\nu_0 \sqrt{\frac{\ln 2}{g_0 L - \ln 2}} \qquad (8.10)$$

Macroscopic bandwidth of the amplifier B_0 is smaller than the local gain bandwidth $\Delta\nu_0$.

8.1.2 Gain saturation

We will now analyse the gain when signal power has large value and saturation effects are becoming important. Assume that $\omega = \omega_0$. Substituting Eq. (8.2) into (8.6), one obtains

$$\frac{dP}{dz} = \frac{g_0 P}{1 + P/P_{sat}} \qquad (8.11)$$

We introduce new variable $u = P/P_{sat}$ and use separation of variables method to integrate the above equation. One obtains

$$\int_{u_{in}}^{u_{out}} \frac{1+u}{u} du = \int_0^L g_0\, dz$$

where $u_{in} = P_{in}/P_{sat}$, $u_{out} = P_{out}/P_{sat}$ and P_{in}, P_{out} are input and output powers, respectively. L is the length of an amplifier. Gain G is defined as

$$G = \frac{P_{out}}{P_{in}} \qquad (8.12)$$

One obtains

$$G = G_0 \exp\left(-\frac{G-1}{G}\frac{P_{out}}{P_S}\right) \qquad (8.13)$$

where $G_0 = \exp(g_0 \cdot L)$.

In Fig. 8.2 we have plotted saturation gain dependence from Eq. (8.13). MATLAB code and its description are provided in Appendix, Listings 8A.1.1, 8A.1.2 and 8A.1.3.

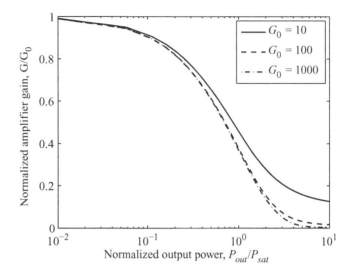

Fig. 8.2 Saturated normalized amplifier gain G/G_0 as a function of the normalized output power for three values of the unsaturated amplifier gain G_0.

8.1.3 Amplifier noise

Signal-to-noise ratio (SNR) of optical amplifiers is degraded because spontaneous emission adds to the signal during its amplification. Amplifier noise figure F_n is defined as [7]

$$F_n = \frac{(SNR)_{in}}{(SNR)_{out}} \quad (8.14)$$

SNR refers to the electrical power generated when the signal is converted to electrical current by using a photodetector. We model F_n by considering an ideal detector limited only by a shot noise

$$(SNR)_{in} = \frac{\langle I \rangle^2}{\sigma_s^2} \quad (8.15)$$

where $\langle I \rangle = RP_{in}$ is the average photocurrent, $R = \frac{q}{h\nu}$ is the responsivity of an ideal detector with unit quantum efficiency, $\sigma_s^2 = 2q(RP_{in})\Delta f$ is the variance from shot noise and Δf is the detector bandwidth. At the output, we should add spontaneous emission to the receiver noise

$$S_{sp}(\nu) = (G-1)n_{sp}h\nu \quad (8.16)$$

Here S_{sp} is the spectral density of the noise induced by spontaneous emission, ν is optical frequency and n_{sp} is the spontaneous-emission factor or population inversion factor. The value of n_{sp} is $n_{sp} = 1$ for amplifiers with complete population inversion (all atoms in the upper state) and $n_{sp} > 1$ for incomplete population inversion.

For a two-level system

$$n_{sp} = \frac{N_2}{N_2 - N_1} \quad (8.17)$$

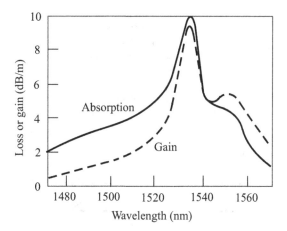

Fig. 8.3 Absorption and gain spectra of Ge silicate amplifier fibre. Copyright (1991) IEEE. Reprinted, with permission from C. R. Giles and E. Desurvire, *J. Lightwave Technol.*, **9**, 271 (1991).

where N_1 and N_2 are the atomic populations in the lower and upper states, respectively.

Total variance of the shot noise plus spontaneous emission noise is thus

$$\sigma^2 = 2q\,(RGP_{in})\,\Delta f + 4\,(GRP_{in})\,(RS_{sp})\,\Delta f \tag{8.18}$$

All other contributions to the receiver noise are neglected. At the output, the SNR of the amplified signal is

$$(SNR)_{out} = \frac{\langle I \rangle^2}{\sigma^2} = \frac{(RGP_{in})^2}{\sigma^2} \approx \frac{GP_{in}}{4S_{sp}\Delta f} \tag{8.19}$$

assuming $G \gg 1$ and we neglected first term in (8.18).

Using definition of F_n, one finds

$$F_n = 2n_{sp}\frac{G-1}{G} \approx 2n_{sp} \tag{8.20}$$

It shows that even for an ideal amplifier ($n_{sp} = 1$), amplified signal is degraded by a factor of 2 (3 dB). In practice, F_n is in the range 6–8 dB.

8.2 Erbium-doped fibre amplifiers (EDFA)

Experimental data on absorption and gain of an EDFA whose core was codoped with Ge to increase the refractive index are shown in Fig. 8.3, after Giles and Desurvire [8].

An optical fibre amplifier consists of an optical fibre where amplification takes place which is doped with a rare-earth element, and a pumping light supply system for supplying pumping light to the optical fibre for amplification, see Fig. 8.4. The pumping light supply system usually includes a semiconductor laser and an optical coupler for guiding the pumping light into the optical fibre for amplification.

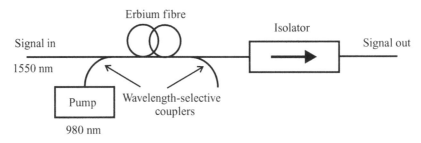

Fig. 8.4 Schematic of EDFA.

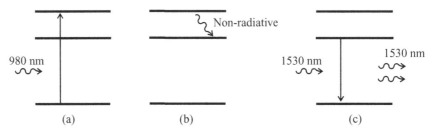

Fig. 8.5 Illustration of amplification in EDFA using a three-level model.

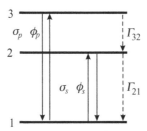

Fig. 8.6 Illustration of notation for a three-level system.

Erbium-doped amplifiers are made by doping a segment of the fibre with erbium and then exciting the erbium atoms to a high energy level through the introduction of pumping light. The energy is transferred gradually to signal light passing through the fibre segment during excitation, resulting in an amplification of the signal light upon exit from the amplifier. Fibre optic amplifiers can amplify signal light including one or more wavelengths within a predetermined wavelength band without converting them into an electrical signal.

Illustration of the amplification process in EDFA based on the three-level model is shown in Fig. 8.5 for the case of a 980 nm pump. A pump photon at a 980 nm wavelength is absorbed by an erbium ion in the ground state and jumps into the highest energy level. Then, through a non-radiative decay the ion loses its energy and arrives into a metastable state. Once there, a photon having wavelength of 1530 nm can force a stimulated transition of the ion into its ground state, creating one additional photon and an amplification.

In this section, we will describe EDFA using a three-level model. We follow Becker [3]. The model is shown in Fig. 8.6. Level 1 is a ground state, level 2 is known as a metastable one (it has a long lifetime) and level 3 is an intermediate state. The population of levels are introduced as N_1, N_2, N_3. The spontaneous transition rates of the ion (transition probabilities) which include radiative and also non-radiative contributions are denoted as Γ_{32} and Γ_{21} and correspond to transitions between levels $3 \rightarrow 2$ and $2 \rightarrow 1$, respectively. σ_p is the pump absorption cross section and σ_s is the signal emission cross section. The incident light intensity fluxes of pump and signal are denoted by ϕ_p and ϕ_s, respectively. They are defined as number of photons per unit time per unit area.

This three-level model represents energy level structure of Er^+ which is involved in the amplification process. To obtain amplification, we need a population inversion between levels 1 and 2. Here, we only consider a one-dimensional model where we assume that the pump and signal intensities and also distribution of Er ions are constant in the transverse direction.

Based on the above observations, the rate equations for the changes of populations for all levels are postulated as

$$\frac{dN_1}{dt} = \Gamma_{21} N_2 - (N_1 - N_3) \sigma_p \phi_p + (N_2 - N_1) \sigma_s \phi_s \tag{8.21}$$

$$\frac{dN_2}{dt} = -\Gamma_{21} N_2 + \Gamma_{32} N_3 - (N_2 - N_1) \sigma_s \phi_s \tag{8.22}$$

$$\frac{dN_3}{dt} = -\Gamma_{32} N_3 + (N_1 - N_3) \sigma_p \phi_p \tag{8.23}$$

8.2.1 Steady-state analysis

In the steady-state conditions

$$\frac{dN_1}{dt} = \frac{dN_2}{dt} = \frac{dN_3}{dt} = 0 \tag{8.24}$$

Also, total population N is assumed to be constant:

$$N = N_1 + N_2 + N_3 \tag{8.25}$$

From Eq. (8.23) one obtains

$$N_3 = N_1 \frac{1}{1 + \frac{\Gamma_{32}}{\sigma_p \phi_p}} \tag{8.26}$$

In what follows we will assume fast decay from level 3 to level 2; i.e. decay from level 3 is dominant compared to pump rate. Mathematically, the assumption corresponds to the following condition $\Gamma_{32} \gg \sigma_p \phi_p$. Lifetime of level 3, τ_{32} is related to transition probability as $\tau_{32} = 1/\Gamma_{32}$. In this limit there is almost no population of level 3, and therefore $N_3 \approx 0$. With those assumptions the system can be effectively considered as consisting of two levels only which are described by the following equations:

$$\frac{\sigma_p \phi_p + \sigma_s \phi_s}{\Gamma_{21} + \sigma_s \phi_s} N_1 - N_2 = 0 \tag{8.27}$$

$$N_1 + N_2 = N \tag{8.28}$$

Solutions by standard methods give for population inversion

$$N_2 - N_1 = N \frac{\sigma_p \phi_p - \Gamma_{21}}{\Gamma_{21} + 2\sigma_s \phi_s + \sigma_p \phi_p} \tag{8.29}$$

8.2.2 Effective two-level approach

Keeping the above assumption, i.e. $\Gamma_{32} \gg \sigma_p \phi_p$ which allows us to neglect level 3, from Eq. (8.28) one has

$$\frac{dN_1}{dt} = -\frac{dN_2}{dt}$$

It is therefore enough to consider only one equation, say for N_2; the other population N_1 can be found from $N_1 = N - N_2$. Following Pedersen et al. [9], to describe the system we postulate the following equations:

$$\frac{dN_2}{dt} = -\Gamma_{21} N_2 + \left[\sigma_s^{(a)} N_1 - \sigma_s^{(e)} N_2 \right] \phi_s - \left[\sigma_p^{(e)} N_2 - \sigma_p^{(a)} N_1 \right] \phi_p \tag{8.30}$$

$$\frac{dN_1}{dt} = \Gamma_{21} N_2 + \left[\sigma_s^{(e)} N_2 - \sigma_s^{(a)} N_1 \right] \phi_s - \left[\sigma_p^{(a)} N_1 - \sigma_p^{(e)} N_2 \right] \phi_p \tag{8.31}$$

where $\sigma_s^{(a)}, \sigma_s^{(e)}, \sigma_p^{(a)}, \sigma_p^{(e)}$ represent signal and pump cross sections for absorption and emission, respectively. Assume steady-state and from Eq. (8.30) determine N_2:

$$N_2 = N \frac{\sigma_s^{(a)} \phi_s + \sigma_p^{(a)} \phi_p}{\frac{1}{\tau} + \left[\sigma_s^{(a)} + \sigma_s^{(e)} \right] \phi_s + \left[\sigma_p^{(a)} + \sigma_p^{(e)} \right] \phi_p}$$

where we have introduced $\tau = 1/\Gamma_{21}$. Further, introduce signal I_s and pump I_p intensities as

$$\phi_s = \frac{I_s}{h\nu_s}, \quad \phi_p = \frac{I_p}{h\nu_p}$$

where h is the Planck constant and ν_s, ν_p are frequencies of the signal and pump, respectively. This allows us to write

$$N_2 = N \frac{\tau \frac{\sigma_s^{(a)}}{h\nu_s} I_s(z) + \tau \frac{\sigma_p^{(a)}}{h\nu_p} I_p(z)}{\tau \frac{\sigma_s^{(a)} + \sigma_s^{(e)}}{h\nu_s} I_s(z) + \tau \frac{\sigma_p^{(a)} + \sigma_p^{(e)}}{h\nu_p} I_p(z) + 1}$$

We further assume that N is independent of distance along fibre z. The variation of signal and pump intensities is described as

$$\frac{dI_s(z)}{dz} = \left[\sigma_s^{(e)} N_2 - \sigma_s^{(a)} N_1 \right] I_s(z)$$

$$\frac{dI_p(z)}{dz} = \left[\sigma_p^{(e)} N_2 - \sigma_p^{(a)} N_1 \right] I_p(z)$$

Table 8.2 Typical parameters of a low-noise in-line EDFA.

Parameter	Symbol	Value	Unit
Signal mode area	πw^2	1.3×10^{-11}	m^2
Erbium concentration	N_{tot}	5.4×10^{24}	m^{-3}
Signal overlapping integral	Γ_s	0.4	
Pump overlapping integral	Γ_p	0.4	
Signal emission cross section	s_{se}	5.3×10^{-25}	m^2
Signal absorption cross section	s_{sa}	3.5×10^{-25}	m^2
Pump absorption cross section	s_p	3.2×10^{-25}	m^2
Signal local saturation power	P_{ss}	1.3	mW
Pump local saturation power	P_{sp}	1.6	mW

8.3 Gain characteristics of erbium-doped fibre amplifiers

We outline a basic approach to evaluate the performance of EDFA. Neglecting amplified spontaneous emission (ASE) and assuming copropagation configuration, the equations describing steady-state are

$$\frac{dI_s(z)}{dz} = 2\pi \Gamma_s \left\{ \sigma_{se} N_{me}(z) I_s(z) - \sigma_{sa} N_{gr}(z) I_s(z) \right\} \tag{8.32}$$

$$\frac{dI_p(z)}{dz} = 2\pi \Gamma_p \sigma_{sa} N_{gr}(z) I_p(z) \tag{8.33}$$

$$N_{me}(z) = N_{tot} \frac{I_s(z)/I_{ss} + I_p(z)/I_{sp}}{1 + I_p(z)/I_{sp} + 2I_s(z)/I_{ss}} \tag{8.34}$$

$$N_{me}(z) + N_{gr}(z) = N_{tot} \tag{8.35}$$

In the previous equations $N_{gr}(z)$ is the population of the ground state, $N_{me}(z)$ is the population of the metastable state, $I_p(z)$ is the intensity of the pump wave propagating at the wavelength λ_p and $I_s(z)$ is the intensity of the signal wave propagating at the wavelength λ_s in the positive z-direction.

Finally, the overall amplifier gain G is obtained using the following relation:

$$G = \frac{I_s(L)}{I_s(0)} \tag{8.36}$$

where L is the doped fibre length.

8.3.1 Typical EDFA characteristics

The above equations were solved numerically using parameter values from Table 8.2 [8], [10], [11]. MATLAB codes are provided in Listings 8A.2.1–8A.4.2.

The variation of gain with fibre length was determined first and the results are summarized in Fig. 8.7 for different values of pump power. MATLAB code is shown in the Appendix, Listings 8A.2.1–8A.2.4. Constant signal input power of 10 µW and constant erbium doping

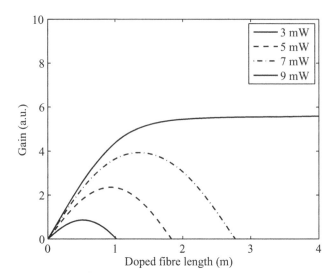

Fig. 8.7 Variation of gain versus fibre length for four different values of pump power.

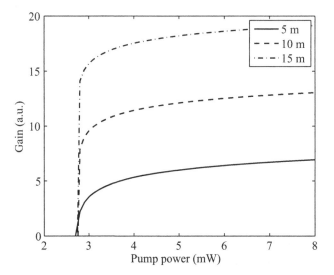

Fig. 8.8 Variation of gain versus pump power for three different values of fibre length.

density have been assumed. Gain was evaluated for four different pump power levels equal to 3, 5, 7 and 9 mW. As it is shown, gain increases up to a certain fibre's length and then begins to decrease after reaching a maximum point. Similar results were reported in the literature, compare [10] and [11].

The optimum fibre length (the one which corresponds to a maximum gain) is a few metres and it increases with the pump power. The reason for the decrease in gain is insufficient population inversion due to excessive pump depletion.

In Fig. 8.8 we show the variation of gain with pump power for three different values of fibre lengths of 5, 10 and 15 m and a constant signal input power equal to 1 mW. MATLAB code is shown in the Appendix, Listings 8A.3.1 and 8A.3.2. Constant Er doping density was also assumed. As it is seen, gain of the EDFA increases with the increasing pump power and then saturates after a certain level of pump power. The pump saturation effect occurs for input powers in the range 3–6 mW. Consult literature for more discussion of the obtained results [10] and [11].

8.4 Problems

1. An EDFA has 20 dB of gain. If the power of incident signal is 150 µW, what is the output power?
2. An optical amplifier amplifies an incident signal of 1 µW to 1 mW. Assuming the saturation power of 20 mW, what will be the output power of 100 µW incident signal on the same amplifier?
3. An optical amplifier produces power gain of 1000 over 30 m with a 10 mW pump. (a) What is the gain exponential coefficient α in nepers per metre? (b) What should be the length of that amplifier to produce gain of 1500?
4. Using steady-state amplifier equations and appropriate approximations, derive an analytical expression for an optimum length of fibre which gives maximum amplification.

8.5 Projects

1. Perform a literature survey on EDFA with noise. Write a MATLAB program to simulate EDFA with realistic noise.
2. Perform a literature survey on Raman amplifiers. Write MATLAB code to simulate Raman amplifier. Compare Raman amplifier and EDFA.
3. Formulate transient model of EDFA with the saturation effect of the forward and backward amplified spontaneous emission [12]. Integrate numerically the resulting equations. Analyse gain saturation and recovery times of an EDFA and the effects on the amplification of optical pulses.

Appendix 8A: MATLAB listings

In Table 8.3 we provide a list of MATLAB files created for Chapter 8 and a short description of each function.

Table 8.3 List of MATLAB functions for Chapter 8.

Listing	Function name	Description
8A.1.1	*sat_gain.m*	Creates data for saturated gain
8A.1.2	*f_sat_gain.m*	Function used by *sat_gain.m*
8A.1.3	*sat_plot.m*	Function plots saturated gain
8A.2.1	*gain_variable_length.m*	Generates date for plots gain vs fibre length for fixed gain
8A.2.2	*edfa_param.m*	Parameters for EDFA
8A.2.3	*edfa_eqs.m*	Equations for EDFA
8A.2.4	*variable_length_plot.m*	Plots gain using data generated by *gain_variable_length.m*
8A.3.1	*gain_variable_pump.m*	Generates data for plots gain vs power for fixed fibre length
8A.3.2	*variable_gain_plot.m*	Plots gain using data generated by *gain_variable_pump.m*

Listing 8A.1.1 Function sat_gain.m. MATLAB program which generates data for saturated normalized amplifier gain G_0. To generate the required data, it must be run three times for three values of G_0. Therefore one should appropriately comment inputs for G_0. Each time a different output file must be specified; i.e. the output file name should be changed.

```
% File name: sat_gain.m
% Generates data for plots of saturated gain G as a function of
% the output power for three values of the unsaturated amplifier gain G_0
% File must be run three times for three values of G_0
clear all
%
G_0 = 10;
%G_0 = 100;
%G_0 = 1000;
u_init = 1d-2;
Delta_u = 0.5d-1;
u_final = 1d1;
uu = u_init;
yy = fzero(@(y) f_sat_gain(y, u_init, G_0), 0.5);
for u = (u_init + Delta_u):Delta_u:u_final
    uu = [uu, u];
    y = fzero(@(y) f_sat_gain(y, u, G_0), 0.5);
    yy = [yy, y];
end
%
d_u = length(uu);
fid = fopen('gain_data_10.dat', 'wt');               % Open the file.
for ii = 1:d_u
```

```
fprintf (fid, ' %11.4f %11.4f\n', uu(ii),yy(ii));   % Print the data
end
status = fclose(fid);                               % Close the file.
```

Listing 8A.1.2 Function f_sat_gain.m. MATLAB function used by a program which plots saturated gain.

```
% defines function used in plot of saturated gain of optical amplifier
function fun = f_sat_gain(y, u, G_0)
fun = y - exp(-(G_0*y-1)*u/(G_0*y));
```

Listing 8A.1.3 Function sat_plot.m. MATLAB function used to plot saturated gain from files. It was not possible here to create one MATLAB program which will perform all the calculations and also to plot all the results. Instead, we had to run *sat_gain.m* three times for three different values of G_0, output the results to data files and then plot final results using a separate program.

```
% File_name: sat_plot.m
% Plots graph of saturated gain using data generated by sat_gain.m
clear all
fid = fopen('gain_data_10.dat');
a_10 = fscanf(fid,'%f %f',[2 inf]); % It has two rows
fclose(fid);
%
fid = fopen('gain_data_100.dat');
a_100 = fscanf(fid,'%f %f',[2 inf]); % It has two rows
fclose(fid);
%
fid = fopen('gain_data_1000.dat');
a_1000 = fscanf(fid,'%f %f',[2 inf]); % It has two rows
fclose(fid);
%
semilogx(a_10(1,:),a_10(2,:),'k-', a_100(1,:),a_100(2,:),'k--',...
    a_1000(1,:),a_1000(2,:),'k-.','LineWidth',1.5)
%
xlabel('Normalized output power, P_{out}/P_{sat}','FontSize',14);
ylabel('Normalized amplifier gain, G/G_0','FontSize',14);
legend('G_0 = 10','G_0 = 100','G_0 = 1000')
set(gca,'FontSize',14);                 % size of tick marks on both axes
pause
close all
```

Listing 8A.2.1 Function gain_variable_length.m. MATLAB function used to compute gain versus fibre length for a fixed value of pump power.

```matlab
% File name: gain_variable_length.m
% Computation of gain vs fibre length for fixed value of pump power
% User selects pump power which is controlled by P_p
% Calculations are repeated for P_p = 3,5,7,9 d-3 Watts
% User should also appropriately rename output file 'gain_P_3.dat'
% based on Iannone-book, p.86
% P_s  == y(1)   signal power
% P_p  == y(2)   pump power
%
clear all
gain = 0.0;
gain_temp = 0.0;
P_s = 100d-7;                          % signal power [Watts]
P_p = 9d-3;     % CHANGE               % pump power [Watts]
%
for d_L = 0.01:0.01:10
    span = [0 d_L];
    y0 = [P_s P_p];                    % initial values of [P_s P_p]
    [z,y] = ode45('edfa_eqs',span,y0); % z - distance
    len = length(z);
    gain_temp = y(len)/y(1);
    gain = [gain, gain_temp];
end
gain_log = log(gain);
d_L_plot = 0.01:0.01:10;
d_L_plot = [0, d_L_plot];
% plot(d_L_plot, gain_log)   % uncomment if you want to see plot when run
% axis([0 4 0 5])
% pause
% close all
%
d_u = length(d_L_plot);
fid = fopen('gain_P_9.dat', 'wt');  % CHANGE     % Open the file.
for ii = 1:d_u
fprintf (fid, ' %11.4f %11.4f\n', d_L_plot(ii),gain_log(ii));
end
status = fclose(fid);                       % Close the file
```

Listing 8A.2.2 Function edfa_param.m. MATLAB function contains parameters for a model of EDFA.

```
% File name: edfa_param.m
%-----------------------------------------------------------------
% Purpose:
% Contains parameters for model of EDFA based on Table 3.2
% Iannone-book
N_tot   = 5.4d24;        % Erbium concentration (m^-3)
Gamma_s = 0.4;           % Signal overlapping integral (dimensionless)
Gamma_p = 0.4;           % Pump overlapping integral (dimensionless)
s_se    = 5.3d-25;       % Signal emission cross-section (m^2)
s_sa    = 3.5d-25;       % Signal absorption cross-section (m^2)
s_p     = 3.2d-25;       % Pump absorption cross-section
P_ss    = 1.3d-3;        % Signal local saturation power (W)
P_sp    = 1.6d-3;        % Pump local saturation power (W)
```

Listing 8A.2.3 Function edfa_eqs.m. MATLAB function contains equations for EDFA.

```
function y_z = edfa_eqs(z,y)
% Purpose:
% To establish equations for EDFA following Iannone-book, p.86
% P_s  == y(1)   signal power
% P_p  == y(2)   pump power
%
edfa_param                          % input of needed parameters
num = y(1)/P_ss + y(2)/P_sp;
denom = y(2)/P_sp + 2*y(1)/P_ss + 1.0;
N_me = N_tot*num/denom;
N_gr = N_tot - N_me;
%
y_z(1) = 2*pi*Gamma_s*(s_se*N_me*y(1) - s_sa*N_gr*y(1));  % derivative
y_z(2) = -2*pi*Gamma_p*s_p*N_gr*y(2);
y_z = y_z';                         % must return column vector
end
```

Listing 8A.2.4 Function variable_length_plot.m. MATLAB function used to plot graph of gain using data generated by *gain_variable_length.m*.

```
% File_name: variable_length_plot.m
% Plots graph of gain using data generated by 'gain_variable_length.m'
clear all
% Open files for 4 values of length of device: 5 m,10 m,15 m
fid = fopen('gain_P_3.dat');
a_3 = fscanf(fid,'%f %f',[2 inf]); % It has two rows
```

```
fclose(fid);
%
fid = fopen('gain_P_5.dat');
a_5 = fscanf(fid,'%f %f',[2 inf]); % It has two rows
fclose(fid);
%
fid = fopen('gain_P_7.dat');
a_7 = fscanf(fid,'%f %f',[2 inf]); % It has two rows
fclose(fid);
%
fid = fopen('gain_P_9.dat');
a_9 = fscanf(fid,'%f %f',[2 inf]); % It has two rows
fclose(fid);
%
plot(a_3(1,:),a_3(2,:),'k-', a_5(1,:),a_5(2,:),'k--',...
    a_7(1,:),a_7(2,:),'k-.',a_9(1,:),a_9(2,:),'k-','LineWidth',1.5)
axis([0 4 0 10])
xlabel('Doped fiber length (m)','FontSize',14);
ylabel('Gain (a.u)','FontSize',14);
legend('3 mW','5 mW','7 mW','9 mW')
set(gca,'FontSize',14);                        % size of tick marks on both axes
pause
close all
```

Listing 8A.3.1 Function gain_variable_pump.m. MATLAB function used to compute gain versus pump power for fixed value of fibre length.

```
% File name: gain_variable_pump.m
% Computation of gain vs power pump for fixed value of fiber length
% User selects fiber length which is controlled by d_L
% Calculations are repeated for d_L = 5,10,15 meters
% User should also appropriately rename output file 'gain_L_5.dat'
% Based on Iannone-book, p.86
% P_s  == y(1)   signal power
% P_p  == y(2)   pump power
%
clear all
gain = 0.0;
gain_temp = 0.0;
P_s = 100d-5;                                  % signal power [Watts]
d_L = 5;         % CHANGE                      % fiber length [meters]
span = [0 d_L];                                % region of integration [meters]
for P_p = 0.1d-3:0.1d-3:10d-3  % pump power [Watts]
    y0 = [P_s P_p];                            % initial values of [P_s P_p]
    [z,y] = ode45('edfa_eqs',span,y0);         % z - distance
    len = length(z);
```

```
        gain_temp = y(len)/y(1);
        gain = [gain, gain_temp];
end
gain_log = log(gain);
P_p_plot = 0.1:0.1:10;
P_p_plot = [0, P_p_plot];
% plot(P_p_plot, gain_log)    % uncomment if you want to see plot when run
% axis([0 10 0 30])
% pause
% close all
%
d_u = length(P_p_plot);
fid = fopen('gain_L_5.dat', 'wt');    % CHANGE    % Open the file.
for ii = 1:d_u
    fprintf (fid, ' %11.4f %11.4f\n', P_p_plot(ii),gain_log(ii));
end
status = fclose(fid);                             % Close the file
```

Listing 8A.3.2 Function variable_pump_plot.m. MATLAB function used to plot graph of gain using data generated by *gain_variable_pump.m*.

```
% File_name: variable_pump_plot.m
% Plots graph of gain using data generated by 'gain_variable_pump.m'
clear all
% Open files for 4 values of length of device: 5 m,10 m,15 m
fid = fopen('gain_L_5.dat');
a_5 = fscanf(fid,'%f %f',[2 inf]); % It has two rows
fclose(fid);
%
fid = fopen('gain_L_10.dat');
a_10 = fscanf(fid,'%f %f',[2 inf]); % It has two rows
fclose(fid);
%
fid = fopen('gain_L_15.dat');
a_15 = fscanf(fid,'%f %f',[2 inf]); % It has two rows
fclose(fid);
%
plot(a_5(1,:),a_5(2,:),'k-', a_10(1,:),a_10(2,:),'k--',...
    a_15(1,:),a_15(2,:),'k-.','LineWidth',1.5)
axis([2 8 0 20])
xlabel('Pump power (mW)','FontSize',14);
ylabel('Gain (a.u)','FontSize',14);
legend('5 m','10 m','15 m')
set(gca,'FontSize',14);                 % size of tick marks on both axes
pause
close all
```

References

[1] A. Yariv and P. Yeh. *Photonics. Optical Electronics in Modern Communications. Sixth Edition*. Oxford University Press, New York, Oxford, 2007.

[2] E. Desurvire. *Erbium-Doped Fiber Amplifiers, Principles and Applications*. Wiley, New York, 2002.

[3] P. C. Becker, N. A. Olsson, and J. R. Simpson. *Erbium-Doped Fiber Amplifiers. Fundamentals and Technology*. Academic Press, San Diego, 1999.

[4] A. Bjarklev. *Optical Fiber Amplifiers: Design and System Applications*. Artech House, Boston, 1993.

[5] A. Yariv. *Quantum Electronics. Third Edition*. Wiley, New York, 1989.

[6] G. P. Agrawal. *Fiber-Optic Communication Systems. Second Edition*. Wiley, New York, 1997.

[7] G. P. Agrawal. Semiconductor optical amplifiers. In G. P. Agrawal, ed., *Semiconductor Lasers. Past, Present, and Future*, pages 243–83. American Institute of Physics Press, Woodbury, New York, 1995.

[8] C. R. Giles and E. Desurvire. Modeling erbium-doped fiber amplifiers. *J. Lightwave Technol.*, **9**:271–83, 1991.

[9] B. Pedersen, A. Bjarklev, O. Lumholt, and J. H. Povlsen. Detailed design analysis of erbium-doped fiber amplifiers. *IEEE Photon. Technol. Lett.*, **3**:548–50, 1991.

[10] E. Iannone, F. Matera, A. Mocozzi, and M. Settembre. *Nonlinear Optical Communication Networks*. Wiley, New York, 1998.

[11] R. Sabella and P. Lugli. *High Speed Optical Communications*. Kluwer Academic Publishers, Dordrecht, 1999.

[12] K. Y. Ko, M. S. Demokan, and H. Y. Tam. Transient analysis of erbium-doped fiber amplifiers. *IEEE Photon. Technol. Lett.*, **6**:1436–8, 1994.

9 Semiconductor optical amplifiers (SOA)

The field of semiconductor optical amplifiers (SOA) is one of the fast growing in recent years. Many new applications of SOA were proposed. For comprehensive summaries, see books by Connolly [1] and Dutta and Wang [2].

In this chapter we will discuss the following topics:

- general concepts
- amplifier equations
- influence of cavity (FP amplifiers)
- gain dependence on polarization and temperature
- some applications.

Typical use of SOA in an amplifier configuration is shown in Fig. 9.1. Here, a signal from transmission fibre is imputed on SOA where it is amplified. After amplification it is redirected again to the fibre. Coupling between fibre and SOA is provided by an appropriate optical element.

9.1 General discussion

SOA is very similar to a semiconductor laser. There are two categories of SOA (see Fig. 9.2): (a) Fabry-Perot (FP) amplifier and (b) travelling-wave amplifier (TWA). The FP amplifier displays high gain but has a non-uniform gain spectrum, whereas TWA has broadband gain but requires very low facet reflectivities. The FP amplifier has large reflectivities at both ends which results in resonant amplification, and also has large gain at the wavelength corresponding to longitudinal modes of the FP cavity.

TWA has very small reflectivities, achieved by AR (anti-reflection) coating; its gain spectrum is broad but small ripples exist in gain spectrum, resulting from residual facet reflectivity. It is more suitable for system applications but the gain must be polarization independent. The phenomenological expression for gain of SOA is written as

$$g_m = a(n - n_0) - a_2(\lambda - \lambda_p)^2 \tag{9.1}$$

and the wavelength peak value as

$$\lambda_p = \lambda_0 + a_3(n - n_0) \tag{9.2}$$

Typical values of the parameters which appear in the above formulas are given in Table 9.1, from [3].

Table 9.1 Basic parameters of SOA.		
Description	Symbol	Value
Differential gain	a	2.7×10^{-16} cm^2
Gain coefficient	a_2	0.15 cm^{-1} nm^{-2}
Gain coefficient	a_3	2.7×10^{-17} nm · cm^{-3}
Transparency density	n_0	1.1×10^{18} cm^{-3}
Amplifier length	L	350 μm
Density at threshold	n_{th}	1.8×10^{18} cm^{-3}

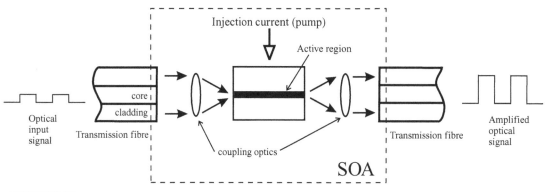

Fig. 9.1 SOA in an amplifier configuration.

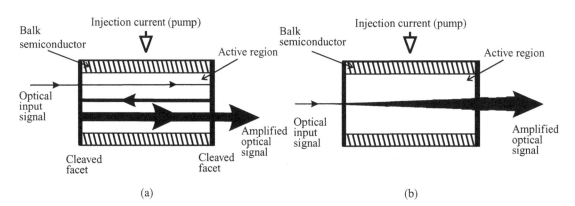

Fig. 9.2 Types of SOA: Fabry-Perot (left) and travelling-wave (right).

From the previous equations, the 3 dB gain bandwidth is determined as

$$2\Delta\lambda = 2\sqrt{\frac{a(n - n_0)}{2n_2}} \tag{9.3}$$

Substituting typical values from Table 9.1 gives SOA bandwidth equal to 54 nm.

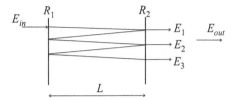

Fig. 9.3 Basic model of a Fabry-Perot amplifier.

9.1.1 Gain formula for SOA with facet reflectivities

Consider a typical Fabry-Perot ethalon with a gain medium in-between, see Fig. 9.3. Total output electric field E_{out} consists of all transmitted contributions:

$$E_{out} = E_1 + E_2 + E_3 + \cdots$$

Here R_1 and R_2 are coefficients of internal reflections for electric field, and $1 - R_1$ and $1 - R_2$ corresponding coefficients of transmission. The single-pass gain G_s for field is

$$G_s = e^{\Gamma(g-\alpha)\cdot L} \tag{9.4}$$

where Γ is optical confinement factor, g is gain coefficient and α are internal losses. The longitudinal propagation constant β_z is

$$\beta_z = k_0 \cdot n_g \tag{9.5}$$

where $k_0 = \frac{2\pi}{\lambda}$ and n_g effective group index

$$n_g = \frac{c}{v_g} \tag{9.6}$$

Here v_g is the group velocity. Applying the standing wave condition to an electromagnetic wave of wavelength λ in a resonator of length L gives

$$L = m\frac{\lambda}{2} \tag{9.7}$$

Expressions for transmitted components are

$$E_1 = E_0 e^{-j\beta_z L} \sqrt{G_s}\sqrt{1-R_1}\sqrt{1-R_2}$$
$$E_2 = E_0 e^{-j\beta_z \cdot 3L} \left(\sqrt{G_s}\right)^3 \sqrt{1-R_1}\sqrt{R_2}\sqrt{R_1}\sqrt{1-R_2}$$
$$E_3 = E_0 e^{-j\beta_z \cdot 5L} \left(\sqrt{G_s}\right)^5 \sqrt{1-R_1}R_2 R_1 \sqrt{1-R_2}$$

Total field, which is determined as a sum of all the above components, is therefore

$$E_{out} = E_0 \sqrt{(1-R_1)(1-R_2)G_s}\, e^{-j\beta_z L}\{1 + G_s\sqrt{R_1 R_2}\, e^{-j\beta_z \cdot 2L} + G_s^2 R_1 R_2 e^{-j\beta_z \cdot 4L} + \cdots\}$$

Summing geometrical series, one finally obtains

$$E_{out} = \frac{\sqrt{1-R_1}\sqrt{1-R_2}\, G_s E_0 e^{-j\beta_z L}}{1 - G_s \sqrt{R_1 R_2}\, e^{-2j\beta_z \cdot L}}$$

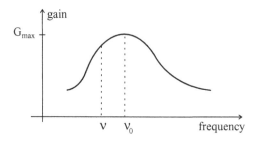

Fig. 9.4 Gain spectra showing frequency corresponding to gain peak.

Finally, gain of SOA with facet reflectivities R_1 and R_2 is

$$G = \frac{(1-R_1)(1-R_2)G_S}{\left(1-\sqrt{R_1 R_2}G_S\right)^2 + 4\sqrt{R_1 R_2}G_S \sin^2 \phi} \quad (9.8)$$

Phase shift ϕ is obtained by assuming that the phase of the incident wave is taken as zero. The relation for phase change ϕ at the output is thus

$$\phi = \frac{2\pi}{\lambda} L \quad (9.9)$$

Using (9.6), we can write the above relation for frequency as

$$\nu = \phi \frac{v_g}{2\pi L} \quad (9.10)$$

The above relation can be interpreted on gain spectrum graph as shown in Fig. 9.4 where frequency ν_0 corresponds to gain peak $G_{\max}$, which is obtained when $\sin \phi = 0$, or $\phi = k\pi$. Using an expression for phase (9.9), one finds its value corresponding to ν_0 (using relation $1/\lambda_0 = \nu_0/v_g$)

$$\phi = \frac{2\pi}{\lambda_0} L = \frac{2\pi}{v_g} \nu_0 L$$

or

$$\nu_0 = \frac{v_g}{2L} \cdot k$$
$$= \frac{v_g}{2L}$$

assuming $k = 1$. The difference in frequencies $\nu - \nu_0$ is determined as

$$\nu - \nu_0 = \frac{v_g}{2\pi L}\phi - \frac{v_g}{2L} = \frac{v_g}{2L}\left(\frac{\phi}{\pi} - 1\right)$$

or

$$\frac{2\pi(\nu - \nu_0)L}{v_g} = \pi\left(\frac{\phi}{\pi} - 1\right) = \phi - \pi$$

Finally

$$\phi = \frac{2\pi(\nu - \nu_0)L}{v_g} \quad (9.11)$$

where we have neglected π since $\sin \pi = 0$.

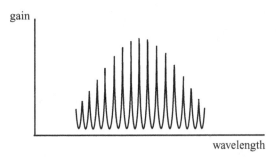

Fig. 9.5 Typical gain spectra of Fabry-Perot amplifier showing gain ripples.

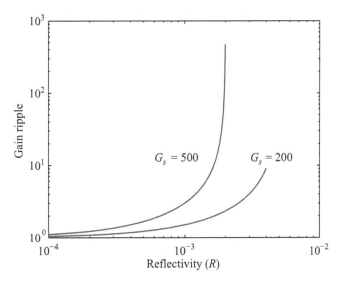

Fig. 9.6 Gain ripple as a function of reflectivity for two values of gain.

9.1.2 The effect of facet reflectivities

An uncoated SOA has facet reflectivities (due to FP reflections) determined by taking the values of refractive indices of a typical semiconductor and the air and is approximately equal to 0.32. Even with the AR coatings, there are some residual reflectivities which result in the appearance of the so-called gain ripples, see Fig. 9.5. Ripples are superimposed on gain spectrum. The peak-to-valley ratio between the resonant and non-resonant gains is known as the amplifier gain ripple G_r. From Eq. (9.8) one obtains

$$G_r = \frac{1 + G_s\sqrt{R_1 R_2}}{1 - G_s\sqrt{R_1 R_2}} \tag{9.12}$$

where G_s is the single-pass amplifier gain.

For an ideal TWA both R_1 and R_2 are zero. In this case $G_r = 1$; i.e. no ripple occurs at the cavity mode frequencies. The quantity G_r is plotted in Fig. 9.6 as a function of reflectivity

R (assuming that $R_1 = R_2$) for two values of gain. One observes that gain ripple increases with increasing gain and also increasing facet reflectivity. The MATLAB code is provided in Appendix, Listing 9A.1.

9.2 SOA rate equations for pulse propagation

In this section we concentrate on basic rate equations describing pulse propagation in SOA [2], [4]. Discussion involves electromagnetic field and carriers. We start with the development of electromagnetic equation.

Pulse propagation inside SOA is described by the wave equation

$$\nabla^2 \mathbf{E}(\mathbf{r}, t) - \frac{\varepsilon}{c^2} \frac{\partial^2 \mathbf{E}(\mathbf{r}, t)}{\partial t^2} = 0 \tag{9.13}$$

where $\mathbf{E}(\mathbf{r}, t)$ is the electric field vector, c is the light velocity and ε is the dielectric constant of the amplifier medium which is expressed as

$$\varepsilon = \overline{n}_b^2 + \chi \tag{9.14}$$

Here $\overline{n}_b$ is the background refractive index of the semiconductor and susceptibility χ which represents the effect of charges inside an active region in the phenomenological model is

$$\chi = -\frac{c\overline{n}}{\omega_0}(\alpha_H + i) g(n) \tag{9.15}$$

where $\overline{n}$ is the effective mode index, α_H is the linewidth enhancement factor (Henry factor) and $g(n)$ is the optical gain approximated here (for bulk devices) as

$$g(n) = a(n - n_0) \tag{9.16}$$

where a is defferential gain, n is the injected carrier density and n_0 is the carrier density at transparency.

In the following we assume a travelling wave semiconductor wave amplifier which supports single mode propagation with a perpendicular electric profile described by $F(x, y)$. The electric field $\mathbf{E}(\mathbf{r}, t)$ which obeys wave equation (9.13) is expressed as

$$\mathbf{E}(\mathbf{r}, t) = \widehat{n} \frac{1}{2} \left\{ F(x, y) A(z, t) e^{i(k_0 z - \omega_0 t)} \right\} \tag{9.17}$$

where $\widehat{n}$ is the polarization vector, $k_0 = \overline{n}\omega_0/c$ and $A(z, t)$ is the slowly varying amplitude of the propagating wave. We introduce the following notation:

$$\nabla^2 = \nabla_\perp^2 + \frac{\partial^2}{\partial z^2}, \quad \text{where } \nabla_\perp^2 = \frac{\partial^2}{\partial x^2} + \frac{\partial^2}{\partial y^2}$$

and evaluate derivatives

$$\nabla^2 \mathbf{E} \sim (\nabla_\perp^2 F) A e^{ik_0 z} + F \frac{\partial^2 A}{\partial z^2} e^{ik_0 z} + 2F \frac{\partial A}{\partial z} i k_0 e^{ik_0 z} + F A (i k_0)^2 e^{ik_0 z} \tag{9.18}$$

In the slowly varying envelope approximation (SVEA) the term with second derivative with respect to z and t is neglected. Substituting (9.18) into (9.13), applying SVEA with respect to z and t and integrating over the transverse direction gives

$$\nabla_\perp^2 F + \frac{\omega_0^2}{c^2}\left(\bar{n}_b^2 - \bar{n}^2\right) F = 0 \qquad (9.19)$$

and

$$\frac{\partial A}{\partial z} + \frac{1}{v_g}\frac{\partial A}{\partial t} = \frac{i\omega_0 \Gamma}{2c\bar{n}}\chi A - \frac{1}{2}\alpha_{loss} A \qquad (9.20)$$

where we have accounted for losses described by α_{loss}. The group velocity is defined as $v_g = c/\bar{n}_g$ and group index $\bar{n}_g$ is

$$\bar{n}_g = \bar{n} + \omega_0 \frac{\partial \bar{n}}{\partial \omega} \qquad (9.21)$$

The confinement factor Γ is

$$\Gamma = \frac{\int_0^w dx \int_0^d dy |F(x,y)|^2}{\int_{-\infty}^{+\infty} dx \int_{-\infty}^{+\infty} dy |F(x,y)|^2} \qquad (9.22)$$

where w and d are the width and thickness of the amplifier active region. At this stage, one simplifies Eq. (9.20) by introducing transformation to a reference frame moving with pulse as

$$\tau = t - \frac{z}{v_g}$$
$$z' = z$$

In the new reference frame Eq. (9.20) takes the form

$$\frac{\partial A}{\partial z} = \frac{i\omega_0 \Gamma}{2c\bar{n}}\chi A - \frac{1}{2}\alpha_{loss} A \qquad (9.23)$$

Rate equation for carrier density n is in the following form [2], [4]:

$$\frac{dn}{dt} = \frac{I}{qV} - \frac{n}{\tau_c} - \frac{\Gamma g(n)}{\hbar \omega_0 \sigma_m}|A|^2 \qquad (9.24)$$

where I is the injection current, V is the active volume, q is the electron charge, τ_c is the carrier lifetime and σ_m is the cross section of the active region. In the new reference frame, the above equation is

$$\frac{dn}{d\tau} = \frac{I}{qV} - \frac{n}{\tau_c} - \frac{\Gamma g(n)}{\hbar \omega_0 \sigma_m}|A|^2 \qquad (9.25)$$

Slowly varying amplitude $A(z,t)$ of the propagating wave is expressed as

$$A = \sqrt{P} e^{i\phi} \qquad (9.26)$$

where $P(z, \tau)$ and $\phi(z, \tau)$ are the instantaneous power and the phase of the propagating pulse. Using the above equations, one obtains [2]

$$\frac{\partial A}{\partial z} = \frac{1}{2}(1 + i\alpha_H)g \cdot A \qquad (9.27)$$

$$\frac{dg}{d\tau} = -\frac{g - g_0}{\tau_c} - \frac{gP}{E_{sat}} \qquad (9.28)$$

$$\frac{\partial P}{\partial z} = (g - \alpha_{loss})P \qquad (9.29)$$

$$\frac{\partial \phi}{\partial z} = -\frac{1}{2}\alpha_H g \qquad (9.30)$$

The quantity E_{sat} is defined as $E_{sat} = \tau_c P_s$, where P_s is the saturation power of the amplifier

$$P_s = \frac{\hbar\omega_0 \sigma_m}{a\Gamma\tau_c} \qquad (9.31)$$

In the above g_0 is the small signal gain

$$g_0 = \Gamma a \left(\frac{I\tau_c}{eV} - n_0\right) \qquad (9.32)$$

Finally, the cross section σ_m of the active region is $\sigma_m = wd$.

Pulse amplification

Using previously defined equations, we will now analyse pulse propagation assuming zero losses, i.e. $\alpha_{loss} = 0$, and also assuming that $\tau_p \ll \tau_c$, where τ_p is the width of the input pulse. Under this approximation pulse is so short that gain has no time to recover. Observing that $\tau_c = 0.2$–0.3 ns for typical SOA [4], this approximation works for τ_p equal to about 50 ps. Under the above approximations, one can obtain the analytical solution of the amplifier equations. One first integrates Eq. (9.29) to obtain output power as

$$P_{out}(\tau) = P_{in}(\tau)e^{h(\tau)} \equiv P_{in}(\tau)G(\tau) \qquad (9.33)$$

where $P_{in}(\tau)$ is the input power, and

$$h(\tau) = \int_0^L g(z, \tau)dz \qquad (9.34)$$

Quantity $h(\tau)$ is known as the total integrated net gain. Replacing the last term in Eq. (9.28) using (9.29) gives

$$\frac{dg}{d\tau} = \frac{g_0 - g}{\tau_c} - \frac{1}{E_{sat}}\frac{dP}{dz}$$

Integrating the above over amplifier length and using (9.33) gives

$$\frac{dh(\tau)}{d\tau} = \frac{g_0 L - h(\tau)}{\tau_c} - \frac{1}{E_{sat}}P_{in}(\tau)[G(\tau) - 1] \qquad (9.35)$$

The solution of the previous equation is [2], [4]

$$G(\tau) = e^{h(\tau)} = \frac{G_0}{G_0 - (G_0 - 1)\exp\left[-E_0(\tau)/E_{sat}\right]} \quad (9.36)$$

where G_0 is the unsaturated gain of the amplifier and the quantity $E_0(\tau)$ is given by

$$E_0(\tau) = \int_{-\infty}^{\tau} P_{in}(\tau')\tau' \quad (9.37)$$

$E_0(\tau)$ represents the fraction of the pulse energy contained in the leading part of the pulse up to $\tau' \leq \tau$.

The above solution shows that due to time dependence of the gain, different parts of input pulse experience different amplification which leads to a modification of pulse shape after being amplified by SOA.

In the following we will restrict our analysis to the Gaussian input pulse

$$P_{in}(\tau) = P_0 \exp\left(-\tau^2\right) \quad (9.38)$$

For the Gaussian input pulse, the expression for the quantity $E_0(\tau)$ can be found in a closed form as

$$E_0(\tau) = P_0 \tau_0 \frac{1}{2}\sqrt{\pi}\left[1 + erf(\tau)\right] \quad (9.39)$$

where $erf(\tau)$ is the error function defined as

$$erf(\tau) = \frac{2}{\sqrt{\pi}} \int_0^{\tau} e^{-x^2} dx \quad (9.40)$$

We have also used the following property of error function [5]:

$$1 - erf(\tau) = \frac{2}{\sqrt{\pi}} \int_{\tau}^{\infty} e^{-x^2} dx \quad (9.41)$$

In Fig. 9.7 we plotted pulse shape for several values of unsaturated gain G_0. One can observe that the amplified pulse becomes asymmetric; i.e. its leading edge is sharper compared to its trailing edge.

9.3 Design of SOA

Here, we briefly describe the main effects which must be considered when designing SOA with proper characteristics. Out of many issues related to the design of SOA, we briefly concentrate here on two: suppression of cavity resonances and polarization insensitivity. Consult the literature [1], [2] and [6] for more information.

Suppression of cavity resonance

To fabricate a travelling wave SOA, the Fabry-Perot cavity resonances must be suppressed. To accomplish this, the reflectivities at both facets must be reduced. Three approaches were

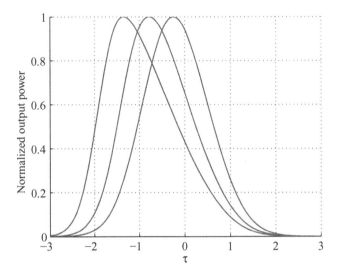

Fig. 9.7 Output pulse shapes for several values of the unsaturated gain $G_0 = 10, 100, 1000$ (increasing values to the left).

used to achieve this goal (see Fig. 9.8): (a) to put anti-reflection (AR) coating at both facets, (b) to tilt the active region and (c) to use transparent window regions.

The principles of AR coating were discussed in Chapter 3. AR coating only works for a particular wavelength and it is not suitable for a wide bandwidth. The analysis [7] shows that with appropriate combination of the previously discussed methods it is possible to obtain an effective facet reflectivity of less than 10^{-4}. Multilayer coatings can broaden wavelength range where there is low reflectivity.

Polarization insensitive structures

The state of polarization of an electric field during propagation in optical fibre changes randomly. Therefore, after propagation when signal is ready for amplification (say, in SOA), its state of polarization is unknown. For amplifications of such signals it is therefore desirable that SOA has polarization independent amplification. The main factor responsible for polarization sensitivity is the difference between confinement factors for TE and TM modes [1]. Proper design of polarization insensitive SOA involves several techniques which are summarized in [1].

Additionally, in modern SOAs which are based on quantum well (QW) designs instead of bulk structures, due to their significantly reduced threshold current and increased efficiency there exist additional polarization related effects associated with QW. QW significantly suffers from polarization sensitivity, that is, a significant difference in gain between the transverse-electric (TE) and transverse-magnetic (TM) polarization modes. This is a major concern as the polarization of a signal cannot be always controlled. Several approaches were discussed to the design of polarization insensitive SOA [8], [9], [10]. The issue is of significant importance for SOA built using quantum wells.

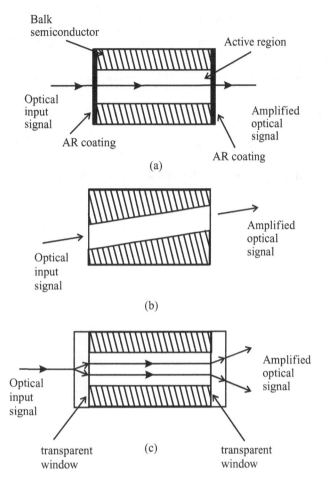

Fig. 9.8 Main ways to make travelling wave SOA.

9.4 Some applications of SOA

9.4.1 Wavelength conversion

One of the important applications of SOA is for wavelength conversion (WC) [11], see also [1]. Three main methods of WC have been analysed: cross-gain modulation (XGM), cross-phase modulation (XPM) and four-wave mixing (FWM). XGM will be discussed in detail in the following section.

Other methods (not discussed here) include those based on the nonlinear optical loop mirror (NOLM) with the nonlinearity achieved by using fibre or SOA (see recent review [12]).

Fig. 9.9 Wavelength conversion using XGM in SOA in copropagation configuration.

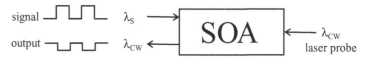

Fig. 9.10 Wavelength conversion using XGM in SOA in a counter-propagation configuration.

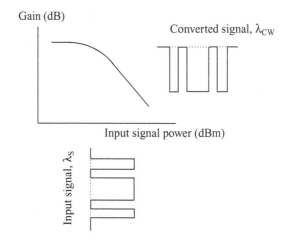

Fig. 9.11 Principle of wavelength conversion in SOA using XGM.

Cross-gain modulation

The reduction of gain, known as gain saturation, typically occurs for input powers of the order of 100 μW or higher. To understand this effect, one must remember that the amplification in SOA is the result of stimulated emission. The rate of stimulated emission in turn depends on the optical input power. At high optical injection, the carrier concentration in the active region is depleted through stimulated emission to such an extent that the gain of SOA is reduced.

WC based on XGM can operate in either copropagation or counter-propagation configurations. Counter-propagation configuration does not require optical filtering of the target wavelength. However, counter-propagation suffers from speed limitations [13].

The schematic representation of the XGM used as wavelength conversion is shown in Fig. 9.9 (copropagation configuration) and in Fig. 9.10 (counter-propagation configuration). The principle of wavelength conversion employing nonlinear characteristics of SOA is explained in Fig. 9.11. Two optical signals enter a single SOA. One of the signals (known as the probe beam) at wavelength λ_{CW} is injected continuously (CW); the other (known as the signal beam) injected at wavelength λ_S is carrying digital information.

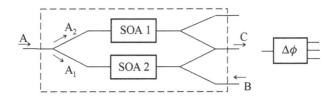

Fig. 9.12 The MZ interferometric configuration and its equivalent symbol.

If the peak optical power in the modulated signal is near the saturation power of the SOA, the gain will be modulated synchronously with the power. When the data signal is at high level (a logical ONE), the gain is depleted and vice versa. This gain modulation is imposed on the unmodulated (CW) input probe beam. Thus, an inverted replica of the input data is created at the probe beam wavelength. This form of wavelength conversion is one of the simplest all-optical wavelength conversion mechanisms available today.

Very fast wavelength conversion can be achieved with speeds in the order of 40 Gb s^{-1} [14] with a small bit error ratio penalty. Previously, it was thought that the speed of WC is limited by the intrinsic carrier lifetime, which is around 0.5 ns. This is because the effective carrier lifetime can be decreased by the use of high optical injection to values as low as 10 ps.

Another effect associated with pulse propagation in long SOA (>1 mm) is that pulse distortion of the input data due to gain saturation effects can tend to sharpen the leading edge of the data pulse at the probe wavelength.

9.4.2 All-optical logic based on interferometric principles

For the future high-speed optical networks, several all-optical signal processing functionalities will be required to avoid cumbersome and power consuming electro-optic conversion. The central role in this process is played by all-optical high-speed logic gates.

In recent years several schemes of implementation have been investigated. Some of those schemes exploit gain saturation of SOA, and the other methods employ the interferometric configurations.

Optical logic devices are constructed using the Mach-Zehnder (MZI) interferometer, see Fig. 9.12 for an introduction to the notation and symbols. $\Delta\phi$ is the phase difference between signals propagating in the upper and lower arms of MZI. Each arm of the interferometer contains SOA where the effect of cross-phase modulation (XPM) is utilized to change phase of the transmitted light. The XPM effect is based on the physical effect of the dependence of refractive index on the carrier density in the active region. Depending on operating conditions which are controlled by driving currents within each SOA and direction and intensity of external light, this configuration can perform like an all-optical gain. In what follows, for several logic gates we schematically show operating conditions of MZI, the resulting phase difference and the equivalent logical table [15].

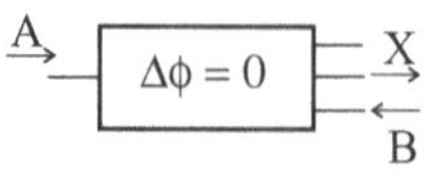

A	B	X
0	0	0
0	1	0
1	0	1
1	1	0

Fig. 9.13 A block diagram of A NOT B and a truth table.

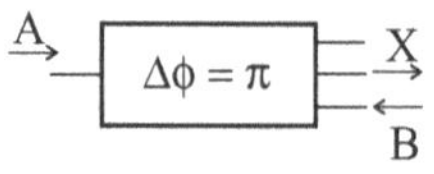

A	B	X
0	0	0
0	1	0
1	0	0
1	1	1

Fig. 9.14 A block diagram of A AND B and a truth table.

A and NOT B

Here, bias conditions are set in such way that, when a second signal B is input in a counter-propagating scheme, the phase difference is zero. The configuration and logical table are shown in Fig. 9.13.

A AND B

Here, both SOA are biased such that in the absence of counterpropagating signal B, a phase difference of π exists at the output X. The relevant configuration and the truth table are shown in Fig. 9.14.

9.5 Problem

1. Derive an expression for gain ripple G_r.

9.6 Project

1. Use software developed by Connelly [1] to analyse optical logical elements.

Appendix 9A: MATLAB listings

In Table 9.2 we provide a list of MATLAB files created for Chapter 9 and a short description of each function.

Table 9.2 List of MATLAB functions for Chapter 9.		
Listing	Function name	Description
9A.1	*gain_ripple.m*	Determines gain ripples as a function of reflectivity
9A.1	*pulse_shape.m*	Determines output pulse shape for various values of G_0

Listing 9A.1 Program gain_ripple.m. MATLAB program which determines gain ripples for SOA.

```
% File name: gain_ripple.m
% Calculates gain ripple as a function of reflectivity
clear all
N_max = 401;                        % number of points for plot
R = linspace(1d-4,4d-3,N_max);      % creation of theta arguments
%
for G_s = [200 500]       % gain coefficient
    G_r = (1 + G_s.*R)./(1 - G_s.*R);
    loglog(R,G_r,'LineWidth',1.5)
    hold on
end
% Redefine figure properties
ylabel('Gain ripple','FontSize',14)
xlabel('Reflectivity (R)','FontSize',14)
set(gca,'FontSize',14);             % size of tick marks on both axes
text(3d-3, 12, 'G_s = 200','Fontsize',14)
text(6d-4, 12, 'G_s = 500','Fontsize',14)
text(8.5, 0.1, 'r = 0.8','Fontsize',14)
pause
close all
```

Listing 9A.2 Program pulse_shape.m. MATLAB program which determines output pulse shape for three values of the unsaturated gain G_0.

```
% File name: pulse_shape.m
% Determines output pulse shapes in soa
% Basic equations after Agrawal and Olsson, JQE25, 2297 (1989)
clear all
P_0 = 1;
w = 0.1;                            % defined as P_0*tau_0/E_sat
tau = -3:0.01:3;
hold on
for G_0 = [10 100 1000]
    x = (1/2)*sqrt(pi)*w*(1 + erf(tau));
```

```
        G = G_0./(G_0 -(G_0-1)*exp(-x));
        y = exp(-tau.^2).*G;
        y_max = max(y);
        plot(tau,y/y_max,'LineWidth',1.5)
end
xlabel('\tau','FontSize',14);
ylabel('Normalized output power','FontSize',14);
set(gca,'FontSize',14);              % size of tick marks on both axes
grid on
pause
close all
```

References

[1] M. J. Connelly. *Semiconductor Optical Amplifiers*. Kluwer Academic Publishers, Boston, 2002.

[2] N. K. Dutta and Q. Wang. *Semiconductor Optical Amplifier*. World Scientific, Singapore, 2006.

[3] M. J. O'Mahony. Semiconductor laser optical amplifiers for use in future fiber systems. *J. Lightwave Technol.*, **6**:531–44, 1988.

[4] G. P. Agrawal and N. A. Olsson. Self-phase modulation and spectral broadening of optical pulses in semiconductor laser amplifiers. *IEEE J. Quantum Electron.*, **25**:2297–306, 1989.

[5] M. R. Spiegel. *Mathematical Handbook of Formulas and Tables*. McGraw-Hill, New York, 1968.

[6] G. P. Agrawal and N. K. Dutta. *Semiconductor Lasers. Second Edition*. Kluwer Academic Publishers, Boston/Dordrecht/London, 2000.

[7] T. Saitoh, T. Mukai, and O. Mikame. Theoretical analysis and fabrication of antireflection coatings on laser-diode facets. *J. Lightwave Technol.*, **3**:288–93, 1985.

[8] Y. Zhang and P. P. Ruden. $1.3-\mu m$ polarization-insensitive optical amplifier structure based on coupled quantum wells. *IEEE J. Quantum Electron.*, **35**:1509–14, 1999.

[9] W. P. Wong and K. S. Chiang. Design of waveguide structures for polarization-insensitive optical amplification. *IEEE J. Quantum Electron.*, **36**:1243–50, 2000.

[10] M. S. Wartak and P. Weetman. The effect of thickness of delta-strained layers in the design of polarization-insensitive semiconductor optical amplifiers. *IEEE Photon. Technol. Lett.*, **16**:996–8, 2004.

[11] T. Durhuus, B. Mikkelsen, C. Joergensen, S. L. Danielsen, and K. E. Stubkjaer. All-optical wavelength conversion by semiconductor optical ampliers. *J. Lightwave Technol.*, **14**:942–54, 1996.

[12] I. Glesk, B. C. Wang, L. Xu, V. Baby, and P. R. Prucnal. Ultra-fast all-optical switching in optical networks. In E. Wolf, ed., *Progress in Optics*, volume **45**, pp. 53–117. Elsevier, Amsterdam, 2003.

[13] D. Nesset, T. Kelly, and D. Marcenac. All-optical wavelength converion using SOA nonlinearities. *IEEE Communications Magazine*, **December**:56–61, 1998.

[14] D. Wolfson, A. Kloch, T. Fjelde, C. Janz, B. Dagens, and M. Renaud. 40-gb/s all-optical wavelength conversion, regeneration, and demultiplexing in and SOA-based all-active Mach-Zehnder interferometer. *IEEE Photon. Technol. Lett.*, **12**:332–4, 2000.

[15] K. Roberts-Byron, J. E. A. Whiteaway, and M. Tait. Optical logic devices and methods. United States Patent, Number 5,999,283. Dec. 7, 1999.

10 Optical receivers

In this chapter we summarize the operation of an optical receiver, which is an important part of an optical communication system. An overview of design principles for receivers used in optical communication systems is provided by Alexander [1]. The aim of a receiver is the recovery of the transmitted data. The process involves two steps [2]:

1. the recovery of the bit clock,
2. the recovery of the transmitted bit within each bit interval.

A block diagram of an optical receiver is shown in Fig. 10.1 [3], [4].

The receiver consists of a photodetector which converts the optical signal into electrical current. A good light detector should generate a large photocurrent at a given incident light power. They should also respond fast to the input changes and add minimal noise to the output signal. This last requirement is of crucial importance since the received signal is typically very weak. In digital optical communication systems the detection process is often conducted with a PIN photodiode.

There are generally two types of detection [4]: direct detection (also called incoherent detection) and coherent detection.

Direct detection detects only the intensity of the incident light. It is used mainly for intensity or amplitude modulation schemes. It can only detect an amplitude modulated (AM) signal.

Coherent detection can detect both the power and phase of the incident light. It is therefore used when phase modulation (PM) or frequency modulation (FM) is preferred. Coherent detection is also important in applications such as WDM.

The coherent detection requires a local oscillator to coherently down-convert the modulated signal from optical frequency to intermediate frequency (IF) [4]. Coherent detection is not discussed here, see [4] for the extensive discussion. The incoherent detection which dominates in currently deployed systems is based on square-law envelope detection of the optical signals.

After detection, the electrical current is often amplified (not shown in Fig. 10.1) and then passes through an electrical filter which is normally of the Bessel type. At that point, electrical eye diagrams are typically observed for the assessment of signal quality. Next, sampling of electrically filtered received signal is performed. The received electrical signal is corrupted with noise of various origins. The noise sources will be discussed in the following sections.

Performance evaluation of an optical transmission system is done by evaluating optical signal-to-noise ratio, eye opening and bit error rate (BER) which is the ultimate indicator of the system's performance.

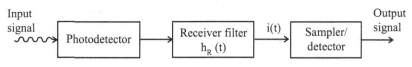

Fig. 10.1 Block diagram of an ideal optical receiver.

10.1 Main characteristics

Before we enter a more specialized discussion of optical receivers, let us summarize their main characteristics [5], which are: receiver sensitivity, dynamic range, bit-rate transparency and bit-pattern independency.

10.1.1 Receiver sensitivity

This property is a measure of the minimum level of optical power P_{sens} at the receiver required for a reliable operation. Specifically, one expects that the BER is smaller than a specified level. Typically, that level is established to be equal to 10^{-9}.

The receiver sensitivity is a function of both the signal and also the noise parameters of photodetector and a preamplifier. It is a measure of the operating limit of the optical receiver, which, however, rarely operates close to that limit. There always exists a possibility of degradation of the system (temperature, ageing, etc.), so a typical margin (normally 3–6 dB) is established.

Receiver sensitivity is a fundamental parameter of an optical receiver. It is directly responsible for the spacing between two points in any optical link, e.g. between transmitter and receiver or between repeaters.

10.1.2 Dynamic range

Dynamic range (expressed in dB) is the difference between the maximum allowable power and minimum power determined by receiver sensitivity. The maximum allowable power on the receiver is determined by nonlinearity and saturation.

Large dynamic range is important because it allows for more flexibility in the design of an optical network. The design of every network should take into account the wide range of possible changes of received optical powers due to changes in temperature, ageing or various types of losses (in the fibre, connectors, etc.).

10.1.3 Bit-rate transparency

It refers to the ability of the optical receiver to operate over a range of bit rates. It describes the ability of the same receiver to be used for several networks operating at different bit rates.

10.1.4 Bit-pattern independency

This is the property of an optical receiver determining its operation for various data formats. The main constraint is imposed by non-return-to-zero (NRZ) code.

10.2 Photodetectors

At the end of travel through a optical fibre, the optical signal reaches a photodetector (PD) where it is converted into an electrical signal. In PD photon of energy $h\nu$ is absorbed and produces photocurrent i_P:

$$i_P = \eta \frac{P}{h\nu} e \qquad (10.1)$$

where P is the power of the incoming light, η is the quantum efficiency, h is Planck's constant and $\nu = c/\lambda$ is the frequency of the absorbed light. PD is similar to a semiconductor diode polarized in the reverse direction. Therefore, PD can be modelled as a current source.

Another very popular detector is a human eye. It is the most popular natural detector of light but it has several disadvantages: it is slow, has bad sensitivity for low-level signals, has no natural connection to electronic amplifiers and its spectral response is limited to the 0.4–0.7 μm range.

Artificial (human made) optical detectors are based on two physical mechanisms: external photoelectric effect and internal photoelectric effect. In the external photoelectric effect, electrons are removed from the metal surface of electrode known as a cathode by absorbing energy from incident light. Then, under an electric field due to the potential difference between both electrodes, they travel to another electrode known as an anode, thus producing electrical current. Vacuum photodiodes and photomultiplier tubes operate on that principle.

Main choices for photodetectors are the PIN (P-type Intrinsic N-type) photodiode and avalanche photodiode (APD). APD provides gain which increases system sensitivity but introduces more noise.

In this section we will discuss only the internal photoelectric effect and its applications in light detection. In this effect, physical processes take place inside semiconductor junction devices. There, free carriers (electrons and holes) are generated by absorption of incoming photons, and as a result an electrical current is produced. These devices can be viewed as the inverse of a light emitting diode (LED).

10.2.1 Principles of photo detection

Photodetection using semiconductors is possible because of optical absorption. When light is incident on the semiconductor surface, it may or may not be absorbed depending on a wavelength. Absorbed optical power is described as

$$\frac{dP}{dx} = \alpha(\lambda) P \qquad (10.2)$$

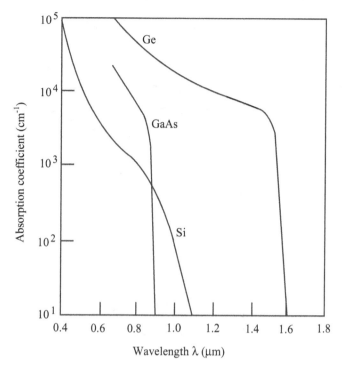

Fig. 10.2 Optical absorption coefficient as a function of wavelength for typical semiconductors. Reprinted from H. Melchior, *Journal of Luminescence*, **7**, 390 (1973). Copyright (1973), with permission from Elsevier.

where $\alpha(\lambda)$ is the absorption coefficient which is wavelength dependent. The spectra of optical absorption for several semiconductors (including compounds) are schematically shown in Fig. 10.2. As can be seen, the absorption coefficient $\alpha(\lambda)$ strongly depends on wavelength.

Integration of Eq. (10.2) gives the absorbed power at the distance x from the surface in terms of the incident optical power P_0:

$$P(x) = P_0 \left(1 - e^{-\alpha(\lambda) \cdot x}\right) \qquad (10.3)$$

A schematic of a photodetector is illustrated in Fig. 10.3. Incident light penetrates the semiconductor surface and generates electrical current I. The device structure consists of p and n regions separated by wide intrinsic (very lightly doped) region of width w under large reverse bias voltage, see Fig. 10.4.

When a photon is absorbed by the solid (and if it has enough energy, i.e. large enough frequency), its energy causes an electron to move from valence band (leaving one hole behind) to the conduction band and to create one extra electron there. The change of energy of an electron should be (at least) E_g, which is known as the energy gap. Photons with smaller energies are usually not able to create electron-hole pairs, see Fig. 10.5. Therefore, photon energy $E_p = h\nu$ must be

$$E_p \geq E_g \quad \text{or} \quad h\nu \geq E_g \quad \text{or} \quad \lambda_c = \frac{ch}{E_g}$$

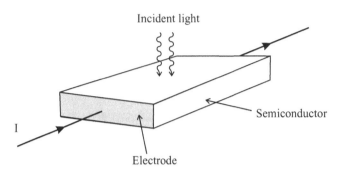

Fig. 10.3 Schematic of a photodetector.

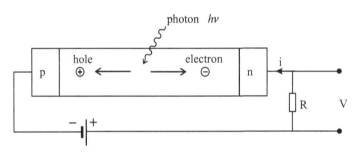

Fig. 10.4 PIN photodetector in external electrical circuit.

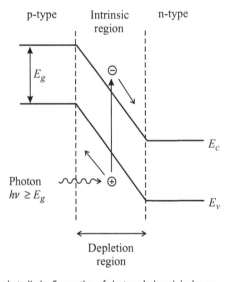

Fig. 10.5 Energy band diagram of a PIN photodiode. Generation of electron-hole pair is shown.

Table 10.1 Bandgaps and cutoff wavelengths for different semiconductors and their compounds.

Semiconductor	$\Delta E (eV)$	$\lambda_c (\mu m)$
C	7	0.18
Si	1.1	1.13
Ge	0.72	1.72
Sn	0.08	15.5
$Ga_xIn_{1-x}As$	1.43–0.36	0.87–3.44
$Ga_xIn_{1-x}As_yP_{1-y}$	1.36–0.36	0.92–3.44

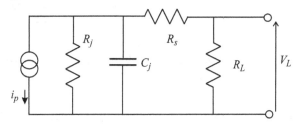

Fig. 10.6 Equivalent circuit model of a PIN photodetector.

where λ_c is known as cutoff and it defines a usable spectrum. The practical formula is

$$\lambda_c = \frac{1.24}{E_g}, \quad \lambda_c \text{ is in } \mu m \text{ and } E_g \text{ in eV}$$

The applied high electric field causes electrons and holes to separate. They subsequently flow to n (electrons) and p (holes) regions and produce flow of current in the external circuit known as photocurrent.

Different semiconductor materials have different bandgaps and therefore different cutoff wavelengths. They are used as detectors in different parts of the electromagnetic spectrum. Bandgaps and cutoff wavelengths for some semiconductors and compounds are summarized in Table 10.1.

An equivalent circuit of PD is shown in Fig. 10.6 [1]. It consists of a current source, parallel junction resistance R_j, which is equivalent to the differential resistance of the diode. R_j is typically large, around 10^6 Ω, and usually it is ignored in the analysis. C_j is the junction capacitance; R_s is the series resistance due to bulk and contact resistance and typically is a few ohms. R_L is the load resistance. In a typical analysis R_j and R_s are neglected.

The spectral function $V_L(\nu)$ of the output voltage is related to the spectral function of the photocurrent $i_P(\nu)$ as

$$V_L(\nu) = i_P(\nu) \frac{R_L}{1 + j2\pi \nu R_L C_j} \tag{10.4}$$

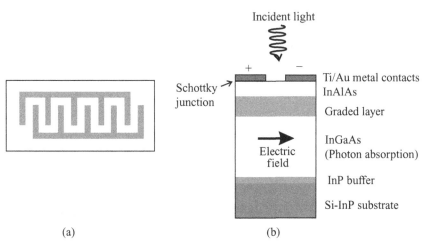

Fig. 10.7 Schematic of a metal-semiconductor-metal (MSM) photodetector. Top view of the structure (a) and its cross-section view (b).

The previous equation is obtained by solving Kirchoff's loop equation. Equivalent impedance is

$$\frac{1}{Z_e} = \frac{1}{R_L} + \frac{1}{\frac{1}{j\omega C_j}} = \frac{1}{R_L} + j\omega C_j = \frac{1 + j\omega C_j R_L}{R_L}$$

Using $V_L(\nu) = i_P(\nu) \cdot Z_e$, Eq. (10.4) follows. Frequency bandwidth is therefore determined by the time constant $C_j R_L$. One wants to keep this time constant small in order to achieve a large bandwidth.

We finish this section with a brief discussion of an integrated optical receiver using a metal-semiconductor-metal (MSM) photodetector, which is shown in Fig. 10.7. The MSM detector structure consists of an interdigital pattern of metal fingers deposited on a semiconductor substrate and a typical vertical PIN detector [6]. Electric potential is applied between alternate fingers creating an electric field which sweeps photogenerated electrons and holes to the positive and negative electrodes, respectively.

This structure shows several improvements compared to traditional designs, like a significant improvement in sensitivity. Most of the improvements result from the lateral design.

10.2.2 Performance parameters of photodetectors

Main parameters which determine characteristics of PD are [7]: dark current, spectral response, quantum efficiency, noise, detectivity, linearity and dynamic range and speed and frequency response. We will now discuss them in more detail.

Dark current

Dark current is the current flowing in the PD without incoming light. Typical values for popular semiconductors are shown in Table 10.2.

Table 10.2 Dark currents for different semiconductors and their compounds.	
Semiconductor	Dark current (nA)
Si	0.1–1
Ge	100
InGaAsP	1–10

Quantum efficiency

Quantum efficiency is a measure of the efficiency of the generation of the electron-hole pairs (EHP). It is defined as [8]

$$\eta = \frac{\text{number of free EHP generated}}{\text{number of incident photons}} = \frac{I_{ph}/e}{P_0/h\nu} \qquad (10.5)$$

Here I_{ph} is the photocurrent flowing in the external circuit, e is the electron's charge, P_0 is the incident optical power and $h\nu$ is the energy of a single photon.

Responsivity

It is also known as spectral responsivity. It is defined as [8]

$$R = \frac{I_{ph}}{P_0} \qquad (10.6)$$

Responsivity specifies the photocurrent generated per unit optical power. It can be expressed as

$$R = \eta \frac{e}{h\nu} = \eta \frac{e\lambda}{hc} \qquad (10.7)$$

Ideal responsivity is therefore a linear function of wavelength λ, see Fig. 10.8.

Speed of response

It determines how PD responds to an optical signal. A typical response to a pulse is shown in Fig. 10.9.

Speed of response is determined by the RC time constant. In terms of parameters of previously introduced equivalent circuit, the rise time τ_r is given by

$$\tau_r = 2.19 \cdot R_L \cdot C_j$$

The evaluation of τ_r is left as an exercise. Time response is directly related to the frequency response. The 3-dB bandwidth is [9]

$$f_{s-dB} = \frac{1}{2\pi R_L C_j}$$

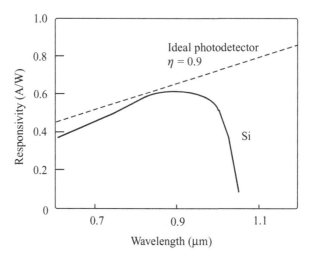

Fig. 10.8 Typical current responsivity versus wavelength for Si. Also shown is the responsivity of an ideal photodiode with quantum efficiency $\eta = 0.9$.

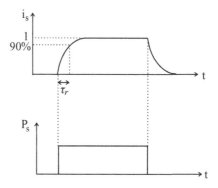

Fig. 10.9 Typical response of a photodetector to a square-pulse signal.

10.2.3 Photodetector noise

Typical optical signals arriving at the receiver front end are very weak. They are, therefore, significantly affected by various noise sources.

There are several physical processes which contribute to noise. For example, the APD generates noise in the avalanche process and optical amplifiers produce noise due to amplified spontaneous emission (ASE). In addition, there is always a quantum noise which exists in all devices and sets a fundamental lower limit on noise power.

The working parameter characterizing the photodetector is the signal-to-noise ratio (SNR), which is defined as [10]

$$\frac{S}{N} = \frac{\text{signal power from photocurrent}}{\text{photodetector noise power} + \text{amplifier noise power}} \qquad (10.8)$$

Noise appears as random fluctuations in a signal. A typical measure of noise is associated with the variance or the mean square deviation of the signal s, which is defined as [7]

$$\sigma_s^2 = \overline{(s - \bar{s})^2} = \overline{s^2} - \bar{s}^2$$

The mean value (average) $\bar{s}$ is defined as

$$\bar{s} = \sum_s p(s)\, s$$

where $p(s)$ is the probability of the measured signal having a value s and the sum is evaluated over all possible values obtained from measuring the signal.

Noise in a signal s can be represented by random variable s_n:

$$s_n = s - \bar{s}$$

If in the device (or system) there are two or more simultaneously present noise sources $s_{n1}, s_{n2}, \ldots$ which are independent, their combined effect is found by adding their mean square values (or their powers):

$$\overline{s_n^2} = \overline{s_{n1}^2} + \overline{s_{n2}^2} + \cdots$$

In terms of the above quantities, the SNR is expressed as

$$SNR = \frac{\overline{s^2}}{\overline{s_n^2}} = \frac{\overline{s^2}}{\sigma_s^2} \quad \text{or} \quad SNR = 10 \log \frac{\overline{s^2}}{\sigma_s^2} [dB]$$

Main types of noise arising in a photodetector will now be summarized.

Shot noise

Shot noise is due to random distribution of electrons generated in the photodetector. It is associated with the quantum nature of photons arriving at the photodetector which generates carriers. Photons arrive at the photodetector randomly in time due to their quantum-mechanical nature. Their randomness is described by Poisson statistics. The resulting expression for shot noise of current in the photodetector is [4], [7]

$$\overline{i_s^2} = 2eB\bar{i_s} \tag{10.9}$$

where e is the electron's charge, B is the bandwidth and $\bar{i_s}$ is the average value of signal current.

Thermal noise

Thermal noise or Johnson noise is due to the random motion of electrons in the resistor R. It is modelled as a Gaussian random process with zero mean and autocorrelation function given by [4]

$$\frac{4k_B T}{R} \delta(\tau)$$

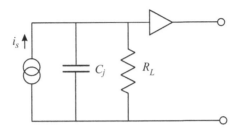

Fig. 10.10 Equivalent circuit model of a PIN photodetector with an amplifier.

where $\delta(\tau)$ is the Dirac delta function. Thermal noise power in a bandwidth B is [7]

$$p_{s,th} = 4k_B T\, B \tag{10.10}$$

where $k_B = 1.38 \times 10^{-23}$ J/K is the Boltzmann's constant and T is the absolute temperature. Thermal noise can be expressed as a current source [4], [7]

$$\overline{i_{s,th}^2} = \frac{4k_B T}{R} B \equiv I_T^2\, B \tag{10.11}$$

Here I_T is the parameter used to specify standard deviation in units of pA/$\sqrt{\text{Hz}}$. Its typical value is 1pA/$\sqrt{\text{Hz}}$ [2].

10.2.4 Detector design

In this section we perform a simple analysis of a PIN photodetector in the presence of noise [11]. Using results 10.6 and 10.7, the signal current of the detector is

$$i_s = \eta \frac{e\lambda}{hc} P \tag{10.12}$$

where P is the optical power coming to the detector at wavelength λ. Equivalent circuit of a PIN photodetector with an amplifier is shown in Fig. 10.10. It is a simple modification of the circuit shown in Fig. 10.6.

Noise generated in the amplifier is essential for the operation. Amplifier noise is represented [11] as Johnson noise originating in the load resistor R_L. Using Eq. (10.11), the signal-to-noise ratio is

$$\frac{S}{N} = \frac{i_s^2 \cdot R_L}{i_{s,yh}^2 \cdot R_L} = \frac{i_s^2 \cdot R_L}{4k_B \cdot T \cdot B} \tag{10.13}$$

where B is the bandwidth and T is the resistor's temperature. The $\frac{S}{N}$ ratio cannot increase indefinitely by increasing value of R_L because of the effect of junction capacitance C_j. In the typical situation, the bandwidth is determined by $(R_L \cdot C_j)^{-1}$. Using the expression for the bandwidth $B = (2\pi \cdot R_L \cdot C_j)^{-1}$ (see problem), Eq. (10.13) takes the form

$$\frac{S}{N} = \frac{i_s^2}{8k_B \cdot T \cdot B^2 \cdot C_j} \tag{10.14}$$

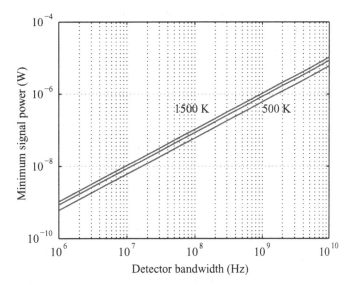

Fig. 10.11 Plot of P_{min} versus bandwidth for PIN detector for three values of temperature (1500 K, 1000 K and 500 K).

Minimum detectable signal $i_{s,\min}$ can be determined from the above by setting $\frac{S}{N} = 1$. One finds

$$i_{s,\min} = 2 \cdot B \cdot \left(2\pi \cdot k_B \cdot T \cdot C_j\right)^{1/2} \tag{10.15}$$

Minimum detectable optical power is determined from Eq. (10.12) as

$$P_{\min} = \frac{2hc}{\eta e \lambda} \cdot B \cdot \left(2\pi \cdot k_B \cdot T \cdot C_j\right)^{1/2} \tag{10.16}$$

The plot of $P_{\min}$ versus bandwidth B is shown in Fig. 10.11. The following parameters were assumed: $C_j = 1 pF$, $\eta = 1$, $T = 1000$ K, $\lambda = 0.85$ μm. MATLAB code is provided in an Appendix, Listing 10A.1.

10.3 Receiver analysis

Optical detection theory is extensively discussed, e.g. by Einarsson [12]. The problem involves detection of signals with noise. In optical systems the information is transmitted using light which consists of photons. Due to their statistical nature, the transmitted information will always show random fluctuations. Those fluctuations determine the lowest limit on transmitted power. In addition, there exist other noise contributions which originate from various processes. Some of them have already been discussed.

Digital signal under consideration operates at a bit rate B. The time slot (or bit interval) T is

$$T = \frac{1}{B}$$

and it is the inverse of the bit rate. The input data sequence in the communication system is denoted by $\{b_k\}$. The optical power $p(t)$ falling on the photodetector is the sequence of pulses and it is written as

$$p(t) = \sum_{k=-\infty}^{k=+\infty} b_k \, h_p(t - k \cdot T) \qquad (10.17)$$

where k is a parameter denoting the k-th time slot and $h_p(t)$ represents the pulse shape of an isolated optical pulse at the photodetector input.

It is assumed that [5]

$$\frac{1}{T} \int_{-\infty}^{+\infty} h_p(t) dt = 1 \qquad (10.18)$$

so that b_k represents the received optical power in the k-th time slot.

Eq. (10.17) is based on the assumption that the system consisting of transmitter and optical fibre is a linear one and time-invariant. b_k in Eq. (10.17) can take two values b_0 and b_1, which correspond to logical values in the k-th time slot being ZERO or ONE. In an ideal case one would expect the b_0 to be zero so no optical power is transmitted for the logical ZERO. However, semiconductor lasers always operate at the nonzero bias current, so there is always some small optical power transmitted.

10.3.1 BER of an ideal optical receiver

There are several ways to measure bit error rate (BER), that is the rate of error occurrence in a digital data stream. It is equal to number of errors occurring over some time interval divided by total number of pulses (both ones and zeros).

The simplest method to define BER is [13]

$$BER = \frac{N_e}{N_p} = \frac{N_e}{B \cdot t} \qquad (10.19)$$

where N_e is the number of errors appearing over time interval t, N_p is the number of pulses transmitted during that interval, $B = 1/T$ is the bit rate and T is the bit interval.

The required BER for high-speed optical communication systems today is typically 10^{-12}, which means that on average one bit error is allowed for every terabit of data transmitted. BER depends on the various signal-to-noise ratio (SNR) of the fibre system, like the receiver noise level.

Direct optical detection is a process of determining the presence or absence of light during a bit interval. No light is interpreted as logical ZERO; some of the light present signals is logical ONE.

In a real life, the detection process is not so simple because of the random nature of photons arriving at the receiver. Their arrival is modelled as a Poisson random process. The random process, in time arrivals of photons at the photodetector, is shown in Fig. 10.12.

For an ideal optical receiver we will assume that there are no noise sources in the system. The average number of photons arriving at the photodetector, with hv_c being the energy of

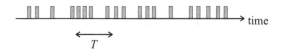

Fig. 10.12 Random arrivals of photons at photodetector are described by the Poisson process. Each photon is represented by a box and they all have the same amplitude.

a single photon, is thus

$$N = \frac{p(t)}{h\nu_c} \qquad (10.20)$$

where $p(t)$ is power of light signal, h is Planck's constant and ν_c is the carrier frequency. The optical power impinging on the photodetector is expressed by Eq. (10.17).

A simple expression for BER for ideal receiver (no noise) can be obtained as follows (following Ramaswani and Sivarajan [2]).

The probability that n photons are received during a bit interval T is

$$e^{-\frac{N}{B}} \frac{\left(\frac{N}{B}\right)^n}{n!}$$

where N is the average number of photons given by Eq. (10.20). Probability of not receiving any photons ($n = 0$) is $\exp(-N/B)$. Assume equal probabilities of receiving ZERO and ONE. The BER of an ideal receiver is thus

$$BER = \frac{1}{2}e^{-N/B} \equiv \frac{1}{2}e^{-M} \qquad (10.21)$$

where $M = \frac{N}{B} = \frac{p}{h\nu_c B}$ represents the average number of photons received during one bit.

Expression (10.21) represents BER for an ideal receiver and it is called the quantum limit. To get a typical bit rate of 10^{-12}, the average number of photons is $M = 27$ per one bit.

10.3.2 Error probability in the receiver

Assume that a binary signal current at the photodetector is [4]

$$\begin{aligned} i_{tot}(t) &= \sum_{n=-\infty}^{+\infty} B_n \cdot h_p(t - nT_B) + I_D + i_{noise}(t) \\ &= i_{ph}(t) + I_D + i_{noise}(t) \end{aligned}$$

where $i_{ph}(t)$ is due to the amplitude modulated pulse signal, I_D is the dark current of the photodiode and $i_{noise}(t)$ is the noise current. B_n takes two values B_H and B_L corresponding to logical ONE and logical ZERO, respectively.

Assume (after Liu [4]) that $H(\omega)$ is the combined transfer function of the front-end receiver and equalizer. The output at the equalizer is

$$y_{out}(t) = i_{tot}(t) \otimes h(t) \qquad (10.22)$$

where $\otimes$ denotes convolution and $h(t)$ corresponds to the transfer function $H(\omega)$. The signal component at the output of the equalizer is

$$y_s(t) = i_{ph}(t) \otimes h(t) = \sum_{n=-\infty}^{+\infty} B_n \cdot h_p(t-nT_B) \otimes h(t)$$

$$\equiv \sum_{n=-\infty}^{+\infty} B_n \cdot h_p(t-nT_B)$$

The signal output of the equalizer is sampled at the bit rate and compared with a threshold. This process allows for detection of the amplitudes B_n. The output signal after equalizer without (constant) dark current is

$$y_{out}(t) = \sum_{n=-\infty}^{+\infty} B_n \cdot p_p(t-nT_B)$$

where $p_p(t)$ contains signal and noise. The above is sampled and the output at time $nT_B + \tau\, (0 < \tau < T_B)$ is

$$y_{out,n}(t) \equiv y_{out}(nT_B+\tau) = \sum_{k} B_k \cdot p_p([n-k]T_B+\tau) + y_{noise,n}$$

$$\equiv B_n \cdot p_p[0] + ISI_n + y_{noise,n} \quad (10.23)$$

with the following definitions:

$$p_p[n] = p_p(nT_B+\tau)$$

$$ISI_n = \sum_{k \neq n} B_k \cdot p_p([n-k])$$

and

$$y_{noise,n} = y_{noise,out}(nT_B+\tau)$$

is the sampled noise term. ISI_n refers to the intersymbol interference.

In a practical receiver, the decision as which bit (ZERO or ONE) was transmitted in each bit interval is done by sampling current after equalizer. With the presence of noise, there is nonzero probability of an error. The errors are typically described by BER (bit error rate).

Introduce y_{th} as a threshold used in error detection. From Eq. (10.23), one can observe that errors occur when

$$B_k = B_H \quad \text{when} \quad y_{out,n} < y_{th}$$

or

$$B_k = B_L \quad \text{when} \quad y_{out,n} > y_{th}$$

Therefore, the probability of an error detection is [14]

$$P_{error} = p_0 P\left(y_{out,n} > y_{th} | B_k = B_L\right) + p_1 P\left(y_{out,n} < y_{th} | B_k = B_H\right)$$

Here p_0 and p_1 are a priori probabilities for bits ZERO and ONE.

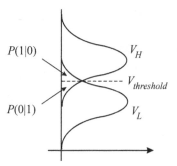

Fig. 10.13 Determination of bit error rate for a Gaussian process.

The decision taken involves comparison of the measured photocurrent I with a threshold I_{th}. If $I > I_{th}$, one decides that a ONE bit was transmitted; otherwise for $I < I_{th}$ a bit ZERO was transmitted.

10.3.3 BER and Gaussian noise

In optical digital communication the transmitted signal is never perfectly recovered. Noise, adjacent channel interference, amplified spontaneous emission, all result in creation of errors. Additionally, pulses are broadened due to fibre dispersion, device nonlinearity or slow photodiode response. Created signal distortion is known as an intersymbol interference (ISI) and it arises when adjacent bits are corrupted. It results in possible detection errors for the transmitted bits. ISI is random because the binary bits in a digital signal are random. To minimize ISI, an equalizer is generally used.

We analyse the performance of real receiver with noise [15]. We want to establish connection between the signal-to-noise ratio (SNR) and BER. By referring to Fig. 10.13, we assume that V_H (high voltage at the output of the receiver) and V_L (low voltage at the output of the receiver) are random variables described by the Gaussian probability density function (pdf) as

$$p(x) = \frac{1}{\sqrt{2\pi}} e^{-\frac{x^2}{2}} \qquad (10.24)$$

BER is defined in terms of two conditional probabilities $P(0|1)$ and $P(1|0)$ as

$$BER = p_0 P(0|1) + p_1 P(1|0) \qquad (10.25)$$

where p_0 and p_1 are a priori probabilities for bits $ZERO$ and ONE. Their values are $p_0 = p_1 = \frac{1}{2}$. Conditional probability $P(A|B)$ for two events A and B is defined as

$$P(A|B) = \frac{P(A \cdot B)}{P(B)}$$

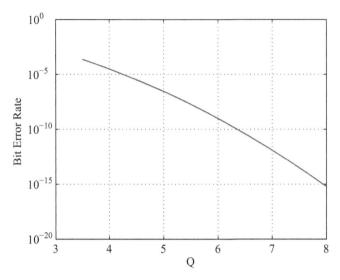

Fig. 10.14 Plot of bit error rate versus Q.

The conditional probabilities $P(0|1)$ and $P(1|0)$ are shown in Fig. 10.13. Assuming Gaussian pdf, they are determined as

$$P(0|1) = \frac{1}{\sqrt{2\pi}} \int_{Q_1}^{\infty} e^{-\frac{x^2}{2}} dx$$

$$P(1|0) = \frac{1}{\sqrt{2\pi}} \int_{Q_2}^{\infty} e^{-\frac{x^2}{2}} dx$$

where functions $Q_1 = \frac{V_{th} - V_L}{\sigma_{noise}}$ and $Q_1 = \frac{V_H - V_{th}}{\sigma_{noise}}$ are integration limits. We assume that distributions of $ZERO$ and ONE are identical and that noise variance σ_{noise} is the same for both. It follows that the optimum threshold is $\frac{1}{2}(V_H + V_L)$. Thus

$$Q = Q_1 = Q_2 = \frac{V_H - V_L}{2\sigma_{noise}}$$

Using definition (10.25), the BER is evaluated as

$$BER = \frac{1}{2} P(0|1) + \frac{1}{2} P(1|0)$$

$$= \frac{1}{\sqrt{2\pi}} \int_{Q}^{\infty} e^{-\frac{x^2}{2}} dx$$

The above function relates BER to the signal-to-noise ratio Q. It is plotted in Fig. 10.14. For a typical BER of 10^{-6}, $Q = 6$. MATLAB code used to create this figure is shown in Appendix, Listing 10A.2.

The average incident input power $\overline{P}$ is defined as

$$\overline{P} = \frac{1}{2}(P_H + P_L) \tag{10.26}$$

where P_H and P_L are optical powers corresponding to logical *ONE* and *ZERO* (high and low), respectively. Those powers are related to voltages V_H and V_L introduced earlier as

$$V_H = R \cdot P_H \cdot y_{s,\max}(t) \tag{10.27}$$
$$V_L = R \cdot P_L \cdot y_{s,\max}(t) \tag{10.28}$$

where R is the responsivity of the detector and $y_{s,\max}(t)$ is the maximum output voltage of the receiver during a given bit period. In terms of the above quantities, the average power $\overline{P}$ is evaluated as follows:

$$\begin{aligned}\overline{P} &= \frac{1}{2}(P_H + P_L) = \frac{1}{2}\frac{P_H + P_L}{P_H - P_L}(P_H - P_L) \\ &= \frac{1}{2} \cdot \frac{1 + \frac{P_L}{P_H}}{1 - \frac{P_L}{P_H}} \cdot \frac{1}{R} \cdot \frac{1}{y_{s,\max}(t)} \cdot (V_H - V_L) \\ &= \frac{1}{2} \cdot \frac{1+r}{1-r} \cdot \frac{1}{R} \cdot Q \cdot \frac{\sqrt{n_{out}^2}}{y_{s,\max}(t)}\end{aligned} \tag{10.29}$$

The above expression gives the average incident input power level necessary to obtain a given BER. In the above derivation, we have used $\sigma_{noise}^2 = n_{out}^2$, where n_{out}^2 is the variance of the output noise. We also defined extinction ratio r as

$$r = \frac{P_L}{P_H} \tag{10.30}$$

For a PIN photodiode, the receiver sensitivity takes the form

$$\eta \overline{P} = \frac{h\nu}{q} \cdot \frac{1+r}{1-r} \cdot Q \cdot \frac{\sqrt{n_{out}^2}}{y_{s,\max}(t)} \tag{10.31}$$

Ideal sensitivity is obtained where no power is transmitted ($r = 0$) as

$$\eta \overline{P} = \frac{h\nu}{q} \cdot Q \cdot \frac{\sqrt{n_{out}^2}}{y_{s,\max}(t)} \tag{10.32}$$

Using the above results, one can analyse receiver sensitivity for various types of noise filters. We refer to the literature for a more detailed discussion [15], [14], [4].

10.4 Modelling of a photoelectric receiver

A schematic of the receiver is shown in Fig. 10.15. It is modelled by a filter function $H(\nu)$, see Geckeler [16]. As a front end it contains an avalanche photodiode, with M being the multiplication factor of the photocurrent and G the gain factor of an electronic amplifier.

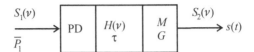

Fig. 10.15 Block scheme of the receiver which is described by a filter function $H(\nu)$ with parameter τ.

Both M and G are assumed to be frequency-independent. Frequency dependence of the receiver is described by a low-pass filter function $H(\nu)$, which we assume to be a real function (not complex). Possible phase distortions which will result in function $H(\nu)$ being complex can be compensated by an equalizer.

In Fig. 10.15 $S_1(\nu)$ is the frequency spectrum of the input, $S_2(\nu)$ is the output frequency spectrum, $s(t)$ is the temporal output signal and $\overline{P}_1$ is the average power arriving at the photoelectric receiver. Parameter τ characterizes filter function $H(\nu)$.

10.5 Problems

1. Determine rise time τ_r for a simple RC circuit. Relate it to 3 dB bandwidth.
2. Analyse noise in an avalanche photodiode (APD). Determine signal-to-noise ratio in terms of gain of APD.
3. Analyse signal limitations for an analogue transmission. Determine signal power in terms of S/N ratio.

10.6 Projects

1. Conduct small-signal equivalent circuit analysis of a photodetector [1]. Write a MATLAB program. Determine frequency dependence of the small signal impedance of a photodiode.
2. Analyse frequency response of a photodiode [17]. Determine the upper limit of the frequency response of *GaAs* p-n junction photodiode assuming data provided in [17].

Appendix 10A: MATLAB listings

In Table 10.3 we provide a list of MATLAB files created for Chapter 10 and a short description of each function.

Appendix 10A

Table 10.3 List of MATLAB functions for Chapter 10.

Listing	Function name	Description
10A.1	min_power.m	Determines minimum signal power
10A.2	berQ.m	Plots bit error rate versus Q

Listing 10A.1 Function min_power.m. MATLAB function which determines minimum signal power for a PIN photodetector for three values of temperature.

```
% File name: min_power.m
% Determines minimum signal power required to give S/N ratio of one
% for a PIN detector
clear all
h_Planck = 6.6261d-34;      % Planck's constant (J s)
c_light=299.80d6;           % speed of light (m/s)
k_B= 1.3807d-23;            % Boltzmann constant (J/K)
e=1.602d-19;                % electron's charge (C)
%
C_j = 1d-12;                % junction capacitance (1pF)
eta = 1;                    % quantum efficiency
lambda = 0.85d-6;           % wavelength (0.85 microns)
B = 1d6:1d5:1d10;           % range of detector bandwidth
%
for T_eff = [500 1000 1500]
    P_min=(2*h_Planck*c_light/(eta*e*lambda))*sqrt(2*pi*k_B*T_eff*C_j)*B;
    loglog(B,P_min,'LineWidth',1.5)
    hold on
end
%
xlabel('Detector bandwidth (Hz)','FontSize',14);
ylabel('Minimum signal power (W)','FontSize',14);
text(1d9, 4d-7, '500K','Fontsize',14)
text(5d7, 4d-7, '1500K','Fontsize',14)
set(gca,'FontSize',14);     % size of tick marks on both axes
grid on
pause
close all
```

Listing 10A.2 Program berQ.m. MATLAB program which plots bit error rate versus Q. Takes one minute to execute.

```
% File name: berQ.m
% Plots bit error rate versus Q
syms x                              % defines symbolic variable
```

```
f=(1/sqrt(2*pi))*exp(-x^2/2);        % defines function for plot
Q_i = 3.5; Q_f = 8; Q_step = 0.5;    % defines range and step
yy = double(int(f,x,Q_i,inf));
for Q = Q_i+Q_step:Q_step:Q_f
    y = double(int(f,x,Q,inf));
    yy = [yy,y];
end
Qplot = Q_i:Q_step:Q_f;              % defines Q for plot
semilogy(Qplot,yy,'LineWidth',1.5)
xlabel('Q','FontSize',14);
ylabel('Bit Error Rate','FontSize',14);
set(gca,'FontSize',14);              % size of tick marks on both axes
grid on
pause
close all
```

References

[1] S. B. Alexander. *Optical Communication Receiver Design*. SPIE Optical Engineering Press and Institution of Electrical Engineers, Bellingham, Washington, and London, 1997.

[2] R. Ramaswami and K. N. Sivarajan. *Optical Networks: A Practical Perspective*. Morgan Kaufmann Publishers, San Francisco, 1998.

[3] J. C. Cartledge and G. S. Burley. The effect of laser chirping on lightwave system performance. *J. Lightwave Technol.*, **7**:568–73, 1989.

[4] M. M.-K. Liu. *Principles and Applications of Optical Communications*. Irwin, Chicago, 1996.

[5] T. V. Muoi. Receiver design of optical-fiber systems. In E. E. Bert Basch, ed., *Optical-Fiber Transmission*, pp. 375–425. Howard W. Sams & Co., Indianapolis, 1986.

[6] D. L. Rogers. Integrated optical receivers using MSM detectors. *J. Lightwave Technol.*, **9**:1635–8, 1991.

[7] J.-M. Liu. *Photonic Devices*. Cambridge University Press, Cambridge, 2005.

[8] S. O. Kasap. *Optoelectronics and Photonics: Principles and Practices*. Prentice Hall, Upper Saddle River, NJ, 2001.

[9] J. C. Palais. *Fiber Optic Communications. Fifth Edition*. Prentice Hall, Upper Saddle River, NJ, 2005.

[10] G. Keiser. *Optical Fiber Communications. Fourth Edition*. McGraw-Hill, Boston, 2011.

[11] J. Wilson and J. F. B. Hawkes. *Optoelectronics. An Introduction*. Prentice Hall, Englewood Cliffs, 1983.

[12] G. Einarsson. *Principles of Lightwave Communications*. Wiley, Chichester, 1996.

[13] G. Keiser. *Local Area Networks. Second Edition*. McGraw Hill, Boston, 2002.

[14] W. van Etten and J. van der Plaats. *Fundamentals of Optical Fiber Communications*. Prentice Hall, New York, 1991.

[15] A. E. Stevens. *An integrate-and-dump receiver for fiber optic networks*. PhD thesis, Columbia University, 1995.

[16] S. Geckeler. *Optical Fiber Transmission System*. Artech House, Inc., Norwood, MA, 1987.

[17] A. Yariv and P. Yeh. *Photonics. Optical Electronics in Modern Communications. Sixth Edition*. Oxford University Press, New York, Oxford, 2007.

11 Finite difference time domain (FDTD) formulation

The finite-difference time-domain (FDTD) method is a widely used numerical scheme for approximate description of propagation of electromagnetic waves. It can be used to study such phenomena as pulse propagation in various media. It is a simple scheme and can be mastered relatively fast.

Definite reference on the subject is a book by Taflove and Hagness [1]. Several other books can be cited: Kunz and Luebbers [2], Sullivan [3], Yu et al. [4], Elsherbeni and Demir [5]. Also, recently published books by Bondeson et al. [6] and by Garg [7] have instructive chapters on the FDTD method.

11.1 General formulation

To develop a general formulation, one starts from Maxwell's equations which we cast in the form suitable for 1D, 2D and 3D discretizations. In the following, we assume that $J = 0$ and $\rho = 0$; i.e there is no flow of current and no free charges. Assume also nonmagnetic material and therefore set $\mu = \mu_0$. General Maxwell's equations are

$$\nabla \times \mathbf{H} = \frac{\partial \mathbf{D}}{\partial t} \tag{11.1}$$

$$\nabla \times \mathbf{E} = -\mu_0 \frac{\partial \mathbf{H}}{\partial t} \tag{11.2}$$

Here **E** and **H** are, respectively, electric and magnetic fields and **D** is the electric displacement flux density which is related to electric field **E**. The relation between **E** and **D** will be summarized in later sections for more detailed models.

First we write the above equations in terms of components. The following cases are typically considered:

1. 1D (one-dimensional) case: infinite medium in y-direction and z-direction and no change in those directions and therefore $\frac{\partial}{\partial y} = \frac{\partial}{\partial z} = 0$.
2. 2D (two-dimensional) case: infinite medium in z-direction and no change in z-direction and therefore $\frac{\partial}{\partial z} = 0$.
3. 3D (three-dimensional) case: no restriction on variations in all three directions.

In this chapter we will not discuss the 3D case in detail since it requires significant hardware resources for typical calculations.

11.1.1 Three-dimensional formulation

Use general mathematical formula

$$\nabla \times \mathbf{H} = \begin{vmatrix} \mathbf{a}_x & \mathbf{a}_y & \mathbf{a}_z \\ \frac{\partial}{\partial x} & \frac{\partial}{\partial y} & \frac{\partial}{\partial z} \\ H_x & H_y & H_z \end{vmatrix}$$

$$= \mathbf{a}_x \left(\frac{\partial H_z}{\partial y} - \frac{\partial H_y}{\partial z} \right) - \mathbf{a}_y \left(\frac{\partial H_z}{\partial x} - \frac{\partial H_x}{\partial z} \right) + \mathbf{a}_z \left(\frac{\partial H_y}{\partial x} - \frac{\partial H_x}{\partial y} \right)$$

By applying the above expression, Maxwell's equations can be written as

$$\begin{aligned}
\frac{\partial D_x}{\partial t} &= \frac{\partial H_z}{\partial y} - \frac{\partial H_y}{\partial z} \\
-\frac{\partial D_y}{\partial t} &= \frac{\partial H_z}{\partial x} - \frac{\partial H_x}{\partial z} \\
\frac{\partial D_z}{\partial t} &= \frac{\partial H_y}{\partial x} - \frac{\partial H_x}{\partial y} \\
-\mu_0 \frac{\partial H_x}{\partial t} &= \frac{\partial E_z}{\partial y} - \frac{\partial E_y}{\partial z} \\
\mu_0 \frac{\partial H_y}{\partial t} &= \frac{\partial E_z}{\partial x} - \frac{\partial E_x}{\partial z} \\
-\mu_0 \frac{\partial H_z}{\partial t} &= \frac{\partial E_y}{\partial x} - \frac{\partial E_x}{\partial y}
\end{aligned} \quad (11.3)$$

These are general equations used as a starting point for 3D implementation, see e.g. [1].

11.1.2 Two-dimensional formulation

In the 2D model we assume that $\frac{\partial}{\partial z} = 0$ and evaluate

$$\nabla \times \mathbf{H} = \begin{vmatrix} \mathbf{a}_x & \mathbf{a}_y & \mathbf{a}_z \\ \frac{\partial}{\partial x} & \frac{\partial}{\partial y} & 0 \\ H_x & H_y & H_z \end{vmatrix} = \mathbf{a}_x \frac{\partial H_z}{\partial y} - \mathbf{a}_y \frac{\partial H_z}{\partial x} + \mathbf{a}_z \left(\frac{\partial H_y}{\partial x} - \frac{\partial H_x}{\partial y} \right)$$

where $\mathbf{a}_x, \mathbf{a}_y, \mathbf{a}_z$ are (as before) unit vectors defining coordinate system. Using the above expression, Maxwell's equations in terms of components become

$$\mathbf{a}_x \frac{\partial H_z}{\partial y} - \mathbf{a}_y \frac{\partial H_z}{\partial x} + \mathbf{a}_z \left(\frac{\partial H_y}{\partial x} - \frac{\partial H_x}{\partial y} \right) = \mathbf{a}_x \frac{\partial D_x}{\partial t} + \mathbf{a}_y \frac{\partial D_y}{\partial t} + \mathbf{a}_z \frac{\partial D_z}{\partial t}$$

$$\mathbf{a}_x \frac{\partial E_z}{\partial y} - \mathbf{a}_y \frac{\partial E_z}{\partial x} + \mathbf{a}_z \left(\frac{\partial E_y}{\partial x} - \frac{\partial E_x}{\partial y} \right) = -\mu_0 \left(\mathbf{a}_x \frac{\partial H_x}{\partial t} + \mathbf{a}_y \frac{\partial H_y}{\partial t} + \mathbf{a}_z \frac{\partial H_z}{\partial t} \right)$$

The above equations are then usually split into two groups, see [1] and [8].

Group 1:

$$\frac{\partial D_x}{\partial t} = \frac{\partial H_z}{\partial y} \qquad (11.4)$$

$$\frac{\partial D_y}{\partial t} = -\frac{\partial E_z}{\partial x} \qquad (11.5)$$

$$\frac{\partial H_z}{\partial t} = -\mu_0 \left(\frac{\partial E_y}{\partial x} - \frac{\partial E_x}{\partial y} \right) \qquad (11.6)$$

Group 2:

$$\frac{\partial H_x}{\partial t} = -\frac{1}{\mu_0} \frac{\partial H_z}{\partial y} \qquad (11.7)$$

$$\frac{\partial H_y}{\partial t} = \frac{1}{\mu_0} \frac{\partial E_z}{\partial x} \qquad (11.8)$$

$$\frac{\partial D_z}{\partial t} = \frac{\partial H_y}{\partial x} - \frac{\partial H_x}{\partial y} \qquad (11.9)$$

The first group is called TE_x mode by Taflove and TE to z polarization by Umashankar, and the group 2 is called TM_z mode by Taflove and TM to z polarization by Umashankar.

11.1.3 One-dimensional model

In one dimension, the medium extends to infinity in the y-direction and z-direction. Therefore $\frac{\partial}{\partial y} = \frac{\partial}{\partial z} = 0$. Using the following result,

$$\nabla \times E = \begin{vmatrix} \mathbf{a}_x & \mathbf{a}_y & \mathbf{a}_z \\ \frac{\partial}{\partial x} & 0 & 0 \\ H_x & H_y & H_z \end{vmatrix} = -\mathbf{a}_y \frac{\partial H_z}{\partial x} + \mathbf{a}_z \frac{\partial H_y}{\partial x}$$

one can write Maxwell's equations in two groups according to field vector components.

First group is

$$\frac{\partial H_y}{\partial t} = \frac{1}{\mu_0} \frac{\partial E_z}{\partial x} \qquad (11.10)$$

$$\frac{\partial D_z}{\partial t} = \frac{\partial H_y}{\partial x} \qquad (11.11)$$

Taflove [1] classifies this group as the x-directed, z-polarized TEM mode. It contains D_y and H_y.

The second group is

$$\frac{\partial H_z}{\partial t} = -\frac{1}{\mu_0} \frac{\partial E_y}{\partial x} \qquad (11.12)$$

$$\frac{\partial D_y}{\partial t} = -\frac{\partial H_z}{\partial x} \qquad (11.13)$$

This group is referred to as the x-directed, y-polarized TEM mode. It contains D_y and H_z. In the following, we will concentrate on group 2. The equations describe wave propagating

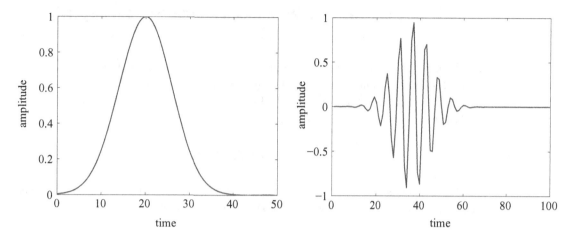

Fig. 11.1 Gaussian pulse (left), and modulated Gaussian pulse (right).

in the x-direction with electric field oriented in the y-direction and magnetic field oriented in the z-direction.

We finish this section with a discussion of the Gaussian pulse and modulated Gaussian pulse.

11.1.4 Gaussian pulse and modulated Gaussian pulse

In practice one often analyses propagation of the Gaussian pulse or modulated Gaussian pulse. A Gaussian pulse centred at t_0 is defined as

$$g(t) = \exp\left[-\frac{(t-t_0)^2}{w^2}\right] \quad (11.14)$$

where w is the width of the pulse in space. Define also the modulation signal as

$$m(t) = \sin(2\pi f_0 \cdot t) \quad (11.15)$$

where $f_0 = c/\lambda_0 = 1/T_0$. Here T_0 is the period of the modulation signal $m(t)$. A continuous Gaussian pulse is created by multiplying Eqs. (11.14) and (11.15):

$$g(t) = \sin(2\pi f_0 \cdot t) \cdot \exp\left[-\frac{(t-t_0)^2}{w^2}\right] \quad (11.16)$$

Its discretized version is

$$g(n) = \sin(2\pi f_0 \cdot n \cdot \Delta t) \cdot \exp\left[-\frac{(n \cdot \Delta t - t_0)^2}{w^2}\right] \quad (11.17)$$

In Fig. 11.1 we plot Gaussian and modulated Gaussian pulses. The results were obtained with MATLAB code presented in Appendix, Listings 11A.1 and 11A.2.

11.2 One-dimensional Yee implementation without dispersion

In this section, we discuss the basic implementation of FDTD method in 1D. It is known as the 1D-Yee algorithm since it was first proposed by Yee [9] for the 3D case. We also introduce the concept of a staggered grid [3], [7]. To illustrate basic principles of FDTD method, let us develop first the 1D lossless case.

11.2.1 Lossless case

Our formulation is based on the second group of Eqs. (11.12) and (11.13). We supplement those equations with the relation between D_y and E_y, which for the free space is

$$D_y = \varepsilon_0 E_y \qquad (11.18)$$

The equations take the form

$$\frac{\partial H_z}{\partial t} = -\frac{1}{\mu_0}\frac{\partial E_y}{\partial x} \qquad (11.19)$$

$$\frac{\partial E_y}{\partial t} = -\frac{1}{\varepsilon_0}\frac{\partial H_z}{\partial x} \qquad (11.20)$$

In order to introduce 1D Yee discretization, let us introduce the following notation:

$$E_y(i \cdot \Delta x, n \cdot \Delta t) \equiv (E_y)_i^n$$
$$H_z(i \cdot \Delta x, n \cdot \Delta t) \equiv (H_y)_i^n$$

To discretize Eqs. (11.19) and (11.20), we use central-difference approximation for space and time derivatives. They are second-order accurate in the space and time increments:

$$\frac{\partial E_y}{\partial x}(i \cdot \Delta x, n \cdot \Delta t) = \frac{(E_y)_{i+1/2}^n - (E_y)_{i-1/2}^n}{\Delta x} + O(\Delta x)^2$$

$$\frac{\partial E_y}{\partial t}(i \cdot \Delta x, n \cdot \Delta t) = \frac{(E_y)_i^{n+1/2} - (E_y)_i^{n-1/2}}{\Delta t} + O(\Delta t)^2$$

In the Yee algorithm [9] **E** and **H** components are interleaved in the space lattice at intervals of $\Delta x/2$. Such an approximation is called the staggered grid approximation. They are also interleaved in time at intervals $\Delta t/2$, see Fig. 11.2.

At time $t = 0$, values of E_y are placed at $x = i \cdot \Delta x$, with $i = 0, 1, 2, \ldots N$. (Total $N + 1$ components).

At time $t = \frac{1}{2}\Delta t$, values of H_z are placed at $x = (i - \frac{1}{2}) \cdot \Delta x$, with $i = 1, 2, \ldots N$. (Total N components).

In general

$$(E_y)_i^n = E_y((i-1) \cdot \Delta x, n \cdot \Delta t), \qquad i = 1, 2, \ldots N+1; n = 0, 1, 2 \ldots$$

$$(H_z)_i^n = H_z\left(\left(i - \frac{1}{2}\right) \cdot \Delta x, \left(n - \frac{1}{2}\right) \cdot \Delta t\right), \qquad i = 1, 2, \ldots N; n = 0, 1, 2 \ldots$$

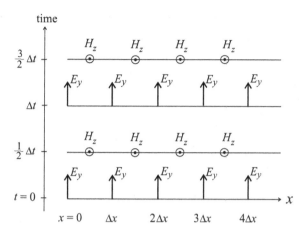

Fig. 11.2 One-dimensional formulation of the FDTD method. Orientations of H_z and E_y fields are shown.

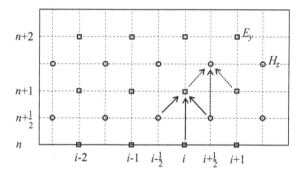

Fig. 11.3 Visual illustration of numerical dependencies in the 1D FDTD method.

Initial values (corresponding to $n = 0$), i.e. $E(\ldots, t = 0)$ and $H(\ldots, t = 0)$, in most cases are set to zero.

The discretized equations are

$$\frac{(E_y)_i^{n+1/2} - (E_y)_i^{n-1/2}}{\Delta t} = -\frac{1}{\varepsilon_0} \frac{(H_z)_{i+1/2}^n - (H_z)_{i-1/2}^n}{\Delta x}$$

$$\frac{(H_z)_{i+1/2}^{n+1} - (H_z)_{i+1/2}^n}{\Delta t} = -\frac{1}{\mu_0} \frac{(E_y)_{i+1}^{n+1/2} - (E_y)_{i+1}^{n-1/2}}{\Delta x}$$

The space-time interdependencies of E_y and H_z fields at different grid points are illustrated in Fig. 11.3. Here, squares represent E_y field and circles represent H_z field. One can observe that the value of a field at any point is determined by three previous values: two from two neighbours of opposite field from the previous half time step and one from the same field at a single previous time step.

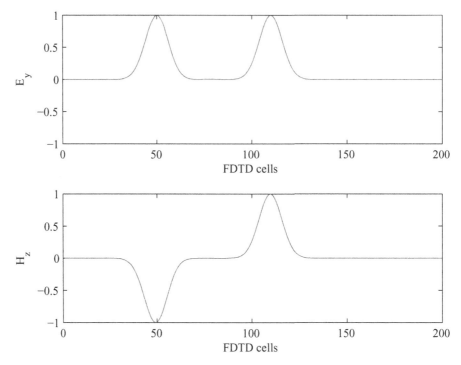

Fig. 11.4 FDTD simulation of a Gaussian pulse after several time steps.

Constants ε_0 and μ_0 appearing in these equations differ by several orders of magnitude. Therefore, one usually introduces a new scaled variable defined as [3]

$$\widetilde{E}_y = \sqrt{\frac{\varepsilon_0}{\mu_0}} E_y$$

Scaled equations are

$$\widetilde{E}_y^{n+1/2}(i) = \widetilde{E}_y^{n-1/2}(i) - \frac{1}{\sqrt{\varepsilon_0 \mu_0}} \frac{\Delta t}{\Delta x} \left[H_z^n(i+1/2) - H_z^n(i-1/2) \right]$$

$$H_z^{n+1}(i+1/2) = H_z^n(i+1/2) - \frac{1}{\sqrt{\varepsilon_0 \mu_0}} \frac{\Delta t}{\Delta x} \left[\widetilde{E}_y^{n+1/2}(i+1) - \widetilde{E}_y^{n+1/2}(i) \right]$$

In our simulations, due to stability reasons we set

$$\frac{1}{\sqrt{\varepsilon_0 \mu_0}} \frac{\Delta x}{\Delta t} = \frac{1}{2} \quad (11.21)$$

The value of $v = \Delta x/\Delta t$ is chosen to be the maximum phase velocity of the wave expected in the medium. The above system has been implemented. The MATLAB program is shown in Appendix, Listing 11A.3. Two versions of MATLAB code have been produced: one of which uses regular loop and one that takes advantage of parallel capabilities of MATLAB. The reader should analyse speed of operation for both versions.

The results of propagation of Gaussian pulse using that code are shown in Fig. 11.4.

11.2.2 Determination of cell size

In one dimension the cell size is just Δx. In 2D it is a square and a cube in 3D. The fundamental question is how to determine size of basic cell.

For a detailed discussion we refer to the book by Kunz and Luebbers [2]. For us, it is enough to notice that cell size should be much less than the smallest wavelength and typical rule is to take it as $\lambda/10$. Such choice should provide accurate results.

11.2.3 Dispersion and stability

Here, we discuss some general properties of the wave equation. It was formulated before. Its discretized version is (here we use index r instead of i to avoid conflict with a symbol $i = \sqrt{-1}$)

$$\frac{E_r^{n+1} - 2E_r^n + E_r^{n-1}}{\Delta t^2} = c^2 \frac{E_{r+1}^n - 2E_r^n + E_{r-1}^n}{\Delta x^2} \quad (11.22)$$

As usual, n refers to time and r is the space index. Explicitly $E_r^n = E(r \cdot \Delta z, n \cdot \Delta t)$. The dispersion relation for Eq. (11.22) is found by assuming

$$E_r^n = e^{i(\omega \cdot n \cdot \Delta t - k \cdot r \cdot \Delta x)}$$

Substituting these assumptions into Eq. (11.22) gives

$$\left[e^{i\omega \cdot \Delta t \cdot (n+1)} - 2e^{i\omega \cdot \Delta t \cdot n} + e^{i\omega \cdot \Delta t \cdot (n-1)} \right] e^{-ik \cdot r \cdot \Delta x}$$

$$= \frac{c^2 \Delta t^2}{\Delta x^2} \left[e^{-ik \cdot \Delta z \cdot (r+1)} - 2e^{-ik \cdot r \cdot \Delta x} + e^{-ik \cdot \Delta x \cdot (r-1)} \right] e^{i\omega \cdot \Delta t \cdot n}$$

Dividing the above by $\exp[i(\omega \cdot n \cdot \Delta t - k \cdot r \cdot \Delta x)]$:

$$e^{i\omega \cdot \Delta t} - 2 + e^{-i\omega \cdot \Delta t} = \frac{c^2 \Delta t^2}{\Delta z^2} \left(e^{-ik \cdot \Delta z} - 2 + e^{ik \cdot \Delta z} \right)$$

The above can be written as a square:

$$\left(e^{i\omega \cdot \Delta t/2} - e^{-i\omega \cdot \Delta t/2} \right)^2 = \frac{c^2 \Delta t^2}{\Delta x^2} \left(e^{ik \cdot \Delta x/2} - e^{-ik \cdot \Delta x/2} \right)^2$$

Finally, using trigonometric relation $e^{i\alpha} - e^{-i\alpha} = 2i \sin \alpha$, one obtains formula for numerical dispersion:

$$\sin \frac{1}{2}\omega \cdot \Delta t = \pm \frac{c\Delta t}{\Delta x} \sin \frac{1}{2} k \cdot \Delta x \quad (11.23)$$

The important case is that of linear dispersion, i.e. when $\omega = c \cdot k$. Eq. (11.23) then holds when $\Delta x = c\Delta t$. For such a linear case different frequency components propagate with the same speed, just c. For nonlinear dispersion relations, different frequencies of a pulse numerically propagate with different velocities and therefore pulse will change shape.

Introduce parameter

$$\alpha = c \frac{\Delta t}{\Delta z} \quad (11.24)$$

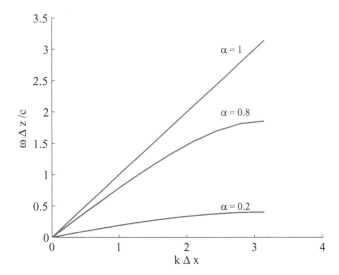

Fig. 11.5 Illustration of numerical dispersion for various values of parameter α.

In Fig. 11.5 we showed numerical dispersions as described by Eq. (11.23) for several values of parameter α. MATLAB code is shown in Appendix, Listing 11A.4. Depending on the value of α, one can distinguish the following cases:

1. $\alpha = 1$. In this case $\Delta t = \frac{\Delta x}{c}$. Numerical dispersion is linear and given by $\omega = \pm c \cdot k$. This choice of time step Δt is called *the magic time step* [10]. In this case wave propagates exactly one cell (or Δx) per time step in both x-directions.
2. $\alpha < 1$. In this case $\Delta t < \frac{\Delta x}{c}$. One observes numerical dispersion which increases when α decreases.
3. $\alpha > 1$. In this case $\Delta t > \frac{\Delta x}{c}$. This inequality implies that numerical front propagates faster than physical speed (given by c). In this case, the scheme becomes unstable.

Because of its importance, we provide next a separate discussion of the stability criterion.

11.2.4 Stability criterion

Stability criterion imposes condition on $\frac{\Delta x}{\Delta t}$ ratio. It is known as Courant-Friedrichs-Levy (CFL) condition [11]. It explains convergence of various schemes.

For the 1D model, the CFL condition is illustrated in Fig. 11.6. Each row of points represents the sampling points in space at a given instant of time (say n). The solution at point A is determined using previous values at the indicated points. In the figure we show characteristics which correspond to stable (left) and unstable (right) situations.

The stability criterion summarizes the physical fact that speed of numerical propagation should not exceed the physical speed. A summary of stability criteria for various dimensions is presented in Table 11.1 [10].

Table 11.1 Summary of stability criteria for various dimensions.

Dimensionality	Criterion
1D medium	$v\, \Delta t \leq \Delta x$
2D medium	$v\, \Delta t \leq \left(\frac{1}{\Delta x^2} + \frac{1}{\Delta y^2}\right)^{-1/2}$
3D medium	$v\, \Delta t \leq \left(\frac{1}{\Delta x^2} + \frac{1}{\Delta y^2} + \frac{1}{\Delta z^2}\right)^{-1/2}$

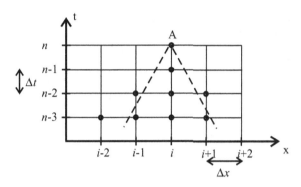
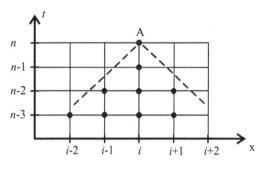

Fig. 11.6 Illustration of the CFL condition. For the parameters chosen within the dashed lines, the propagation is stable (left figure) and becomes unstable (right figure.)

11.2.5 One-dimensional model with losses

When we consider ohmic losses, Maxwell's Eq. (11.1) is modified as

$$\nabla \times \mathbf{H} = \frac{\partial \mathbf{D}}{\partial t} + \sigma \mathbf{E} \tag{11.25}$$

where σ is the conductivity.

The discretization of Eqs. (11.25) and (11.2) with losses is done in a similar way as for the lossless case. The resulting equations after scaling are

$$\frac{(E_y)_i^{n+1/2} - (E_y)_i^{n-1/2}}{\Delta t} = -\frac{1}{\varepsilon_r \sqrt{\varepsilon_0 \mu_0}} \frac{(H_z)_{i+1/2}^n - (H_z)_{i-1/2}^n}{\Delta x} \\ - \frac{\sigma}{2\varepsilon_r \varepsilon_0} \left((E_y)_i^{n+1/2} + (E_y)_i^{n-1/2}\right) \tag{11.26}$$

and

$$\frac{(H_z)_{i+1/2}^{n+1} - (H_z)_{i+1/2}^n}{\Delta t} = -\frac{1}{\mu_0} \frac{(E_y)_{i+1}^{n+1/2} - (E_y)_i^{n+1/2}}{\Delta x} \tag{11.27}$$

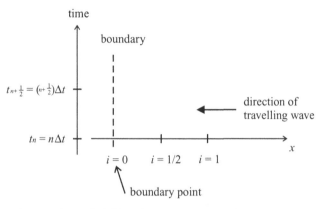

Fig. 11.7 Illustration of propagating wave close to the left boundary.

When selecting the time step given by Eq. (11.21), one arrives at the following equations:

$$(E_y)_i^{n+1/2} - (E_y)_i^{n-1/2} = -\frac{1}{2\varepsilon_r}\left((H_z)_{i+1/2}^n - (H_z)_{i-1/2}^n\right) - \frac{\sigma \Delta t}{2\varepsilon_r \varepsilon_0}\left((E_y)_i^{n+1/2} + (E_y)_i^{n-1/2}\right)$$

$$(H_z)_{i+1/2}^{n+1} - (H_z)_{i+1/2}^n = -\frac{1}{\mu_0}\frac{\Delta t}{\Delta x}\left((E_y)_{i+1}^{n+1/2} - (E_y)_i^{n-1/2}\right)$$

Solving the first equation for $E_y^{n+1/2}(i)$ gives

$$(E_y)_i^{n+1/2} = \frac{1 - \frac{\sigma \Delta t}{2\varepsilon_r \varepsilon_0}}{1 + \frac{\sigma \Delta t}{2\varepsilon_r \varepsilon_0}}(E_y)_i^{n-1/2} - \frac{1}{2\varepsilon_r\left(1 + \frac{\sigma \Delta t}{2\varepsilon_r \varepsilon_0}\right)}\left((H_z)_{i+1/2}^n - (H_z)_{i-1/2}^n\right) \quad (11.28)$$

Implementation and analysis of the above equations is left as a project.

11.3 Boundary conditions in 1D

11.3.1 Mur's first-order absorbing boundary conditions (ABC)

Here, based on intuitive arguments we derive the so-called Mur's first-order absorbing boundary conditions [12]. Consider plane wave $\phi(x, t)$ travelling left at the left boundary, see Fig. 11.7.

Reflections of this wave will be eliminated if [7]

$$\phi(x, t) = \phi(x - \Delta x, t + \Delta t) \quad (11.29)$$

Perform Taylor expansion

$$\phi(x - \Delta x, t + \Delta t) = \phi(x, t) - \frac{\partial \phi}{\partial x}\Delta x + \frac{\partial \phi}{\partial t}\Delta t$$

Velocity of the propagating wave is

$$c = \frac{\Delta x}{\Delta t}$$

Substitute into Eq. (11.29) and have

$$\phi(x,t) = \phi(x,t) - \frac{\partial \phi}{\partial x} c \Delta t + \frac{\partial \phi}{\partial t} \Delta t$$

or

$$\frac{\partial \phi}{\partial x} = \frac{1}{c} \frac{\partial \phi}{\partial t} \qquad (11.30)$$

It describes ABC for a plane wave incident from the right.

Finite difference of Eq. (11.30) close to the left boundary is

$$\frac{\phi_1^{n+1/2} - \phi_0^{n+1/2}}{\Delta x} = \frac{1}{c} \frac{\phi_{1/2}^{n+1} - \phi_{1/2}^n}{\Delta t} \qquad (11.31)$$

The values at half grid points and half time steps are determined as

$$\phi_i^{n+1/2} = \frac{1}{2}\left(\phi_i^{n+1} + \phi_i^n\right)$$

$$\phi_{i+1/2}^n = \frac{1}{2}\left(\phi_{i+1}^n + \phi_i^n\right)$$

Substitute the above into Eq. (11.31) and obtain an expression for ϕ_0^{n+1}:

$$\phi_0^{n+1} = \phi_1^n + \frac{c\Delta t - \Delta x}{c\Delta t + \Delta x}\left(\phi_1^{n+1} + \phi_0^n\right) \qquad (11.32)$$

The formula (11.32) determines the new (updated) field value at the boundary node point at $x = 0$.

The above expression is known as the first-order Mur's absorbing boundary conditions (ABC). A similar derivation can be performed for right boundary. It is left as an exercise.

11.3.2 Second-order boundary conditions in 1D

We will provide a more rigorous discussion of the boundary conditions based on the work by Umashankar [8] and Taflove [10]. It allows for the introduction of the second-order Mur's radiation boundary conditions.

We start with a 1D wave equation for any scalar component of the field

$$\frac{\partial^2 U}{\partial x^2} - \frac{1}{c^2} \frac{\partial^2 U}{\partial t^2} = 0 \qquad (11.33)$$

Introduce operator L defined as

$$L = \frac{\partial^2}{\partial x^2} - \frac{1}{c^2} \frac{\partial^2}{\partial t^2} = D_x^2 - \frac{1}{c^2} D_t^2$$

The wave equation can therefore be written as

$$LU = 0$$

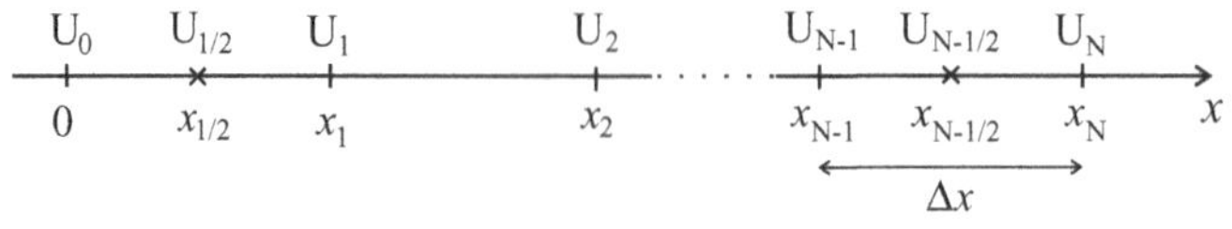

Fig. 11.8 Illustration of derivation of ABC in one dimension.

The above operator can be factored as

$$LU = L^+ L^- U = 0$$

where $L^+ = D_x + \frac{1}{c}D_t$ and $L^- = D_x - \frac{1}{c}D_t$.

Suppose that 1D computational region extends from 0 to h; i.e. one has $0 \leq x \leq h$. Engquist and Majda [13] showed that at $x = 0$ the application of an operator L^- to U exactly absorbs a plane wave propagating towards the boundary. A similar relation holds for an operator L^+ operating at $x = h$.

In light of the above, the PDE that can be numerically implemented as a first-order accurate ABC at $x = 0$ grid boundary is

$$L^- U = \left(D_x - \frac{1}{c} D_t \right) u = 0$$

Multiplying the above by an operator D_t, one finds

$$D_t \left(D_x - \frac{1}{c} D_t \right) U = 0$$

In the full form

$$\frac{\partial^2 U}{\partial t \partial x} - \frac{1}{c} \frac{\partial^2 U}{\partial t^2} = 0 \tag{11.34}$$

which is a second-order ABC at $x = 0$. For the above equation original numerical schemes have been proposed by Mur [12]. Here, we followed an improved scheme as described by Taflove [10].

Let U represent a Cartesian component of **E** or **H** located on the Yee grid, see Fig. 11.8. We use central differences expansion around an auxiliary grid point $x = 1/2$ and time $t_n = n \cdot \Delta t$. The derivatives are approximated as follows:

$$\left. \frac{\partial^2 U}{\partial t \partial x} \right|_{1/2}^n = \frac{1}{2\Delta t} \left[\left. \frac{\partial U}{\partial x} \right|_{1/2}^{n+1} - \left. \frac{\partial U}{\partial x} \right|_{1/2}^{n-1} \right]$$

$$= \frac{1}{2\Delta t} \left[\frac{U_1^{n+1} - U_0^{n+1}}{\Delta x} - \frac{U_1^{n-1} - U_0^{n-1}}{\Delta x} \right]$$

and

$$\left. \frac{\partial^2 U}{\partial t^2} \right|_{1/2}^n = \frac{1}{2} \left[\left. \frac{\partial^2 U}{\partial t^2} \right|_0^n + \left. \frac{\partial^2 U}{\partial t^2} \right|_1^n \right]$$

$$= \frac{1}{2} \left[\frac{U_0^{n+1} - 2U_0^n + U_0^{n-1}}{\Delta t^2} + \frac{U_1^{n+1} - 2U_1^n + U_1^{n-1}}{\Delta t^2} \right]$$

Substitution of the above results into Eq. (11.34) and solving for U_0^{n+1} gives

$$U_0^{n+1} = U_0^{n-1} + \frac{c\Delta t - \Delta x}{c\Delta t + \Delta x}\left(U_1^{n+1} + U_0^{n-1}\right) + \frac{2\Delta x}{c\Delta t + \Delta x}\left(U_1^n + U_0^n\right) \quad (11.35)$$

That expression determines the value of the field $U(x,t)$ at the next time $t_{n+1} = (n+1)\cdot \Delta t$ for $x = 0$.

In exactly the same way one obtains the updated equation at the right boundary. The ABC wave equation is

$$\frac{\partial^2 U}{\partial t \partial x} + \frac{1}{c}\frac{\partial^2 U}{\partial t^2} = 0 \quad (11.36)$$

The derivatives are approximated as follows:

$$\left.\frac{\partial^2 U}{\partial t \partial x}\right|_{N-1/2}^n = \frac{1}{2\Delta t}\left[\left.\frac{\partial U}{\partial x}\right|_{N-1/2}^{n+1} - \left.\frac{\partial U}{\partial x}\right|_{N-1/2}^{n-1}\right]$$

$$= \frac{1}{2\Delta t}\left[\frac{U_N^{n+1} - U_{N-1}^{n+1}}{\Delta x} - \frac{U_N^{n-1} - U_{N-1}^{n-1}}{\Delta x}\right]$$

and

$$\left.\frac{\partial^2 U}{\partial t^2}\right|_{N-1/2}^n = \frac{1}{2}\left[\left.\frac{\partial^2 U}{\partial t^2}\right|_N^n + \left.\frac{\partial^2 U}{\partial t^2}\right|_{N-1}^n\right]$$

$$= \frac{1}{2}\left[\frac{U_N^{n+1} - 2U_N^n + U_N^{n-1}}{\Delta t^2} + \frac{U_{N-1}^{n+1} - 2U_{N-1}^n + U_{N-1}^{n-1}}{\Delta t^2}\right]$$

Substituting the last result into Eq. (11.36) and solving for U_0^{n+1} gives

$$U_N^{n+1} = U_{N-1}^{n-1} + \frac{c\Delta t - \Delta x}{c\Delta t + \Delta x}\left(U_{N-1}^{n+1} + U_N^{n-1}\right) + \frac{2\Delta x}{c\Delta t + \Delta x}\left(U_N^n + U_{N-1}^n\right) \quad (11.37)$$

11.4 Two-dimensional Yee implementation without dispersion

We assume medium with the constitutive relation of the form

$$D_y = \varepsilon_0 \varepsilon_r E_y$$

Using the above constitutive relation, from the group 2 (Eqs. 11.8) (also called TM_z mode) of Maxwell's equations one has ($\varepsilon = \varepsilon_0 \varepsilon_r$)

$$\frac{\partial E_z}{\partial t} = \frac{1}{\varepsilon}\left(\frac{\partial H_y}{\partial x} - \frac{\partial H_x}{\partial y}\right)$$

$$\frac{\partial H_x}{\partial t} = -\frac{1}{\mu_0}\frac{\partial H_z}{\partial y}$$

$$\frac{\partial H_y}{\partial t} = \frac{1}{\mu_0}\frac{\partial E_z}{\partial x}$$

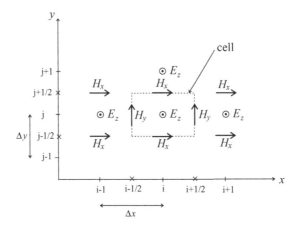

Fig. 11.9 Yee mesh in 2D for the TM_z mode.

For 2D discretization we introduce the Yee mesh, shown in Fig. 11.9. A Yee cell is formed around point (i, j). The electric field distribution within each cell is assumed to be constant and the z-component E_z of electric field is defined at the midpoint of each cell. We also introduce $(i, j) = (i\Delta x, j\Delta y)$ and $F^n(i, j) = F(i\Delta x, j\Delta y, n\Delta t)$. Maxwell's equations are approximated as

$$\frac{E_z^{n+1}(i,j) - E_z^n(i,j)}{\Delta t} = \frac{1}{\varepsilon_0 \varepsilon_r(i,j)} \frac{H_y^{n+1/2}(i+1/2, j) - H_y^{n+1/2}(i-1/2, j)}{\Delta x}$$
$$- \frac{1}{\varepsilon_0 \varepsilon_r(i,j)} \frac{H_x^{n+1/2}(i, j+1/2) - H_x^{n+1/2}(i, j-1/2)}{\Delta y} \quad (11.38)$$

$$\frac{H_x^{n+1/2}(i, j+1/2) - H_x^{n-1/2}(i, j+1/2)}{\Delta t} = -\frac{1}{\mu_0} \frac{E_z^n(i, j+1) - E_z^n(i, j)}{\Delta y} \quad (11.39)$$

$$\frac{H_y^{n+1/2}(i+1/2, j) - H_y^{n-1/2}(i+1/2, j)}{\Delta t} = \frac{1}{\mu_0} \frac{E_z^n(i+1, j) - E_z^n(i, j)}{\Delta x} \quad (11.40)$$

From the above equations one determines the time-stepping numerical algorithm for the interior region. For the same reason as in the 1D case, we need to perform scaling. As before we introduce new scaled variable $\widetilde{E}_z$ as follows:

$$\widetilde{E}_z = \sqrt{\frac{\varepsilon_0}{\mu_0}} E_z \quad (11.41)$$

An algorithm for scaled variables is

$$\widetilde{E}_z^{n+1}(i,j) = \widetilde{E}_z^n(i,j) + \frac{\Delta t}{\varepsilon_0 \varepsilon_r(i,j)} \frac{1}{\sqrt{\varepsilon_0 \mu_0}} \left\{ \frac{H_y^{n+1/2}(i+1/2, j) - H_y^{n+1/2}(i-1/2, j)}{\Delta x} \right.$$
$$\left. - \frac{H_x^{n+1/2}(i, j+1/2) - H_x^{n+1/2}(i, j-1/2)}{\Delta y} \right\} \quad (11.42)$$

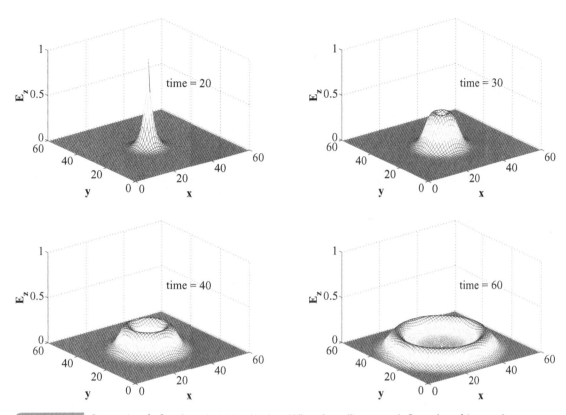

Fig. 11.10 Propagation of a Gaussian pulse initiated in the middle and travelling outwards. Four values of time are shown.

$$H_x^{n+1/2}(i, j+1/2) = H_x^{n-1/2}(i, j+1/2) - \frac{\Delta t}{\Delta y} \frac{1}{\sqrt{\varepsilon_0 \mu_0}} \left[\widetilde{E}_z^n(i, j+1) - \widetilde{E}_z^n(i, j) \right]$$
(11.43)

$$H_y^{n+1/2}(i+1/2, j) = H_y^{n-1/2}(i+1/2, j) - \frac{\Delta t}{\Delta y} \frac{1}{\sqrt{\varepsilon_0 \mu_0}} \left[\widetilde{E}_z^n(i+1, j) - \widetilde{E}_z^n(i, j) \right]$$
(11.44)

The above equations were implemented in MATLAB for the propagation of Gaussian pulse. A MATLAB program implementing 2D propagation of the Gaussian pulse is shown in Appendix, Listing 11A.6.

Propagation of the Gaussian pulse initiated in the middle and travelling outwards is shown in Fig. 11.10. We illustrate pulse shape at four values of time. One can then observe its spread in time.

11.5 Absorbing boundary conditions (ABC) in 2D

Absorbing boundary conditions (ABC) in 2D are derived here. We follow the methodology based on the wave equation used in the 1D case [7] and [10]. The two-dimensional wave

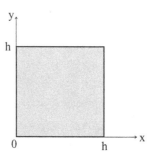

Fig. 11.11 Rectangular region used to illustrate derivations of ABC.

equation for scalar component U is

$$\frac{1}{c^2}\frac{\partial^2 U}{\partial t^2} = \frac{\partial^2 U}{\partial x^2} + \frac{\partial^2 U}{\partial y^2}$$

In a rectangular coordinate system, the plane-wave time-harmonic solution to wave equation is

$$U(x, y, t) = e^{i(\omega t - k_x x - k_y y)}$$

Substituting into a wave equation gives the condition

$$k_x^2 + k_y^2 = \frac{\omega^2}{c^2} \equiv k^2$$

Solution for k_x is therefore

$$\begin{aligned} k_x &= \pm\left(k^2 - k_y^2\right)^{1/2} \\ &= \pm k\left(1 - \frac{k_y^2}{k^2}\right)^{1/2} \\ &\simeq \pm k\left(1 - \frac{1}{2}\frac{k_y^2}{k^2}\right) \end{aligned}$$

The last result can be written as

$$kk_x = \pm k^2 \mp \frac{1}{2}k_y^2 \tag{11.45}$$

From the above relation, we can reconstruct partial differential equations that can be numerically implemented as a second-order accurate ABC. First observe that for a rectangular region shown in Fig. 11.11 near $x = 0$ boundary, the left travelling wave is described by the following equation:

$$\frac{1}{c}\frac{\partial^2 U}{\partial x \partial t} = \frac{1}{c^2}\frac{\partial^2 U}{\partial t^2} - \frac{1}{2}\frac{\partial^2 U}{\partial y^2} \tag{11.46}$$

Analogously, at the $x = h$ boundary

$$\frac{1}{c}\frac{\partial^2 U}{\partial x \partial t} = -\frac{1}{c^2}\frac{\partial^2 U}{\partial t^2} + \frac{1}{2}\frac{\partial^2 U}{\partial y^2}$$

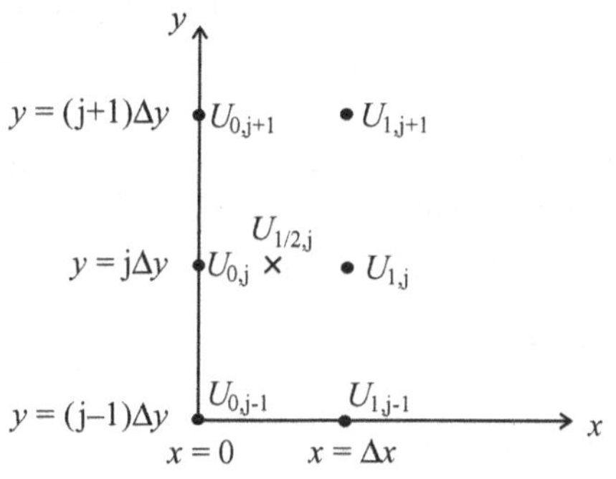

Fig. 11.12 Numerical boundary conditions.

To obtain numerical boundary conditions, we approximate the above derivatives around the central point $(\frac{1}{2}, j)$, see Fig. 11.12.

One obtains

$$\frac{\partial^2 U^n_{1/2,j}}{\partial x \partial t} = \frac{1}{2\Delta t} \left\{ \frac{\partial U^{n+1}_{1/2,j}}{\partial x} - \frac{\partial U^{n-1}_{1/2,j}}{\partial x} \right\}$$

$$= \frac{U^{n+1}_{1,j} - U^{n+1}_{0,j}}{2\Delta t \Delta x} - \frac{U^{n-1}_{1,j} - U^{n-1}_{0,j}}{2\Delta t \Delta x}$$

The second-order derivative with respect to time t is implemented as an average of time derivatives at the adjacement points $(0, j)$ and $(1, j)$ as

$$\frac{\partial^2 U^n_{1/2,j}}{\partial t^2} = \frac{1}{2} \left\{ \frac{\partial^2 U^n_{0,j}}{\partial t^2} + \frac{\partial^2 U^n_{1,j}}{\partial t^2} \right\}$$

$$= \frac{U^{n+1}_{0,j} - 2U^n_{0,j} + U^{n-1}_{0,j}}{2\Delta t^2} + \frac{U^{n+1}_{1,j} - 2U^n_{1,j} + U^{n-1}_{1,j}}{2\Delta t^2}$$

Also, as an average of the y derivatives we implement

$$\frac{\partial^2 U^n_{1/2,j}}{\partial y^2} = \frac{1}{2} \left\{ \frac{\partial^2 U^n_{0,j}}{\partial y^2} + \frac{\partial^2 U^n_{1,j}}{\partial y^2} \right\}$$

$$= \frac{U^n_{0,j+1} - 2U^n_{0,j} + U^n_{0,j-1}}{2\Delta y^2} + \frac{U^n_{1,j+1} - 2U^n_{1,j} + U^n_{1,j-1}}{2\Delta y^2}$$

Substitute the above results into Eq. (11.46) and have

$$\frac{U^{n+1}_{1,j} - U^{n+1}_{0,j}}{2\Delta t \Delta x} - \frac{U^{n-1}_{1,j} - U^{n-1}_{0,j}}{2\Delta t \Delta x}$$
$$= -\frac{1}{c^2 2\Delta t^2} \left(U^{n+1}_{0,j} - 2U^n_{0,j} + U^{n-1}_{0,j} + U^{n+1}_{1,j} - 2U^n_{1,j} + U^{n-1}_{1,j} \right) \quad (11.47)$$
$$= -\frac{1}{4\Delta y^2} \left(U^n_{0,j+1} - 2U^n_{0,j} + U^n_{0,j-1} + U^n_{1,j+1} - 2U^n_{1,j} + U^n_{1,j-1} \right)$$

From the above we can find the time-stepping algorithm for the component of U along $x = 0$ boundary, i.e. $U^{n+1}_{0,j}$.

11.6 Dispersion

Dispersion exists when medium properties are frequency-dependent. Examples include dielectric constant and/or conductivity which can vary with frequency. Generally, the following situations exist [1]:

- linear dispersion
- nonlinearity
- nonlinear dispersion
- gain.

Here, we will only summarize the most important models of material dispersion and briefly outline one of the possible numerical approaches to dispersion.

11.6.1 Material dispersion

Usually, the dependencies are established in frequency domain. The defining equation is

$$D(\omega) = \varepsilon_0 \varepsilon_r(\omega) E(\omega)$$

The physics is contained in $\varepsilon_r(\omega)$. We assume non-conducting media. Main cases are:

- Debye medium (first order)

$$\varepsilon(\omega) = \varepsilon_\infty + \frac{\varepsilon_s - \varepsilon_\infty}{1 + j\omega t_0} \qquad (11.48)$$

- Lorentz medium

$$\varepsilon_r(\omega) = \varepsilon_r + \frac{\varepsilon_1}{1 + 2j\delta_0 \left(\frac{\omega}{\omega_0}\right) - \left(\frac{\omega}{\omega_0}\right)^2}$$

Specific analytical and numerical work with the inclusion of dispersion is left as a project.

11.7 Problems

1. Derive an expression for the first-order Mur's boundary conditions at the right interface.
2. Determine the residual reflection produced by Mur's first-order ABC. Assume the existence of the following wave at the boundary:

$$\phi(x,t) = \exp(i\omega t) \cdot [\exp(ikx) + R\exp(-ikx)] \qquad (11.49)$$

Analyse reflection coefficient R.

Table 11.2 List of MATLAB functions for Chapter 11.

Listing	Function name	Description
11A.1	gauss.m	Plots Gaussian pulse
11A.2	gauss_modul.m	Plots modulated Gaussian pulse
11A.3	fdtd_free.m	Propagation of Gaussian pulse in a free space
11A.4	num_disp.m	Illustrates numerical dispersion
11A.5	gmf.m	Animates Gaussian pulse in a free space (not used in text)
11A.6	fdtd_2D.m	Propagation of Gaussian pulse in 2D

11.8 Projects

1. Implement a 1D FDTD approach with losses. Analyse propagation of Gaussian and super-Gaussian pulses.
2. Implement boundary conditions for 1D and 2D problems.
3. Use a convolution integral approach which relates D and E:

$$D(t) = \varepsilon_\infty \varepsilon_0 E(t) + \varepsilon_0 \int_0^t E(t-t')\chi(t')dt' \quad (11.50)$$

   to develop discretization scheme for the first-order Debye medium as described by Eq. (11.48). Implement in MATLAB the resulting numerical scheme. For more details, consult the book by Kunz and Lubbers [2].
4. Conduct similar analysis for the first-order Drude dispersion; i.e.

$$\varepsilon(\omega) = 1 + \frac{\omega_p^2}{\omega(j\nu_c - \omega)} \quad (11.51)$$

   where ω_p is the plasma frequency and ν_c is the collision frequency.

Appendix 11A: MATLAB listings

In Table 11.2 we provide a list of MATLAB files created for Chapter 11 and a short description of each function.

Listing 11A.1 Program gauss.m. MATLAB program which creates Gaussian pulse.

```
% File name: gauss.m
% Plots Gaussian pulse
clear all;
%
t_zero = 20.0;                      % center of incident pulse
```

```
width = 6.0;                           % width of the incident pulse
%
N      = 300.0;
time = linspace(0,50,N);
pulse = exp(-0.5*((t_zero - time)/width).^2);        % Gaussian pulse
plot(time,pulse,'LineWidth',1.5)
xlabel('time','FontSize',14);
ylabel('amplitude','FontSize',14);
set(gca,'FontSize',14);                % size of tick marks on both axes
pause
close all
```

Listing 11A.2 Program gauss_modul.m. MATLAB program which creates a modulated Gaussian pulse.

```
% File name: gauss_modul.m
% Plots modulated Gaussian pulse
alpha = 0.6;
mu_0=4*pi*1e-7; e_0=8.854e-12;    % fundamental constants
c=1/sqrt(mu_0*e_0);               % velocity of light
dx = 4.2e-2;                      % step in space
dt=alpha*dx/c;                    % time step
f_mod = 2e9;                      % modulating frequency (2 GHz)
t_0 = 3e-09;                      % peak position of Gaussian pulse in time
T = t_0/3;                        % pulse width
N_x = 400;
spec = [0:N_x];
for n = 1:100
    y(n) = exp(-(((n-1)*dt-t_0)/T).^2).*cos(2*pi*f_mod*((n-1)*dt-t_0));
end
plot(y,'LineWidth',1.5)
xlabel('time','FontSize',14); ylabel('amplitude','FontSize',14);
set(gca,'FontSize',14);           % size of tick marks on both axes
pause
close all
```

Listing 11A.3 Function fdtd_free.m MATLAB program which describes propagation of a Gaussian pulse in a free space using the FDTD method.

```
% File name: fdtd_free.m
% Plots E and H components of Gaussian pulse in a free space
% after some number of time steps using FDTD method
% No animation
clear all;
N_x = 200;                        % number of x-cells
```

```
N_time = 100;                  % number of time steps
time = 0.0;
E_y = zeros(N_x,1);            % Initialize fields to zero at all points
H_z = zeros(N_x,1);
t_zero = 40.0;                 % center of incident pulse
width = 12;                    % width of the incident pulse
tic;    % start timer
for istep = 1:N_time           % time loop
    time = time + 1;
    pulse = exp(-0.5*((t_zero - time)/width)^2); % Gaussian pulse
    for i = 2:(N_x)
        E_y(120) = pulse;      % location of initial pulse
        E_y(i) = E_y(i) - 0.5*(H_z(i) - H_z(i-1));
    end
    %
    for i = 1:(N_x -1)
        H_z(i) = H_z(i) - 0.5*(E_y(i+1) - E_y(i));
    end
end
toc     % measure elapsed time since last call to tic
subplot(2,1,1); plot(E_y)
ylabel('E_y'); xlabel('FDTD cells')
axis([0 200 -1 1])
subplot(2,1,2); plot(H_z)
ylabel('H_z'); xlabel('FDTD cells')
axis([0 200 -1 1])
pause
close all
```

Listing 11A.4 Function num_disp.m. MATLAB program used to produce Fig. 11.5 to illustrate numerical dispersion.

```
% File name: num_disp.m
% Illustrates numerical dispersion
N_max = 10;                             % number of points for plot
x = linspace(0,pi,N_max);
hold on
for alpha = [1.0 0.8 0.2]
    y = 2*asin(alpha*sin(x/2));
    plot(x,y,'LineWidth',1.5)
end
xlabel('k \Delta z','FontSize',14)
ylabel('\omega \Delta x /c','FontSize',14)
set(gca,'FontSize',14);                 % size of tick marks on both axes
```

```
text(2.5, 3, '\alpha = 1','Fontsize',12)
text(2.5, 2, '\alpha = 0.8','Fontsize',12)
text(2.5, 0.5, '\alpha = 0.2','Fontsize',12)
pause
close all
```

Listing 11A.5 Function gmf.m. MATLAB program which generates and animates propagation of a Gaussian pulse in a free space.

```
% File name: gmf.m
% gauss, modulated, free space
% Animation of propagation of Gaussian pulse in a free space
% without boundary conditions
% Program not used in the main text
clear all
alpha = 0.5;            % alpha = c*dt/dx
mu0=4*pi*1e-7;          % magnetic permeability
e0=8.854e-12;           % electric permittivity
c=1/sqrt(mu0*e0);       % velocity of light in a vacuum
dx = 4.2e-2;            % space step
dt=alpha*dx/c;          % time step
f_mod = 1e9;            % modulating frequency
lambda = c/f_mod;       % wavelength
x_end = 400;
t_0 = 3e-09;            % peak position of Gaussian pulse in time
dxl = dx/lambda;
index=[0:x_end];
e=zeros(length(index),1); h=e; e1=e; h0=e;
ha = plot(e);           % get the handle to the plot
set(ha,'EraseMode','xor')    % for simulation of wave propagation
%
t_end = 500;            % number of time steps
for n = 1:t_end
    for i=1:400         % simulate for h-nodes
        h(i) = h(i) - (e(i+1)-e(i))*dt/dx/mu0;
    end

    for i=2:400                  % simulate for e-nodes
        e(i) = e(i) - (h(i)-h(i-1))*dt/dx/e0;
    end
% launch Gaussian or modulated Gaussian pulse
    T = t_0/3;          % specify pulse width
    y(n) = exp(-(((n-1)*dt-t_0)/T).^2).*cos(2*pi*f_mod*((n-1)*dt-t_0));
%   y(n) = exp(-(((n-1)*dt-t_0)/T).^2);
    e(1)= y(n);   % Gauss pulse starts here
    drawnow;     %refresh data and plot the current condition of the pulse
```

```
        set(ha,'YData',(e));
end
pause
close all
```

Listing 11A.6 Function fdtd_2D.m. MATLAB program which describes propagation of a Gaussian pulse in 2D.

```
% File name: fdtd_2D.m
% Program for finite-difference time-domain method in 2D
% Propagation of Gaussian pulse
clear all;
N_x = 60;                % number of steps in x-direction
N_y = 60;                % number of steps in y-direction
N_steps = 60;            % number of time steps
%
N_x_middle = N_x/2.0;    % middle point; location of pulse
N_y_middle = N_y/2.0;    %
%
delta_x = 0.01;          % cell size
delta_t = delta_x/6e8;   % time step
epsilon = 8.8e-12;
%
% Initialization
E_z = zeros(N_x,N_y);    % Initialize field to zero at all points
H_x = zeros(N_x,N_y);
H_y = zeros(N_x,N_y);
%
t_zero = 20.0;           % center of incident pulse
width = 6.0;             % width of the incident pulse
time = 0.0;
%
for istep = 1:N_steps    % main loop
    time = time + 1;
% calculate E_z field
    for i = 2:N_x
        for j = 2:N_y
            E_z(i,j)=E_z(i,j)+0.5*(H_y(i,j)-H_y(i-1,j)-H_x(i,j)+H_x(i,j-1));
        end
    end
% Gaussian pulse in the middle
pulse = exp(-0.5*((t_zero - time)/width)^2); % Gaussian pulse
E_z(N_x_middle,N_y_middle) = pulse;
%
% Calculate H_x field
    for i = 1:(N_x-1)
        for j = 1:(N_y-1)
```

```
                H_x(i,j) = H_x(i,j) - 0.5*(E_z(i,j+1) - E_z(i,j));
            end
        end
%
% Calculate H_y field
    for i = 1:(N_x-1)
        for j = 1:(N_y-1)
            H_y(i,j) = H_y(i,j) + 0.5*(E_z(i+1,j) - E_z(i,j));
        end
    end
end
%---------- Plotting----------------------------
[x,y] = meshgrid(0:1:N_x-1,0:1:N_y-1);
mesh(x,y,E_z);
xlabel('\bfx','FontSize',14);
ylabel('\bfy','FontSize',14);
zlabel('\bfE_z','FontSize',14);
set(gca,'FontSize',14);            % size of tick marks on both axes
axis([0 60 0 60 0 1])
text(50,40,0.5, 'time = 60', 'Fontsize', 14)
pause
close all
```

References

[1] A. Taflove and S. C. Hagness. *Computational Electrodynamics. The Finite-Difference Time-Domain Method*. Artech House, Boston, 2005.

[2] K. S. Kunz and R. J. Luebbers. *The Finite Difference Time Domain Method for Electromagnetics*. CRC Press, Boca Raton, 1993.

[3] D. M. Sullivan. *Electromagnetic Simulation using the FDTD Method*. IEEE Press, New York, 2000.

[4] W. Yu, X. Yang, Y. Liu, and R. Mittra. *Electromagnetic Simulation Techniques Based on the FDTD Method*. Wiley, Hoboken, NJ, 2009.

[5] A. Z. Elsherbeni and V. Demir. *The Finite-Difference Time-Domain Method for Electromagnetics with Matlab Simulations*. Scitech Publishing, Inc., Raleigh, NC, 2009.

[6] A. Bondeson, Th. Rylander, and P. Ingelstroem. *Computational Electromagnetics*. Springer, New York, 2005.

[7] R. Garg. *Analytical and Computational Methods in Electromagnetics*. Artech House, Boston, 2008.

[8] K. R. Umashankar. Finite-difference time domain method. In S. M. Rao, ed., *Time Domain Electromagnetics*, pages 151–235. Academic Press, San Diego, 1999.

[9] K. S. Yee. Numerical solution of initial boundary value problems involving Maxwell's equations in isotropic media. *IEEE Trans. Antennas and Propag.*, **17**:585, 1966.

[10] A. Taflove. *Computational Electrodynamics. The Finite-Difference Time-Domain Method*. Artech House, Boston, 1995.

[11] W. H. Press, S. A. Teukolsky, W. T. Vetterling, and B. P. Flannery. *Numerical Recipes in Fortran 90. The Art of Scientific Computing. Second Edition*. Cambridge University Press, Cambridge, 1992.

[12] G. Mur. Absorbing boundary conditions for the finite-difference approximation of the time-domain electromagnetic-field equations. *IEEE Trans. Electromagn. Compat.*, **23**:377–82, 1981.

[13] B. Engquist and A. Majda. Absorbing boundary conditions for the numerical simulation of waves. *Math. Comp.*, **31**:629–51, 1977.

12 Beam propagation method (BPM)

In this chapter we summarize the beam propagation method (BPM). The method is widely used for the numerical solution of the Helmholtz equation and also for the numerical solution to the nonlinear Schroedinger equation (to be discussed in Chapter 15 dealing with solitons). It is the most powerful technique for studying the propagation of light in integrated optics. The method was originally introduced by Feit and Fleck in the late 1970s [1]. The BPM was initially based on FFT algorithm. Later on it has been extended to finite-difference based BPM schemes (FD-BPM) and finite-element BPM (FE-BPM) and many others. The main characteristics [2] of FD-BPM are the ability to simulate structures with large index discontinuity, less memory and time consumption in modelling complex structures, the possibility to incorporate wide-angle and full vector algorithms and the ability to incorporate transparent boundary conditions.

There is a large number of algorithms available in the literature and almost all of them are based on the concept of a propagator. Propagators are mathematical 'objects' which propagate fields from one space coordinate to another. An example of the system where propagation takes place, known as an optical waveguide, is shown in Fig. 12.1. It splits input optical signal into two arms, see also Fig. 12.2. The role of BPM is to determine field profile along the waveguide knowing the distribution of refractive index over the whole waveguide.

There is a huge amount of literature on the principles of BPM and its applications in integrated photonics. We recommend books by Kawano and Kitoh [3], Pollock and Lipson [4], Okamoto [5] and Lifante [6]. We start our discussion by illustrating the principles of BPM in the paraxial approximation.

12.1 Paraxial formulation

12.1.1 Introduction

We start this section by outlining the simplest version of BPM, where one assumes scalar electric field and paraxial approximations which restrict its applicability to the fields propagating at small angles with respect to the axis of the waveguide (guiding axis), which we define as the z axis. The geometry of the waveguide is determined by refractive index $\bar{n}(x, y, z)$. The method consists of the repeated propagation of the electric field from a perpendicular plane at a given position along the waveguide to the next parallel plane. To

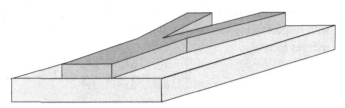

Fig. 12.1 Example of a planar waveguide. Here we show the beam splitter.

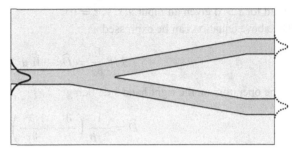

Fig. 12.2 Schematical illustration of a typical problem solved by BPM. Knowing the input electric field on the left, determine fields at the output (shown by dotted lines on the right).

illustrate the method, one starts from the wave equation for a monochromatic wave:

$$\frac{\partial^2 E}{\partial x^2} + \frac{\partial^2 E}{\partial y^2} + \frac{\partial^2 E}{\partial z^2} + k^2(x,y,z)E = 0 \tag{12.1}$$

Here, the spatially dependent wavenumber is given by $k(x,y,z) = k_0 \bar{n}(x,y,z)$ and $k_0 = 2\pi/\lambda$ is the wavenumber in free space. The main assumption is that the phase variation due to propagation along the z axis represents the fastest variation in the field E. This variation is exemplified by introducing slowly varying field u and expressing E as

$$E(x,y,z) = u(x,y,z)e^{-j\beta z} \tag{12.2}$$

where β is a constant that represents the characteristic propagation wave vector, $\beta = \bar{n}_0 \omega / c$. Here $\bar{n}_0$ is a reference refractive index which can be, for example, the refractive index of the substrate or cladding. β represents the average phase variation of the field E and it is known as propagation constant. Substituting Eq. (12.2) into Eq. (12.1) yields the equation equivalent to the exact Helmholtz for the slowly varying field

$$-\frac{\partial^2 u}{\partial z^2} + 2j\beta \frac{\partial u}{\partial z} = \frac{\partial^2 u}{\partial x^2} + \frac{\partial^2 u}{\partial y^2} + (k^2 - \beta^2)u \tag{12.3}$$

At this stage one assumes the slow variation of optical field in the propagation direction (SVEA or slowly varying envelope approximation), which requires

$$\left| \frac{\partial^2 u}{\partial z^2} \right| \ll \left| 2\beta \frac{\partial u}{\partial z} \right| \tag{12.4}$$

This approximation allows us to ignore the first term on the left hand side of Eq. (12.3) with respect to the second one. It is also known as Fresnel approximation. Eq. (12.3) thus reduces to

$$2i\beta \frac{\partial u}{\partial z} = \frac{\partial^2 u}{\partial x^2} + \frac{\partial^2 u}{\partial y^2} + (k^2 - \beta^2)u \quad (12.5)$$

which is known as a Fresnel or paraxial equation. It is a starting point for the description of the evolution of electric (or magnetic) field in inhomogeneous medium, for example an optical waveguide. It does not describe polarization effects. It determines the evolution of the field for $z > 0$ given an input $u(x, y, z = 0)$.

The above equation can be expressed as

$$2j\beta \frac{\partial u}{\partial z} = \widehat{D}u + \widehat{W}u \quad (12.6)$$

with the operators on the right hand side being

$$\widehat{D} = \frac{1}{2j\beta}\left(\frac{\partial^2}{\partial x^2} + \frac{\partial^2}{\partial y^2}\right) \quad (12.7)$$

$$\widehat{W} = \frac{1}{2j\beta}\left(k^2 - \beta^2\right) \quad (12.8)$$

Operator $\widehat{D}$ represents free-space propagation (diffraction) and operator $\widehat{W}$ describes guiding effects.

Assuming that operators are z-independent, the solution is symbolically written as

$$u(x, y, z + \Delta z) = e^{(\widehat{D}+\widehat{W})\Delta z} u(x, y, z)$$

Use the Baker-Hausdorf lemma [7]

$$e^{\widehat{A}} e^{\widehat{B}} = e^{\widehat{A}+\widehat{B}} e^{\frac{1}{2}[\widehat{A},\widehat{B}]}$$

given that $\widehat{A}$ and $\widehat{B}$ each commutes with $[\widehat{A}, \widehat{B}]$. Using it, we can approximate

$$u(x, y, z + \Delta z) \approx e^{\widehat{D}\Delta z} e^{\widehat{W}\Delta z} u(x, y, z)$$

From the above, it follows that actions due to both operators can be considered independently.

The action of operator $\widehat{D}$ is better understood in the spectral domain. It will be described in the next section. The second operator $\widehat{W}$ describes the effect of propagation in the presence of medium inhomogeneities and its action is incorporated in the spatial domain.

12.1.2 Operators $\widehat{D}$ and $\widehat{W}$

Consider an equation containing only operator $\widehat{D}$:

$$2j\beta \frac{\partial u(x, y, z)}{\partial z} = \left(\frac{\partial^2}{\partial x^2} + \frac{\partial^2}{\partial y^2}\right) u(x, y, z) \quad (12.9)$$

Define 2D continuous Fourier transform as

$$\tilde{u}(k_x, k_y, z) = \int_{-\infty}^{+\infty} u(x, y, z) e^{-j(k_x x + k_y y)} dx dy \equiv F_{x,y}\{u(x, y, z)\}$$

and its inverse

$$u(x, y, z) = \int_{-\infty}^{+\infty} \tilde{u}(k_x, k_y, z) e^{j(k_x x + k_y y)} dk_x dk_y F_{x,y}^{-1}\{\tilde{u}(x, y, z)\}$$

Apply the 2D FT to Eq. (12.9) and have for Fourier components

$$2j\beta \frac{\partial}{\partial z}\tilde{u}(x, y, z) = -(k_x^2 + k_y^2)\tilde{u}(x, y, z)$$

Integrate the above from z to $z + \Delta z$ and have

$$\begin{aligned}\tilde{u}(x, y, z + \Delta z) &= e^{-\frac{1}{2j\beta}(k_x^2 + k_y^2)\Delta z}\tilde{u}(x, y, z) \\ &\equiv \hat{H}(k_x, k_y, \Delta z)\tilde{u}(x, y, z)\end{aligned} \quad (12.10)$$

To analyse operator $\hat{W}$, introduce the relation

$$\beta = k_0 n_{eff} \quad (12.11)$$

Write also refractive index $\bar{n}$ as

$$\bar{n} = n_{eff} + \Delta \bar{n} \quad (12.12)$$

Substituting the above into Eq. (12.8) gives an expression for operator $\hat{W}$ to the first order in $\Delta \bar{n}$

$$\hat{W} = -jk_0 \Delta n \quad (12.13)$$

12.1.3 The implementation using the Fourier transform split-step method

A general scheme of FD-BPM is described by the propagator U, which we use to advance field as

$$\tilde{\mathbf{E}}_t(z + \Delta z) = U(\Delta z)\tilde{\mathbf{E}}_t(z)$$

Propagator U can take many forms depending on the chosen BPM technique. Its operation is illustrated in Fig. 12.3. In practice two versions of BPM are used: $1 + 1$ FD-BPM and $1 + 2$ FD-BPM. The nomenclature refers to one dimension along the propagation direction (usually the z-axis) and one or two perpendicular dimensions (along x-axis (1D case) or in x-y plane (2D case)). Finite difference discretization in the $1 + 1$ and $1 + 2$ cases is illustrated in Fig. 12.4.

We illustrate the application of Fourier transform split-step BPM for the propagation of a two-dimensional Gaussian pulse. The guiding structure is divided into a large number of segments. During each segment of length Δz, the optical pulse is propagated as shown in

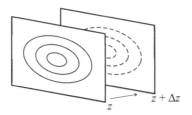

Fig. 12.3 Illustration of the BPM scheme. Field profile is propagated from the position within one plane to a position within another plane along the z-axis.

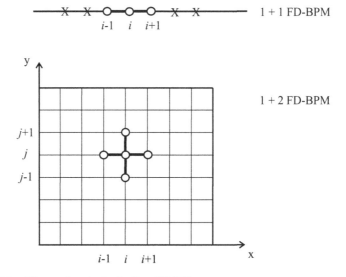

Fig. 12.4 Illustration of finite-difference discretization for 1D and 2D BPM.

the flow diagram, see Fig. 12.5. The accompanying MATLAB script is shown in Appendix, Listing 12B.1. The results of a 3D view of pulse before propagation and after propagation are shown in Fig. 12.6. One can observe spread of pulse width as expected.

12.2 General theory

12.2.1 Introduction

Polarization effects are not included in the previous version of BPM. When one wants to describe polarization effects, the vector wave equation must replace the scalar Helmholtz equation used earlier. In this section we will outline a general approach to systematically handle those effects. General formulation is based on an approach established by W. P. Huang's group [8], [9]. As usual, one starts from Maxwell's equations in the frequency

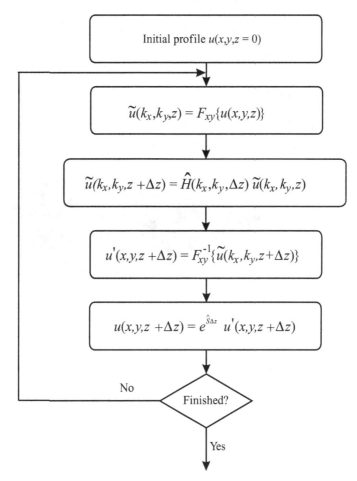

Fig. 12.5 Flow diagram for the BPM split-step method.

domain:

$$\nabla \times \mathbf{E} = -j\omega\mu_0 \mathbf{H} \tag{12.14}$$

$$\nabla \times \mathbf{H} = j\omega\varepsilon_0 \bar{n}^2 \mathbf{E} \tag{12.15}$$

$$\nabla \cdot \bar{n}^2 \mathbf{E} = 0 \tag{12.16}$$

where we have introduced refractive index $\bar{n}$ as $\varepsilon = \varepsilon_0 \bar{n}^2$. We separate all quantities into longitudinal (along z-axis) and transversal components, as follows:

$$\nabla = \nabla_t + \hat{z}\frac{\partial}{\partial z}, \quad \mathbf{E} = \mathbf{E}_t + \hat{z}E_z, \quad \mathbf{H} = \mathbf{H}_t + \hat{z}H_z$$

where

$$\nabla_t = \left[\frac{\partial}{\partial x}, \frac{\partial}{\partial y}, 0\right], \quad \mathbf{E}_t = [E_x, E_y, 0], \quad \mathbf{H}_t = [H_x, H_y, 0]$$

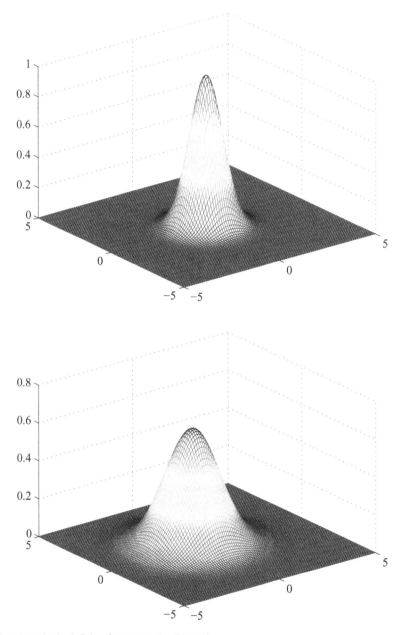

Fig. 12.6 Initial Gaussian pulse (top). Pulse after propagation (bottom).

are the transversal components and $\hat{z} = [0, 0, 1]$ is the unit vector along z-axis. Eq. (12.16) can be rearranged as follows:

$$\nabla \cdot \overline{n}^2 \mathbf{E} = 0$$

or
$$\nabla_t \cdot \overline{n}^2 \mathbf{E}_t + \frac{\partial}{\partial z} \overline{n}^2 E_z = 0$$

or
$$\nabla_t \cdot \overline{n}^2 \mathbf{E}_t + E_z \frac{\partial \overline{n}^2}{\partial z} + \overline{n}^2 \frac{\partial E_z}{\partial z} = 0$$

$$\frac{\partial}{\partial x} \overline{n}^2 E_x + \frac{\partial}{\partial y} \overline{n}^2 E_y + \frac{\partial}{\partial z} \overline{n}^2 E_z = 0$$

The z-derivative is
$$\frac{\partial E_z}{\partial z} = -\frac{1}{\overline{n}^2} E_z \frac{\partial \overline{n}^2}{\partial z} - \frac{1}{\overline{n}^2} \nabla_t \cdot \overline{n}^2 \mathbf{E}_t \qquad (12.17)$$

If the refractive index varies slowly along z-axis, we can set $\frac{\partial \overline{n}^2}{\partial z} \approx 0$ (z-invariant structures [9]) and obtain

$$\frac{\partial E_z}{\partial z} \approx -\frac{1}{\overline{n}^2} \nabla_t \cdot \overline{n}^2 \mathbf{E}_t$$
$$= -\frac{1}{\overline{n}^2} \frac{\partial}{\partial x} \overline{n}^2 E_x - \frac{1}{\overline{n}^2} \frac{\partial}{\partial y} \overline{n}^2 E_y \qquad (12.18)$$

The above relation is exact for z-invariant systems, i.e. when there is no change of refractive index $\overline{n}^2$ along z-axis.

The wave equation is derived by taking $\nabla \times \cdots$ operation on Eq. (12.14). One finds

$$\nabla \times \nabla \times \mathbf{E} = -j\omega\mu_0 \nabla \times \mathbf{H} = \omega^2 \mu_0 \varepsilon_0 \overline{n}^2 \mathbf{E} \qquad (12.19)$$

In the last step we have used Eq. (12.15). Next, we use the following general relation which holds for arbitrary vector

$$\nabla \times \nabla \times \mathbf{E} = \nabla (\nabla \cdot \mathbf{E}) - \nabla^2 \mathbf{E}$$

Applying it to expression (12.19), we obtain wave equation

$$-\nabla^2 \mathbf{E} + \nabla (\nabla \cdot \mathbf{E}) = \overline{n}^2 k_0^2 \mathbf{E} \qquad (12.20)$$

where $\omega^2 \mu_0 \varepsilon_0 = k_0^2$.

Transversal components of the wave equation (12.20) are

$$-\nabla^2 E_x + \frac{\partial}{\partial x} (\nabla \cdot \mathbf{E}) = \overline{n}^2 k_0^2 E_x \qquad (12.21)$$

$$-\nabla^2 E_y + \frac{\partial}{\partial y} (\nabla \cdot \mathbf{E}) = \overline{n}^2 k_0^2 E_y \qquad (12.22)$$

where $\nabla^2 E_x = \frac{\partial^2 E_x}{\partial x^2} + \frac{\partial^2 E_x}{\partial y^2} + \frac{\partial^2 E_x}{\partial z^2}$. We will analyse the behaviour of E_z component. In the above equations, term $\nabla \cdot \mathbf{E}$ is expanded as

$$\nabla \cdot \mathbf{E} = \frac{\partial E_x}{\partial x} + \frac{\partial E_y}{\partial y} + \frac{\partial E_z}{\partial z}$$

and the last term is replaced using relation (12.18). One finds

$$\nabla \cdot \mathbf{E} = \frac{\partial E_x}{\partial x} + \frac{\partial E_y}{\partial y} - \frac{1}{\bar{n}^2} \frac{\partial}{\partial x} \bar{n}^2 E_x - \frac{1}{\bar{n}^2} \frac{\partial}{\partial y} \bar{n}^2 E_y \qquad (12.23)$$

Substituting (12.23) into (12.21), one obtains

$$\frac{\partial}{\partial x} \frac{1}{\bar{n}^2} \frac{\partial}{\partial x} \bar{n}^2 E_x + \frac{\partial^2 E_x}{\partial y^2} + \frac{\partial^2 E_x}{\partial z^2} + \frac{\partial}{\partial x} \frac{1}{\bar{n}^2} \frac{\partial}{\partial y} \bar{n}^2 E_y - \frac{\partial^2 E_y}{\partial x \partial y} + \bar{n}^2 k_0^2 E_x = 0 \qquad (12.24)$$

Similarly, from (12.22) we have

$$\frac{\partial}{\partial y} \frac{1}{\bar{n}^2} \frac{\partial}{\partial x} \bar{n}^2 E_x - \frac{\partial^2 E_x}{\partial y \partial x} + \frac{\partial^2 E_y}{\partial x^2} + \frac{\partial^2 E_y}{\partial z^2} + \frac{\partial}{\partial y} \frac{1}{\bar{n}^2} \frac{\partial}{\partial y} \bar{n}^2 E_y + \bar{n}^2 k_0^2 E_y = 0 \qquad (12.25)$$

The above equations in a matrix form are

$$\begin{bmatrix} \frac{\partial}{\partial x} \frac{1}{\bar{n}^2} \frac{\partial}{\partial x} \bar{n}^2 + \frac{\partial^2}{\partial y^2} + \frac{\partial^2}{\partial z^2} + \bar{n}^2 k_0^2 & \frac{\partial}{\partial x} \frac{1}{\bar{n}^2} \frac{\partial}{\partial y} \bar{n}^2 - \frac{\partial^2}{\partial x \partial y} \\ \frac{\partial}{\partial y} \frac{1}{\bar{n}^2} \frac{\partial}{\partial x} \bar{n}^2 - \frac{\partial^2}{\partial y \partial x} & \frac{\partial}{\partial y} \frac{1}{\bar{n}^2} \frac{\partial}{\partial y} \bar{n}^2 + \frac{\partial^2}{\partial x^2} + \frac{\partial^2}{\partial z^2} + \bar{n}^2 k_0^2 \end{bmatrix} \begin{bmatrix} E_x \\ E_y \end{bmatrix} = 0$$

$$(12.26)$$

Those equations form the starting point for detailed approximations.

12.2.2 Slowly varying envelope approximation (SVEA)

Assume that the transversal field components E_x and E_y are of the form

$$E_x(x, y, z) = u_x(x, y, z) e^{-j\beta z} \qquad (12.27)$$

$$E_y(x, y, z) = u_y(x, y, z) e^{-j\beta z} \qquad (12.28)$$

SVEA requires that

$$\left| \frac{\partial^2 u_i}{\partial z^2} \right| \ll 2\beta \left| \frac{\partial u_i}{\partial z} \right| \qquad (12.29)$$

Applying SVEA to term $\frac{\partial^2}{\partial z^2}$ where $(i = x, y)$, one obtains

$$\frac{\partial^2}{\partial z^2} E_i = \frac{\partial^2}{\partial z^2} u_i e^{-j\beta z} = e^{-j\beta z} \left\{ \frac{\partial^2 u_i}{\partial z^2} - 2j\beta \frac{\partial u_i}{\partial z} - \beta^2 u_i \right\}$$

$$\simeq e^{-j\beta z} \left\{ -2j\beta \frac{\partial u_i}{\partial z} - \beta^2 u_i \right\}$$

Using SVEA in Eqs. (12.24) and (12.25) results in the paraxial wave equations [9]

$$j \frac{\partial u_x}{\partial z} = A_{xx} u_x + A_{xy} u_y \qquad (12.30)$$

$$j \frac{\partial u_y}{\partial z} = A_{yx} u_x + A_{yy} u_y \qquad (12.31)$$

where the differential operators are defined by [9]

$$A_{xx}u_x = \frac{1}{2\beta}\left\{\frac{\partial}{\partial x}\frac{1}{\bar{n}^2}\frac{\partial}{\partial x}\bar{n}_x^2 + \frac{\partial_x^2}{\partial y^2} + \frac{\partial_x^2}{\partial z^2} + \left(\bar{n}^2 k_0^2 - \beta^2\right)\right\}u_x \quad (12.32)$$

$$A_{xy}u_y = \frac{1}{2\beta}\left\{\frac{\partial}{\partial x}\frac{1}{n^2}\frac{\partial}{\partial y}n_y^2 - \frac{\partial_y^2}{\partial x \partial y}\right\}u_y \quad (12.33)$$

$$A_{yx}u_x = \frac{1}{2\beta}\left\{\frac{\partial}{\partial y}\frac{1}{\bar{n}^2}\frac{\partial}{\partial x}\bar{n}_x^2 - \frac{\partial_x^2}{\partial y \partial x} + \frac{\partial_y^2}{\partial x^2}\right\}u_x \quad (12.34)$$

$$A_{yy}u_y = \frac{1}{2\beta}\left\{\frac{\partial_y^2}{\partial x^2} + \frac{\partial_y^2}{\partial z^2} + \frac{\partial}{\partial y}\frac{1}{\bar{n}^2}\frac{\partial}{\partial y}\bar{n}_y^2 + \left(n^2 k_0^2 - \beta^2\right)\right\}u_y \quad (12.35)$$

The above can be written in a matrix form:

$$j\frac{\partial}{\partial z}\begin{bmatrix}u_x \\ u_y\end{bmatrix} = \begin{bmatrix}A_{xx} & A_{xy} \\ A_{yx} & A_{yy}\end{bmatrix}\begin{bmatrix}u_x \\ u_y\end{bmatrix} \quad (12.36)$$

The evolution of electric field described by Eq. (12.36) is known as vectorial BPM [9]. It takes into account the polarization ($A_{xx} \neq A_{yy}$) and includes coupling between E_x and E_y ($A_{xy} \neq A_{yx}$).

12.2.3 Semi-vector BPM

Often the coupling between two polarizations is weak and may be neglected. The two polarizations are decoupled as long as they are not synchronized with some mechanism within the device. The resulting equations which describe semi-vector formulation are

$$j\frac{\partial u_x}{\partial z} = A_{xx}u_x$$

$$j\frac{\partial u_y}{\partial z} = A_{yy}u_y$$

Under the semi-vector approximation, the polarization dependencies of the EM waves are taken into account.

12.2.4 Scalar formulation

If the structure is weakly guiding and/or the polarization dependence is not important, we can neglect it. The resulting scalar approximation is described by the equation

$$j\frac{\partial u}{\partial z} = Au \quad (12.37)$$

where the operator A is

$$A = \frac{1}{2\beta}\left\{\frac{\partial_x^2}{\partial x^2} + \frac{\partial_x^2}{\partial y^2} + \left(\bar{n}^2 k_0^2 - \beta^2\right)\right\}$$

The above reproduces a simple theory summarized in the introductory section.

We finish this section with the remark that similar equations can be derived for the magnetic field, known as the H-formulation (consult [9] for more details).

12.2.5 Finite-difference (FD) approximations

In this section we provide finite-difference discretization of the above equations. Due to its simplicity, the FD is the popular numerical method of solving such problems. One replaces continuous space by a discrete lattice where all fields are determined at the lattice's points. The lattice points are defined by $x_i = i \cdot \Delta x$ and $y_j = j \cdot \Delta y$, see Fig. 12.4. Some details of discretization are provided in Appendix 12A. The final discrete form of the previous equations for E-formulation is [9] (here we made a replacement $\varepsilon = \bar{n}^2$)

$$A_{xx}u_x$$
$$= \frac{1}{2\beta} \left\{ \frac{T_{i,j+1}u_x(i, j+1) - \left[2 - R_{i,j+1} - R_{i-1,j}\right]u_x(i, j) + T_{i-1,j}u_x(i-1, j)}{\Delta x^2} \right.$$
$$\left. + \frac{u_x(i, j+1) - 2u_x(i, j) + u_x(i, j-1)}{\Delta y^2} + \left[\varepsilon_{i,j,k} - \beta^2\right]u_x(i, j) \right\} \quad (12.38)$$

$$A_{yy}u_y$$
$$= \frac{1}{2\beta} \left\{ \frac{T_{i+1,j}u_x(i+1, j) - \left[2 - R_{i+1,j} - R_{i-1,j}\right]u_y(i, j) + T_{i,j-1}u_y(i, j-1)}{\Delta y^2} \right.$$
$$\left. + \frac{u_y(i+1, j) - 2u_y(i, j) + u_y(i-1, j)}{\Delta x^2} + \left[\varepsilon_{i,j,k} - \beta^2\right]k^2 u_y(i, j) \right\} \quad (12.39)$$

$$A_{xy}u_y$$
$$= \frac{1}{8\beta \Delta x \Delta y} \left\{ \left(\frac{\varepsilon_{i+1,j+1,k}}{\varepsilon_{i+1,j,k}} - 1\right) u_x(i+1, j+1) - \left(\frac{\varepsilon_{i+1,j-1,k}}{\varepsilon_{i+1,j,k}} - 1\right) u_x(i+1, j-1) \right.$$
$$\left. - \left(\frac{\varepsilon_{i-1,j+1,k}}{\varepsilon_{i-1,j,k}} - 1\right) u_x(i-1, j+1) + \left(\frac{\varepsilon_{i-1,j-1,k}}{\varepsilon_{i-1,j,k}} - 1\right) u_x(i-1, j-1) \right\} \quad (12.40)$$

$$A_{yx}u_x$$
$$= \frac{1}{8\beta \Delta x \Delta y} \left\{ \left(\frac{\varepsilon_{i+1,j+1,k}}{\varepsilon_{i,j+1,k}} - 1\right) u_x(i+1, j+1) - \left(\frac{\varepsilon_{i-1,j+1,k}}{\varepsilon_{i,j-1,k}} - 1\right) u_x(i-1, j+1) \right.$$
$$\left. - \left(\frac{\varepsilon_{i+1,j-1,k}}{\varepsilon_{i,j+1,k}} - 1\right) u_x(i+1, j-1) + \left(\frac{\varepsilon_{i-1,j-1,k}}{\varepsilon_{i,j-1,k}} - 1\right) u_x(i-1, j-1) \right\} \quad (12.41)$$

In the previous equations

$$T_{i\pm 1,j} = \frac{2\varepsilon_{i\pm 1,j,k}}{\varepsilon_{i\pm 1,j,k} + \varepsilon_{i,j,k}} \quad (12.42)$$

$$R_{i+1,j} = T_{i\pm 1,j} - 1 \quad (12.43)$$

are the transmission and reflection coefficients across index interfaces between points i and $i + 1$. Similarly,

$$T_{i,j\pm 1} = \frac{2\varepsilon_{i,j\pm 1,k}}{\varepsilon_{i,j\pm 1,k} + \varepsilon_{i,j,k}} \quad (12.44)$$

$$R_{i,j\pm 1} = T_{i,j\pm 1} - 1 \quad (12.45)$$

are the transmission and reflection coefficients across index interfaces between points j and $j + 1$.

12.3 The $1 + 1$ dimensional FD-BPM formulation

If we can neglect y-dependence of the refractive index (wide waveguides), the scalar formulation described by Eq. (12.37) can be further simplified to the $1 + 1$ description. Cross-sectional dependence of the refractive index is $\bar{n} = \bar{n}(x, z)$. The relevant Helmholtz equation is [10]

$$2j\beta \frac{\partial u}{\partial z} = \frac{\partial_x^2 u}{\partial x^2} + \left(\bar{n}^2 k_0^2 - \beta^2\right) u \quad (12.46)$$

where u is the only electric field component of the TE mode of the waveguide. This equation will now be analysed using two methods. The second method will be implemented.

12.3.1 Simple approach

Here we discuss the FD-BPM formulation as described by Chung and Dagli [10]. First, replace continuous field $u(z, x)$ by its discrete values as

$$u_i \equiv u(i \cdot \Delta x, z), \quad i = 0, 1, 2 \ldots N - 1$$

Second derivative is approximated as

$$\frac{\partial^2 u}{\partial x^2} = \frac{u_{i-1} - 2u_i + u_{i+1}}{\Delta x^2}$$

The resulting finite difference equation is

$$2j\beta \frac{\partial u_i}{\partial z} = \frac{u_{i-1} - 2u_i + u_{i+1}}{\Delta x^2} + \left(\bar{n}_i^2 k_0^2 - \beta^2\right) u_i \equiv f_i(z)$$

Integrate the equation using trapezoidal rule

$$2j\beta \int_{u_i(z)}^{u_i(z+\Delta z)} du_i = \int_z^{z+\Delta z} f_i(z) dz$$

and have

$$2j\beta \left[u_i(z + \Delta z) - u_i(z)\right] = \frac{1}{2}\Delta z \left[f_i(z + \Delta z) + f_i(z)\right]$$

Using the definition of $f_i(z)$ in the above formula and combining relevant terms, one finds [10]

$$-au_{i+1}(z + \Delta z) + bu_i(z + \Delta z) - au_{i-1}(z + \Delta z) = au_{i+1}(z) + cu_i(z) + au_{i-1}(z) \qquad (12.47)$$

where

$$a = \frac{\Delta z}{2\Delta x^2}$$
$$b = \frac{\Delta z}{\Delta x^2} - \frac{1}{2}\Delta z \left[k_0^2 \bar{n}_i^2(z + \Delta z) - \beta^2\right] + 2j\beta$$
$$c = -\frac{\Delta z}{\Delta x^2} + \frac{1}{2}\Delta z \left[k_0^2 \bar{n}_i^2(z) - \beta^2\right] + 2j\beta$$

The scheme results in a tridiagonal system of linear equations. For more details, see [10].

12.3.2 Propagator approach

Approach 1

When one approximates propagation within the waveguide by a one-dimensional approach using variables (z, x) with z as the propagation direction, the y dependence is irrelevant and one then sets $\partial/\partial y = 0$. The resulting equation is

$$2j\beta \frac{\partial u}{\partial z} = \frac{\partial^2 u}{\partial x^2} + (k^2 - \beta^2)u \qquad (12.48)$$

where $k = \bar{n}k_0$.

The finite-difference discretization of Eq. (12.48) in 1D is the following:

$$j\frac{u_i^{n+1} - u_i^n}{h} = \frac{1}{2\beta} \frac{u_{i+1}^n - 2u_i^n + u_{i-1}^n}{\Delta x^2} + \frac{1}{2\beta}(k_i^2 - \beta^2) \qquad (12.49)$$

where $k_i^2 = k_0^2 \, \bar{n}_i^2(x) \equiv k_0^2 \, \bar{n}^2(x_i)$. With an introduction of the propagation operator P, the above scheme is

$$j\frac{u_i^{n+1} - u_i^n}{h} = \sum_{k=1}^{N} P_{ik} \, u_k^n \qquad (12.50)$$

The matrix elements of the operator P are

$$P_{ik} = \frac{1}{2\beta} \frac{1}{\Delta x^2} (\delta_{i+1,k} - 2\delta_{i,k} + \delta_{i-1,k}) + \frac{1}{2\beta}(k_i^2 \, \delta_{i,k} - \beta^2) \qquad (12.51)$$

The solution of Eq. (12.50) can be written as

$$\vec{u}^{n+1} = \left(\overleftrightarrow{I} - jh\overleftrightarrow{P}\right) \vec{u}^n \qquad (12.52)$$

where $\vec{u}^n$ is a column vector which consists of elements u_i^n, $\overleftrightarrow{I}$ is the identity matrix and $\overleftrightarrow{P}$ is the propagation matrix operator having elements given by Eq. (12.51). The scheme is known as the explicit scheme. It is numerically unstable if the h step is too large.

The improvement is made by applying operator P to the future value of u [11]. From Eq. (12.50)

$$j\frac{u_i^{n+1} - u_i^n}{h} = \sum_{k=1}^{N} P_{ik}\, u_k^{n+1} \qquad (12.53)$$

or

$$\vec{u}^{\,n+1} = \vec{u}^{\,n} - jh\,\overleftrightarrow{P}\,\vec{u}^{\,n+1}$$

or

$$\left(\overleftrightarrow{T} + jh\,\overleftrightarrow{P}\right)\vec{u}^{\,n+1} = \vec{u}^{\,n}$$

The solution is

$$\vec{u}^{\,n+1} = \left(\overleftrightarrow{T} + jh\,\overleftrightarrow{P}\right)^{-1}\vec{u}^{\,n} \qquad (12.54)$$

The scheme is called implicit method, and it is stable, see Problem.

The combination of the above two methods is created by taking the average between implicit and explicit schemes. The new scheme is known as the Crank-Nicolson method [11]. It is both more accurate and stable. Adding Eqs. (12.50) and (12.53), one obtains (the Crank-Nicolson scheme)

$$2j\frac{u_i^{n+1} - u_i^n}{h} = \sum_{k=1}^{N} P_{ik}\left(u_k^n + u_k^{n+1}\right) \qquad (12.55)$$

In matrix form

$$\vec{u}^{\,n+1} = \vec{u}^{\,n} - \frac{1}{2}jh\,\overleftrightarrow{P}\left(u_k^n + u_k^{n+1}\right) \qquad (12.56)$$

The above can be expressed as

$$\left(\overleftrightarrow{T} + \frac{1}{2}jh\,\overleftrightarrow{P}\right)\vec{u}^{\,n+1} = \left(\overleftrightarrow{T} - \frac{1}{2}jh\,\overleftrightarrow{P}\right)\vec{u}^{\,n} \qquad (12.57)$$

or

$$L_+\,\vec{u}^{\,n+1} = L_-\,\vec{u}^{\,n} \qquad (12.58)$$

where

$$L_+ = \overleftrightarrow{T} + \frac{1}{2}jh\,\overleftrightarrow{P} \qquad (12.59)$$

and

$$L_- = \overleftrightarrow{T} - \frac{1}{2}jh\,\overleftrightarrow{P} \qquad (12.60)$$

From that equation, the final form which can be implemented is

$$\vec{u}^{\,n+1} = L_+^{-1}\,L_-\,\vec{u}^{\,n} \qquad (12.61)$$

Approach 2

The above derivation can be made more formal. For that purpose, write Eq. (12.48) as

$$j\frac{\partial u}{\partial z} = P \cdot u \qquad (12.62)$$

where the propagation operator P is defined as

$$P = \frac{1}{2\beta}\left[\partial_x^2 + (k^2 - \beta^2)\right] \qquad (12.63)$$

The solution of Eq. (12.62) is

$$u(z) = e^{-jzP} u(0)$$

or, when one introduces 'small' step h:

$$u(z + h) = e^{-jhP} u(z) \qquad (12.64)$$

For a small step, the solution (12.64) can be approximated as

$$u(z + h) \approx (1 - jhP)\, u(z) \qquad (12.65)$$

One might notice that the solution (12.64) can also be written as

$$e^{jhP} u(z + h) = u(z) \qquad (12.66)$$

or

$$(1 + jhP)\, u(z + h) \approx u(z) \qquad (12.67)$$

or

$$u(z + h) \approx (1 + jhP)^{-1}\, u(z) \qquad (12.68)$$

To solve this problem, a practically important method is known as the Crank-Nicolson scheme which is obtained from the solutions (12.65) and (12.68). One thus makes the following manipulations:

$$\begin{aligned} u(z + h) &= \left(1 + \frac{jh}{2}P\right)^{-1} u\left(z + \frac{h}{2}\right) \\ &= \left(1 + \frac{jh}{2}P\right)^{-1} \left(1 + \frac{jh}{2}P\right) u(z) \end{aligned} \qquad (12.69)$$

using Eq. (12.65) for $h \to h/2$. The implementation of this scheme is shown in Appendix, Listing 12A.2. It is used to propagate a Gaussian pulse in an empty space. In our implementation we assumed that $\beta = \bar{n} \cdot k_0$. With this choice, last term in propagator P vanishes.

Profiles of a Gaussian pulse at various locations along the propagation direction in a free space are shown in Fig. 12.7. One can observe reduction of a peak value and also the effects of reflections from the boundaries of the computational window. Elimination of those reflections will be discussed in the next sections. In Fig. 12.8 we show a three-dimensional view of profiles of Gaussian pulses shown in Fig. 12.7.

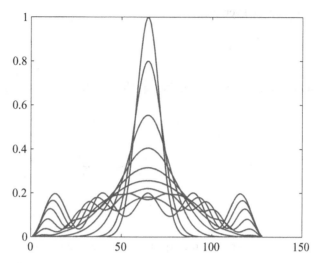

Fig. 12.7 Profiles of Gaussian pulses at various locations along z-axis.

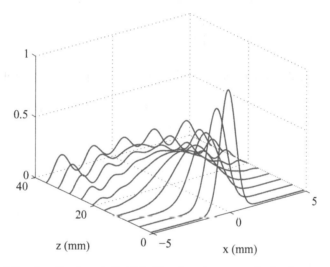

Fig. 12.8 Three-dimensional view of profiles shown in Fig. 12.7 of Gaussian pulses propagating in a free space.

Approach 3

Before starting the discussion of boundary conditions, we want to summarize yet another possible numerical approach to Eq. (12.48). It will then be used to derive details of transparent boundary conditions.

Assuming an equidistant step along x-axis and using standard formulas ($h \equiv \Delta z$)

$$\frac{\partial u}{\partial z} = \frac{u_i^{n+1} - u_i^n}{h} \tag{12.70}$$

$$\frac{\partial^2 u}{\partial x^2} = \frac{u_{i+1} - 2u_i + u_{i-1}}{\Delta x^2} \tag{12.71}$$

the discretized version of Eq. (12.48) is

$$2j\beta\frac{u_i^{n+1} - u_i^n}{h} = \frac{u_{i+1} + u_{i-1}}{\Delta x^2} + \left[-\frac{2}{\Delta x^2} + (k_i^2 - \beta^2)\right]u_i \qquad (12.72)$$

Introduce modifications on the right hand side of Eq. (12.72):

$$u_{i+1} = \frac{1}{2}\left(u_{i+1}^{n+1} + u_{i+1}^n\right) \qquad (12.73)$$

$$u_{i-1} = \frac{1}{2}\left(u_{i-1}^{n+1} + u_{i-1}^n\right) \qquad (12.74)$$

$$k_i^2 = \frac{1}{2}\left[k_i^2(n+1) + k_i^2(n)\right] \qquad (12.75)$$

After new algebraic steps, Eq. (12.72) can be expressed as

$$u_i^{n+1} + \frac{jh}{2}\left\{\frac{1}{2\beta}\frac{1}{\Delta x^2}\left(u_{i+1}^{n+1} - 2u_i^{n+1} + u_{i-1}^{n+1}\right) + \frac{1}{2\beta}\left[k_i^2(n+1) - \beta^2\right]u_i^{n+1}\right\}$$

$$= u_i^n - \frac{jh}{2}\left\{\frac{1}{2\beta}\frac{1}{\Delta x^2}\left(u_{i+1}^n - 2u_i^n + u_{i-1}^n\right) + \frac{1}{2\beta}\left[k_i^2(n) - \beta^2\right]u_i^n\right\} \qquad (12.76)$$

Note the different n dependence in k_i^2. The final form of the above in a condensed form is

$$L_+(n+1)\vec{u}^{n+1} = L_-(n)\vec{u}^n \qquad (12.77)$$

where

$$L_+(n+1)\vec{u}^{n+1} = u_i^{n+1} + \frac{jh}{2}\left\{\frac{1}{2\beta}\frac{1}{\Delta x^2}\left(u_{i+1}^{n+1} - 2u_i^{n+1} + u_{i-1}^{n+1}\right)\right.$$

$$\left. + \frac{1}{2\beta}\left[k_i^2(n+1) - \beta^2\right]u_i^{n+1}\right\} \qquad (12.78)$$

and

$$L_-(n)\vec{u}^n = u_i^n - \frac{jh}{2}\left\{\frac{1}{2\beta}\frac{1}{\Delta x^2}\left(u_{i+1}^n - 2u_i^n + u_{i-1}^n\right) + \frac{1}{2\beta}\left[k_i^2(n) - \beta^2\right]u_i^n\right\} \qquad (12.79)$$

12.3.3 Transparent boundary conditions

To eliminate reflections shown in the previous example, one needs to introduce appropriate boundary conditions outside the computational window. The popular solution is to introduce the so-called transparent boundary conditions (TBC) which were invented by Hadley [12], [13]. TBC have been extensively discussed in the literature [14], [15], [16], [17].

Consider the system with left and right boundaries as shown in Fig. 12.9. Nodes x_0 and x_{N+1} are outside the system; however, they are needed in the implementation of numerical scheme.

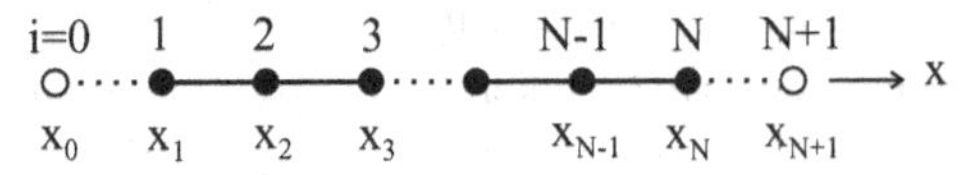

The 1 + 1 dimensional FD-BPM formulation

Fig. 12.9 Numbering of nodes used in deriving transparent boundary conditions.

Left-hand boundary

Analyse the left boundary first. Assume that near the boundary the field is approximated as

$$u(x, z) \approx A(z) e^{ik_x x} \tag{12.80}$$

For the first left points shown in Fig. 12.9, one obtains

$$u_0 = u(x_0) = A(z) e^{ik_x x_0} \tag{12.81}$$

$$u_1 = u(x_1) = A(z) e^{ik_x x_1} \tag{12.82}$$

$$u_2 = u(x_2) = A(z) e^{ik_x x_2} \tag{12.83}$$

From the above equations one finds

$$\frac{u_2}{u_1} = \frac{e^{ik_x x_2}}{e^{ik_x x_1}} = e^{ik_x(x_2 - x_1)} \equiv e^{ik_x \Delta x} \tag{12.84}$$

and

$$\frac{u_1}{u_0} = \frac{e^{ik_x x_1}}{e^{ik_x x_0}} = e^{ik_x(x_1 - x_0)} \equiv e^{ik_x \Delta x} \tag{12.85}$$

assuming uniform grid; i.e. $\Delta x = x_2 - x_1 = x_1 - x_0$. From the above equations, one determines value of the field at the outside point x_0:

$$u_0 = u_1 e^{ik_x \Delta x} \tag{12.86}$$

Wave number k_x is determined from known fields using Eq (12.84):

$$k_x = \frac{1}{i \Delta x} \ln \frac{u_2}{u_1} \tag{12.87}$$

For a field travelling leftward (an outgoing wave), the real part of k_x, $Re(k_x)$ is positive.

The implementation of TBC at the left point for operator L_+^{n+1} will be illustrated now. Consider the first element for $i = 1$, which from Eq. (12.78) is

$$L_+^{n+1}(1) = u_1^{n+1} + \frac{1}{2}jh\frac{1}{2\beta}\frac{1}{\Delta x^2}(-2)u_1^{n+1} + \frac{1}{2}jh\frac{1}{2\beta}\frac{1}{\Delta x^2}u_0^{n+1} \tag{12.88}$$

Replacing u_0^{n+1} using Eq. (12.86) gives

$$\begin{aligned}L_+^{n+1}(1) &= u_1^{n+1} + \frac{1}{2}jh\frac{1}{2\beta}\frac{1}{\Delta x^2}(-2)u_1^{n+1} + \frac{1}{2}jh\frac{1}{2\beta}\frac{1}{\Delta x^2}u_1^{n+1}e^{jk_x\Delta x} \\ &= \left(1 + \frac{1}{2}jh\frac{1}{2\beta}\frac{1}{\Delta x^2}(-2) + \frac{1}{2}jh\frac{1}{2\beta}\frac{1}{\Delta x^2}e^{jk_x\Delta x}\right)u_1^{n+1} \\ &= (\text{old term} + \text{left correction})\, u_1^{n+1}\end{aligned} \tag{12.89}$$

The last expression is directly implemented. For operator L_-^{n+1} there is only a change of sign. Similar steps can be taken for the right boundary, which is left as a problem.

Right-hand boundary

At the right-hand boundary, the point x_{N+1} is outside the system and the field there must be determined. One starts with assuming the following expression for the right-travelling field:

$$u(x) \approx A(z)\, e^{-ik_x x} \tag{12.90}$$

Based on this assumption, we can write fields at three points around the right boundary:

$$u_{N-1} = u(x_{N-1}) = A(z)\, e^{-ik_x x_{N-1}} \tag{12.91}$$

$$u_N = u(x_N) = A(z)\, e^{-ik_x x_N} \tag{12.92}$$

$$u_{N+1} = u(x_{N+1}) = A(z)\, e^{-ik_x x_{N+1}} \tag{12.93}$$

Here u_{N-1} and u_N are fields within the system and u_{N+1} is outside. From the above relations one determines the outside field as

$$u_{N+1} = u_N\, e^{-ik_x \Delta x} \tag{12.94}$$

where the wavenumber is

$$k_x = \frac{1}{i\Delta x} \ln \frac{u_2}{u_1} \tag{12.95}$$

Cross sections are shown in Fig. 12.10. One can observe that reflections were eliminated. The three-dimensional view of an initial Gaussian pulse at various positions is shown in Fig. 12.11. MATLAB code is shown in Appendix, Listings 12B.3 and 12B.3.1.

12.4 Concluding remarks

A few years ago an article was published with the title 'What is the future for beam propagation methods?' [18]. The role played by BPM in the simulations of photonic devices was discussed. The authors also provided an extensive literature summary on the following issues associated with the BPM: transverse discretization, explicit and implicit formulations, vector effects, wide angle effects, reflective schemes and time domain BPM. They concluded with highlighting the outstanding issues (as of 2004) of BPM, which were: flexible meshing, development of wide-angle schemes, improvement of boundary conditions and creation of hybrid BPM schemes, e.g. with time and/or frequency domain methods.

A more recent assessment of BPM (and also Eigenmode Expansion Method and Finite Difference Time Domain) for Photonic CAD is reported by Gallagher [19]. He summarizes the main algorithms used in photonic modelling and discusses their strengths and weaknesses.

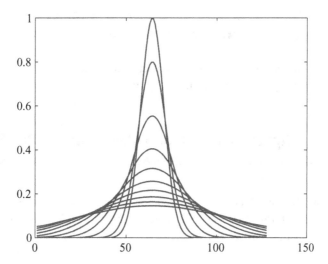

Fig. 12.10 Comparison of field profiles at various locations along z-axis of Gaussian pulse propagating in a free space by BPM with transparent boundary conditions. One observes reduction of its peak value and spreading of pulse. No reflections are observed.

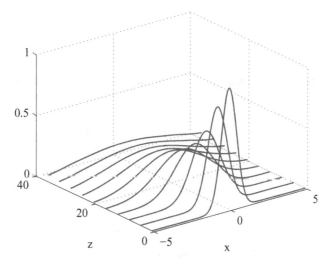

Fig. 12.11 Three-dimensional view of Gaussian pulse shown in Fig. 12.10 propagating in a free space by BPM with transparent boundary conditions.

12.5 Problems

1. Develop H-formulation of BPM.
2. Perform von Neumann stability analysis for the implicit method.
3. Based on the example in the text for TBC for the left boundary, develop expressions for the right TBC.

12.6 Project

1. Implement 1-1 FD-BPM as discussed in Section 12.3.1. Consult work by [10]. Analyse waveguides described in Ref. [10].

Appendix 12A: Details of derivation of the FD-BPM equation

We describe details of discretization of the earlier equations, like Eq. (12.32). We present details for the 1D case only. Generalization to 2D is straightforward. A typical term which needs discretization is

$$K = \frac{\partial}{\partial x} \frac{1}{\varepsilon} \frac{\partial}{\partial x} (\varepsilon u) \equiv \frac{\partial}{\partial x} g(x)$$

Simple approach

First-order derivative is

$$\frac{\partial \phi}{\partial x} = \frac{\phi_{p+1} - \phi_p}{\Delta x}$$

where $\phi = \varepsilon u$. Applying the above formula with difference centred around the points $p + \frac{1}{2}$ and $p - \frac{1}{2}$, one finds

$$K = \frac{1}{\Delta x} \left\{ \frac{1}{\varepsilon_{p+\frac{1}{2}}} \frac{\phi_{p+1} - \phi_p}{\Delta x} - \frac{1}{\varepsilon_{p-\frac{1}{2}}} \frac{\phi_p - \phi_{p-1}}{\Delta x} \right\}$$

where

$$\phi_{p\pm 1} = \varepsilon_{p\pm 1} u_{p\pm 1}$$
$$\phi_p = \varepsilon_p u_p$$

We also use the following intuitive relations (which will be proved in the next section):

$$\varepsilon_{p+\frac{1}{2}} = \frac{\varepsilon_{p+1} + \varepsilon_p}{2}$$
$$\varepsilon_{p-\frac{1}{2}} = \frac{\varepsilon_{p+1} + \varepsilon_{p-1}}{2}$$

Combining the above relations, one obtains

$$K = \frac{1}{\Delta x} \frac{2}{\varepsilon_{p+1} + \varepsilon_p} \frac{\varepsilon_{p+1} u_{p+1} - \varepsilon_p u_p}{\Delta x} - \frac{1}{\Delta x} \frac{2}{\varepsilon_p + \varepsilon_{p-1}} \frac{\varepsilon_p u_p - \varepsilon_{p-1} u_{p-1}}{\Delta x}$$
$$= \frac{2}{\Delta x^2} \left\{ \frac{\varepsilon_{p+1}}{\varepsilon_{p+1} + \varepsilon_p} u_{p+1} - \left[\frac{\varepsilon_p}{\varepsilon_{p+1} + \varepsilon_p} + \frac{\varepsilon_p}{\varepsilon_p + \varepsilon_{p-1}} \right] u_p + \frac{\varepsilon_{p-1}}{\varepsilon_p + \varepsilon_{p-1}} u_{p-1} \right\}$$

Taylor series approach

We differentiate

$$K = \frac{\partial}{\partial x}\frac{1}{\varepsilon}\frac{\partial}{\partial x}(\varepsilon u) = \frac{\partial}{\partial x}\frac{1}{\varepsilon}\left(\frac{\partial \varepsilon}{\partial x}\right)u + \frac{\partial^2 u}{\partial x^2}$$

First term is evaluated as follows. First, we expand around point p:

$$\left(\frac{1}{\varepsilon}\left(\frac{\partial \varepsilon}{\partial x}\right)u\right)_{p\pm\frac{1}{2}} = \left(\frac{1}{\varepsilon}\left(\frac{\partial \varepsilon}{\partial x}\right)u\right)_p \pm \frac{\Delta x}{2}\frac{\partial}{\partial x}\frac{1}{\varepsilon}\left(\frac{\partial \varepsilon}{\partial x}\right)u\bigg|_p$$

Subtracting gives

$$\frac{\partial}{\partial x}\frac{1}{\varepsilon}\left(\frac{\partial \varepsilon}{\partial x}\right)u\bigg|_p = \frac{1}{\Delta x}\left\{\left(\frac{1}{\varepsilon}\left(\frac{\partial \varepsilon}{\partial x}\right)u\right)_{p+\frac{1}{2}} - \left(\frac{1}{\varepsilon}\left(\frac{\partial \varepsilon}{\partial x}\right)u\right)_{p-\frac{1}{2}}\right\} \quad (12.96)$$

From the above, one can observe that the values of u and ε are needed at intermediate points $p+\frac{1}{2}$ and $p-\frac{1}{2}$. Those can be obtained by performing expansions

$$u_{p+1} = u_{p+\frac{1}{2}} + \frac{\partial u}{\partial x}\bigg|_{p+\frac{1}{2}}\frac{\Delta x}{2}$$

and

$$u_p = u_{p+\frac{1}{2}} - \frac{\partial u}{\partial x}\bigg|_{p+\frac{1}{2}}\frac{\Delta x}{2}$$

Adding the above gives

$$u_{p+\frac{1}{2}} = \frac{1}{2}(u_p + u_{p+1}) \quad (12.97)$$

Expansions of ε gives

$$\varepsilon_{p+1} = \varepsilon_{p+\frac{1}{2}} - \frac{\partial \varepsilon}{\partial x}\bigg|_{p+\frac{1}{2}}\frac{\Delta x}{2}$$

and

$$\varepsilon_p = \varepsilon_{p+\frac{1}{2}} - \frac{\partial \varepsilon}{\partial x}\bigg|_{p+\frac{1}{2}}\frac{\Delta x}{2}$$

Adding the above gives

$$\varepsilon_{p+\frac{1}{2}} = \frac{1}{2}(\varepsilon_p + \varepsilon_{p+1}) \quad (12.98)$$

Subtracting gives

$$\frac{\partial \varepsilon}{\partial x}\bigg|_{p+\frac{1}{2}} = \frac{1}{2}(\varepsilon_{p+1} - \varepsilon_p) \quad (12.99)$$

From Eqs. (12.98) and (12.99) we have

$$\left.\frac{1}{\varepsilon}\frac{\partial \varepsilon}{\partial x}\right|_{p+\frac{1}{2}} = \frac{2}{\Delta x}\frac{\varepsilon_{p+1} - \varepsilon_p}{\varepsilon_p + \varepsilon_{p+1}} \quad (12.100)$$

Using Eqs. (12.97) and (12.100), we obtain

$$\left.\frac{1}{\varepsilon}\frac{\partial \varepsilon}{\partial x}u\right|_{p+\frac{1}{2}} = \frac{1}{\Delta x}\frac{\varepsilon_{p+1} - \varepsilon_p}{\varepsilon_p + \varepsilon_{p+1}}(u_p + u_{p+1}) \quad (12.101)$$

In exactly the same way, performing the relevant expansions around point $p - \frac{1}{2}$ one finds

$$\left.\frac{1}{\varepsilon}\frac{\partial \varepsilon}{\partial x}u\right|_{p-\frac{1}{2}} = \frac{1}{\Delta x}\frac{\varepsilon_p - \varepsilon_{p-1}}{\varepsilon_p + \varepsilon_{p-1}}(u_p + u_{p-1}) \quad (12.102)$$

Substituting Eqs. (12.101) and (12.102) into (12.96), we obtain an intermediate result

$$\left.\frac{\partial}{\partial x}\frac{1}{\varepsilon}\left(\frac{\partial \varepsilon}{\partial x}\right)u\right|_p = \frac{1}{\Delta x^2}\left\{\frac{\varepsilon_{p+1} - \varepsilon_p}{\varepsilon_p + \varepsilon_{p+1}}(u_p + u_{p+1}) - \frac{\varepsilon_p - \varepsilon_{p-1}}{\varepsilon_p + \varepsilon_{p-1}}(u_p + u_{p-1})\right\} \quad (12.103)$$

A complete derivative is obtained by combining discretization for a second derivative

$$\frac{\partial^2 u}{\partial x^2} = \frac{1}{\Delta x^2}\left(u_{p+1} - 2u_p + u_{p-1}\right)$$

with expression (12.103). The result is

$$K = \frac{\partial}{\partial x}\frac{1}{\varepsilon}\frac{\partial}{\partial x}(\varepsilon u)$$

$$= \frac{2}{\Delta x^2}\left\{\frac{\varepsilon_{p+1}}{\varepsilon_p + \varepsilon_{p+1}}u_{p+1} - \varepsilon_p\left(\frac{1}{\varepsilon_p + \varepsilon_{p+1}} + \frac{1}{\varepsilon_p + \varepsilon_{p+1}}\right)u_p + \frac{\varepsilon_{p-1}}{\varepsilon_p + \varepsilon_{p-1}}u_{p-1}\right\}$$

The last result is the same as the one obtained by Lidgate [20].

Appendix 12B: MATLAB listings

In Table 12.1 we provide a list of MATLAB files created for Chapter 12 and a short description of each function.

Table 12.1 List of MATLAB functions for Chapter 12.

Listing	Function name	Description
12B.1	*pbpm.m*	Gaussian pulse in free space in the paraxial approximation
12B.2	*fd_bpm_free.m*	Gaussian pulse in free space (Crank-Nicholson)
12B.3	*bpm_tbc.m*	Transparent boundary conditions
12B.3.1	*prop.m*	Function called by *bpm_tbc.m*. Performs single step

Listing 12B.1 Function pbpm.m. Function performs propagation of a two-dimensional Gaussian pulse using the Fourier transform split-step beam propagation method in the paraxial approximation.

```matlab
% File name: pbpm.m
% Propagation of 2D Gaussian pulse by paraxial FT split-step BPM
clear all N= 10; delta_x = 1/N;
x = -5:delta_x:5;                   % creation of space arguments
y = -5:delta_x:5;                   % creation of space arguments
delta_z = 5.0;                      % step size along z-axis
delta_n = 0.4; k_zero = 100;
beta = 20;                          % propagation constant
%
u_init=(exp(-x.^2))'*exp(-y.^2);    % initial Gaussian pulse
mesh(x,y,abs(u_init));              % plots original pulse
pause close all
%
k_x = -5:1/N:5;                     % creation of Fourier variables
k_y = -5:1/N:5;                     % creation of Fourier variables
%
temp = delta_z/(2*beta); H_transfer =
(exp(1i*temp*k_x.^2))'*(exp(1i*temp*k_y.^2)); H_transfer =
fftshift(H_transfer); S_phase = exp(-1i*k_zero*delta_n);
%
for n=1:100
    z = fft2(u_init);
    zz = z.*H_transfer;
    u_prime = ifft2(zz);
    u = S_phase.*u_prime;
    u_init = u;
end
%
mesh(x,y,abs(u_init))               % plots pulse after propagation
pause close all
```

Listing 12B.2 Program fd_bpm_free.m. Illustrates propagation of a Gaussian pulse in a free space.

```matlab
% File name: fd_bpm_free.m
% Propagation of Gaussian pulse in a free space by Crank-Nicholson method
% No boundary conditions are introduced
clear all
L_x=10.0;                           % transversal dimension (along x-axis)
w_0=1.0;                            % width of input Gaussian pulse
lambda = 0.6;                       % wavelength
```

```
n=1.0;                              % refractive index of the medium
k_0=2*pi/lambda;                    % wavenumber
N_x=128;                            % points on x axis
Delta_x=L_x/(N_x-1);                % x axis spacing
h=5*Delta_x;                        % propagation step along z-axis
N_z=100;                            % number of propagation steps
plotting=zeros(N_x,N_z);            % storage for plotting
x=linspace(-0.5*L_x,0.5*L_x,N_x);   % coordinates along x-axis
x = x';
E=exp(-(x/w_0).^2);                 % initial Gaussian field
%
% beta = n*k_0. With this choice, last term in propagator vanishes
prefactor = 1/(2*n*k_0*Delta_x^2); main = ones(N_x,1); above =
ones(N_x-1,1); below = above;
P = prefactor*(diag(above,-1)-2*diag(main,0)+diag(below,1)); % matrix P
%
step_plus = eye(N_x) + 0.5i*h*P;    % step forward
step_minus =eye(N_x)-0.5i*h*P;      % step backward
%
z = 0; z_plot = zeros(N_z); for r=1:N_z
    z = z + h;
    z_plot(r) = z + h;
    plotting(:,r)=abs(E).^2;
    E=step_plus\step_minus*E;
end;
%
for k = 1:N_z/10:N_z                % choosing 2D plots every 10-th  step
    plot(plotting(:,k),'LineWidth',1.5)
    set(gca,'FontSize',14);         % size of tick marks on both axes
    hold on
end pause close all
%
for k = 1:N_z/10:N_z                % choosing 3D plots every 10-th step
    y = z_plot(k)*ones(size(x));    % spread out along y-axis
    plot3(x,y,plotting(:,k),'LineWidth',1.5)
    hold on
end grid on xlabel('x (mm)','FontSize',14)
ylabel('z (mm)','FontSize',14)      % along propagation direction
set(gca,'FontSize',14);             % size of tick marks on both axes
pause close all
```

Listing 12B.3 Program bpm_tbc.m. Illustrates propagation of Gaussian pulse in a free space with transparent boundary conditions.

```
% File name: bpm_tbc.m
% Illustrates propagation of Gaussian pulse in a free space
```

```
% using BPM with transparent boundary conditions
% Operator P is determined in a separate function
clear all
L_x=10.0;                            % transversal dimension (along x-axis)
w_0=1.0;                             % width of input Gaussian pulse
lambda = 0.6;                        % wavelength
n=1.0;                               % refractive index of the medium
k_0=2*pi/lambda;                     % wavenumber
N_x=128;                             % number of points on x axis
Delta_x=L_x/(N_x-1);                 % x axis spacing
h=5*Delta_x;                         % propagation step
N_z=100;                             % number of propagation steps
plotting=zeros(N_x,N_z);             % storage for plotting
x=linspace(-0.5*L_x,0.5*L_x,N_x);    % coordinates along x-axis
x = x';
E=exp(-(x/w_0).^2);                  % initial Gaussian field
%
z = 0;
z_plot = zeros(N_z);
for r=1:N_z                          % BPM stepping
    z = z + h;
    z_plot(r) = z + h;
    plotting(:,r)=abs(E).^2;
    E = step(Delta_x,k_0,h,n,E);     % Propagates pulse over one step
end;
%
for k = 1:N_z/10:N_z                 % choosing 2D plots every 10-th step
    plot(plotting(:,k),'LineWidth',1.5)
    set(gca,'FontSize',14);          % size of tick marks on both axes
    hold on
end
pause
close all
%
for k = 1:N_z/10:N_z                 % choosing 3D plots every 10-th step
    y = z_plot(k)*ones(size(x));     % spread out along y-axis
    plot3(x,y,plotting(:,k),'LineWidth',1.5)
    hold on
end
grid on
xlabel('x','FontSize',14)
ylabel('z','FontSize',14)            % along propagation direction
set(gca,'FontSize',14);              % size of tick marks on both axes
pause
close all
```

Listing 12B.3.1 Function step.m used by *bpm_tbc.m*. Function performs a single step in BPM.

```matlab
% File name: step.m
function E_new = step(Delta_x,k_0,h,n,E_old)
% Function propagates BPM solution along one step

N_x = size(E_old,1);                    % determine size of the system
%--- Defines operator P outside of a boundary
prefactor = 1/(2*n*k_0*Delta_x^2);
main = ones(N_x,1);
above = ones(N_x-1,1);
below = above;
P = prefactor*(diag(above,-1)-2*diag(main,0)+diag(below,1)); % matrix P
%
L_plus = eye(N_x) + 0.5i*h*P;           % step forward
L_minus = eye(N_x)-0.5i*h*P;            % step backward
%
%---- Implementation of boundary conditions
%
pref = 0.5i*h/(2*k_0*Delta_x^2);
k=1i/Delta_x*log(E_old(2)/E_old(1));
if real(k)<0
    k=1i*imag(k);
end;
left = pref*exp(1i*k*Delta_x);          % left correction for next step
L_plus(1) = L_plus(1)+left;
L_minus(1) = L_minus(1)-left;
%
k=-1i/Delta_x*log(E_old(N_x)/E_old(N_x-1));
if real(k)<0
    k=1i*imag(k);
end;
right = pref*exp(1i*k*Delta_x);         % right correction for nest step
L_plus(N_x) = L_plus(N_x) + right;
L_minus(N_x) = L_minus(N_x) - right;
%
E_new = L_minus\L_plus*E_old;           % determine new solution
```

References

[1] M. D. Feit and J. A. Fleck Jr. Light propagation in grated0index optical fibres. *Appl. Opt.*, **17**:3990–8, 1978.

[2] V. P. Tzolov, D. Feng, S. Tanev, and Z. J. Jakubczyk, Modeling tools for integrated and fiber optical devices. In G. C. Righini and S. Iraj Najafi, eds., *Integrated Optics Devices III. Proceedings of SPIE*, volume **3620**, pages 162–73. EMW Publishing, Cambridge, MA, 1999.

[3] K. Kawano and T. Kitoh. *Introduction to Optical Waveguide Analysis. Solving Maxwell's Equations and the Schroedinger Equation*. Wiley, New York, 2001.

[4] C. R. Pollock and M. Lipson. *Integrated Photonics*. Kluwer Academic Publishers, Boston, 2003.

[5] K. Okamoto. *Fundamentals of Optical Waveguides*. Academic Press, Amsterdam, 2006.

[6] G. Lifante. *Integrated Photonics. Fundamentals*. Wiley, Chichester, 2003.

[7] R. L. Liboff. *Introductory Quantum Mechanics*. Addison-Wesley, Reading, MA, 1992.

[8] C. L. Xu and W. P. Huang. Finite-difference beam propagation method for guide-wave optics. In J. A. Kong, ed., *Progress in Electromagnetic Research, PIER 11*, pages 1–49. EMW Publishing, Cambridge, MA, 1995.

[9] W. P. Huang and C. L. Xu. Simulation of three-dimensional optical waveguides by a full-vector beam propagation method. *IEEE J. Quantum Electron.*, **29**:2639–49, 1993.

[10] Y. Chung and N. Dagli. An assessment of finite difference beam propagation method. *IEEE J. Quantum Electron.*, **26**:1335–9, 1990.

[11] A. L. Garcia. *Numerical Methods for Physics. Second Edition*. Prentice Hall, Upper Saddle River, 2000.

[12] G. R. Hadley. Transparent boundary condition for beam propagation. *Opt. Lett.*, **16**:624–6, 1991.

[13] G. R. Hadley. Transparent boundary condition for the beam propagation method. *IEEE J. Quantum Electron.*, **28**:363–70, 1992.

[14] F. Schmidt and P. Deuflhard. Discrete transparent boundary conditions for the numerical solution of Fresnel's equation. *Computers Math. Applic.*, **29**:53–76, 1995.

[15] F. Fogli, G. Bellanca, and P. Bassi. TBC and PML conditions for 2D and 3D BPM: a comparison. *Optical and Quantum Electronics*, **30**:443–56, 1998.

[16] D. Yevick, T. Friese, and F. Schmidt. A comparison of transparent boundary conditions for the Fresnel equation. *Journal of Computational Physics*, **168**:433–44, 2001.

[17] F. Schmidt, T. Friese, and D. Yevick. Transparent boundary conditions for split-step Pade approximations of the one-way Helmholtz equation. *Journal of Computational Physics*, **170**:696–719, 2001.

[18] T. M. Benson, B. B. Hu, and P. Sewell. What is the future for beam propagation methods? In G. Lampropoulos, J. Armitage, and R. Lessard, eds., *Proc. of SPIE Photonics North 2004: Photonic Applications in Telecommunications, Sensors, Software and Lasers*, volume **5579**, pages 351–8. SPIE, Bellingham, WA, 2004.

[19] D. Gallagher. Photonic CAD matures. *IEEE LEOS Newsletter*, **2**:8–14, 2008.

[20] S. Lidgate. *Advanced Finite Difference – Beam Propagation Method Analysis of Complex Components*. PhD thesis, Nottingham, 2004.

13 Some wavelength division multiplexing (WDM) devices

Wavelength division multiplexing (WDM) is a modern practical method of increasing transmission capacity in fibre communication systems. It uses the principle that optical beams with different wavelengths can propagate simultaneously over a single fibre without interfering with one another. In the wavelength range of 1280–1650 nm (like an AllWave fibre [1]) the useable bandwidth of a single mode fibre is about 53 THz. In recent years an improved (denser) WDM system known as DWDM is under development.

In this chapter we discuss some of the WDM devices and also provide some applications of BPM developed earlier to simulate those devices. We start by summarizing the basic WDM system.

13.1 Basics of WDM systems

WDM is the main technique used in the realization of all optical networks. WDM is the technology which combines a number of wavelengths onto the same fibre.

Key features include:

- capacity upgrade
- transparency (each optical channel can carry any transmission format)
- wavelength routing
- wavelength switching

Implementation of a typical WDM system employing N channels is shown in Fig. 13.1. In the shown system *three* wavelengths are multiplexed in one fibre to increase transmission capacity. The light of laser diodes with wavelengths recommended by the ITU is launched into the inputs of a wavelength multiplexer (MUX) where all wavelengths are combined and coupled into a single-mode fibre. When needed, propagating light can be amplified by an optical fibre amplifier and eventually imputed at the wavelength demultiplexer (DMUX) which separates all optical channels and sends them to different outputs.

In order to find the optical bandwidth corresponding to a spectral width in optical region, we use the relation $c = \lambda \cdot \nu$, where λ is wavelength and ν carrier frequency and c velocity of light. Differentiating

$$d\nu = c\frac{d}{d\lambda}\left(\frac{1}{\lambda}\right)d\lambda = -\frac{c}{\lambda^2}d\lambda$$

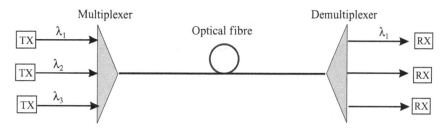

Fig. 13.1 Implementation of a typical WDM link.

or

$$|\Delta \nu| = \frac{c}{\lambda^2} |\Delta \lambda| \qquad (13.1)$$

The above equation describes the frequency change $\Delta \nu$ which corresponds to the wavelength change $\Delta \lambda$ around λ. Using the above formula, we can estimate the usable wavelength range for a standard single-mode fibre. Assuming that telecommunication wavelength range extends from $\lambda_1 = 1280$ nm to $\lambda_2 = 1625$ nm, the ultimate bandwidth of optical fibre is 40 THz [2]. Assuming 50 or 25 GHz channel spacing, there is the possibility to transmit 800–1600 wavelength channels.

In the remainder of this chapter, we will consider some of the basic devices used in WDM systems and provide two applications of the beam propagation method (BPM) to simulate simple structures.

13.2 Basic WDM technologies

In an ideal WDM system where nonlinear effects are neglected, the discrete wavelengths can be optically processed, i.e. routed and/or switched without interfering with each other. Optical processing can be performed using passive or active components.

One refers to passive components where there is no external control of their operation. Typically, they are used to split or combine signals. Examples of passive components are: $N \times N$ couplers, power splitters and star couplers.

Active components, which can be controlled electronically, include tunable optical filters, tunable sources, optical amplifiers. Here, we will concentrate on passive components.

Basic technologies which are used to develop WDM devices are [3]: fibre Bragg grating (FBG), array waveguide grating (AWG), thin-film filter (TFF) and diffraction grating (DG).

Another classification of main WDM components used in fibre optic communication systems is described by Agrawal [4] and includes: tunable optical filters (Fabry-Perot filters, Mach-Zehnder filters, grating-based filters, application-based filters), multiplexers and demultiplexers, add/drop multiplexers and filters, broadcast star couplers, wavelength routers, optical cross-connects, wavelength converters and WDM transmitters and receivers.

In the following, we will discuss basic WDM technologies. We start with fibre Bragg grating.

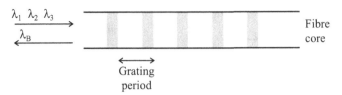

Fig. 13.2 Fibre grating operating as an optical filter.

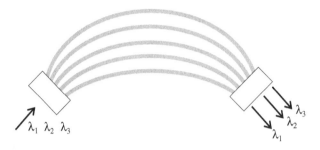

Fig. 13.3 Schematic illustration of a simple array waveguide demultiplexer.

13.2.1 Fibre Bragg grating

Fibre Bragg grating (FBG) devices are based on the principle discovered by Bragg in 1913 and demonstrated in fibre by Hill in 1978 [5]. It is illustrated in Fig. 13.2.

In a fibre's core a periodic change of the refractive index (known as grating) is created using the effect of photosensitivity in germanium-doped optical fibre [5]. When a light signal consisting of several wavelengths travels in optical fibre (here from left to right), the signal with a wavelength which obeys the Bragg condition is reflected. The reflected wavelength centred at λ_B that fulfills the Bragg condition is

$$\lambda_B = 2\Lambda n_{eff} \qquad (13.2)$$

where Λ is the grating period and n_{eff} is the effective group refractive index of the core.

Applications of Bragg grating are numerous and include filters and dispersion compensation. A comprehensive review on FBG including fibre grating lasers and amplifiers was written by Kashyap [6].

13.2.2 Array waveguide grating

Array waveguide grating (AWG) is shown schematically in Fig. 13.3 [2]. It is formed by several waveguides with different lengths. At both ends all waveguides are converging to the same points. Light composed of several wavelenths ($\lambda_1, \lambda_2, \ldots$) enters the device on the left. The length of each waveguide is carefully designed so to provide precise phase difference between neighbouring waveguides at the end of the guiding structure (on the right). At the output, a diffraction pattern is created which allows output waveguide to collect light at a particular wavelength. AWG can be used as a passive optical multiplexer and/or demultiplexer.

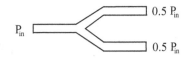

Fig. 13.4 Y-branch power splitter.

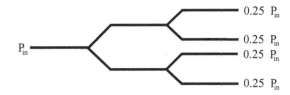

Fig. 13.5 A four-port splitter.

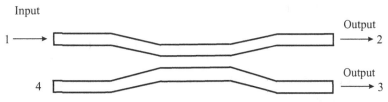

Fig. 13.6 Integrated optics directional coupler.

13.2.3 Couplers and splitters

Optical couplers are passive devices which either split optical signal into multiple paths, or combine several signals into one path. A prime characteristic of couplers is the number of input and output ports, which is typically expressed as an $N \times M$ configuration, where N represents the number of inputs and M represents the number of outputs.

A splitter is a modification of a coupler. As an example, a Y-branch power splitter is shown in Fig. 13.4. Typically, such Y-branch splits power evenly between the two output ports. By combining several Y-branches more outputs can be provided, as shown in Fig. 13.5.

Couplers can be constructed within the waveguide structure as shown in Fig. 13.6. Such a structure can be formed by ion exchange, ion implantation or chemical vapour deposition.

Couplers can also be constructed by twisting two or more fibres together and melting them in a flame. The process creates a fused coupler, a very popular device.

These techniques can be used to make 50–50% or 99–1% couplers. The numbers indicate splitting ratios. The length of the coupling region (fused region) as well as the twisting determines the splitting ratio. Such couplers are simple to construct but fabrication requires significant experience.

In the coupler shown in Fig. 13.6, the optical signal enters port 1. Some of its power exits in port 2 and some in port 3. In an ideal situation, no light reaches port 4 and also no power is lost. In practice, however, a few tenths of a dB are lost and the coupling to port 4 is about 30 dB relative to input power at port 1. The percentage of light coupled to different ports can be varied by changing the coupling length L.

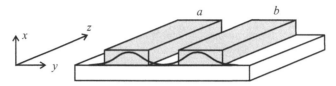

Fig. 13.7 Two coupled optical waveguides. Distribution of electric fields is also shown.

Some of the common applications of couplers and splitters are

- to monitor output of light locally (usually 99−1% couplers are used)
- to distribute an incoming signal to several locations simultaneously. For example, a four-part splitter (Fig. 13.5) allows a signal to drive four receivers.

Next, we provide a summary of the mathematical description of a passive coupler.

13.2.4 Mathematical theory of a passive coupler

Consider two waveguides 'a' and 'b' (Fig. 13.7), see [7]. For a single waveguide, say 'a', one can express the field as

$$\vec{E}(x,y,z) = \vec{E}^{(a)}(x,y)\,a(z)$$
$$\vec{H}(x,y,z) = \vec{H}^{(a)}(x,y)\,a(z)$$

and where

$$a(z) = a_0 e^{i\beta_a z}$$

where $\vec{E}^{(a)}(x,y)$ and $\vec{H}^{(a)}(x,y)$ are modal distributions in (x,y) plane. They are normalized as

$$\frac{1}{2}\mathrm{Re}\int\int \vec{E}^{(a)*}(x,y) \times \vec{H}^{(a)*} dxdy \cdot \hat{z} = 1$$

Also

$$\frac{da(z)}{dz} = i\beta_a a(z)$$

Total guided power

$$P = \frac{1}{2}\int\int \vec{E}^{(a)*}(x,y,z) \times \vec{H}^*(x,y,z) \cdot \hat{z} dxdy$$
$$= |a(z)|^2$$

For two parallel waveguides, fields in each one are written as

$$\vec{E}(x,y,z) = a(z)\vec{E}^{(a)}(x,y) + b(z)\vec{E}^{(b)}(x,y) \tag{13.3}$$
$$\vec{H}(x,y,z) = a(z)\vec{H}^{(a)}(x,y) + b(z)\vec{H}^{(b)}(x,y) \tag{13.4}$$

Amplitudes $a(z)$ and $b(z)$ satisfy the following (coupled-mode) equations:

$$\frac{da(z)}{dz} = i\beta_a a(z) + i\kappa_{ab} b(z) \tag{13.5}$$

$$\frac{db(z)}{dz} = i\kappa_{ba} a(z) + i\beta_b b(z) \tag{13.6}$$

where κ_{ab} and κ_{ba} are coupling coefficients. Guided power is

$$P = s_a |a(z)|^2 + s_b |b(z)|^2 + \text{Re}\left\{a(z) b^*(z) C_{ba} + b(z) a^*(z) C_{ab}\right\} \tag{13.7}$$

where

$$C_{pq} = \frac{1}{2} \int\int_{-\infty}^{+\infty} \vec{E}^{(q)}(x,y) \times \vec{H}^{(p)*}(x,y) \cdot \hat{z}\, dx dy \tag{13.8}$$

In the above, $s_a, s_b = +1$ is for propagation in the $+z$ direction and $s_a, s_b = -1$ is for propagation in the $-z$ direction.

Coupled mode equations can be written in a matrix form as

$$\frac{d}{dz}\begin{bmatrix} a(z) \\ b(z) \end{bmatrix} = i\overleftrightarrow{M} \begin{bmatrix} a(z) \\ b(z) \end{bmatrix} \tag{13.9}$$

where

$$\overleftrightarrow{M} = \begin{bmatrix} \beta_a & \kappa_{ab} \\ \kappa_{ba} & \beta_b \end{bmatrix} \tag{13.10}$$

The solution is assumed to be

$$\begin{bmatrix} a(z) \\ b(z) \end{bmatrix} = \begin{bmatrix} A \\ B \end{bmatrix} e^{i\beta z} \tag{13.11}$$

After substitution into Eq. (13.9), one finds

$$\left[\overleftrightarrow{M} - \beta \overleftrightarrow{1}\right] \begin{bmatrix} A \\ B \end{bmatrix} = 0 \tag{13.12}$$

where $\overleftrightarrow{1}$ is an identity matrix. In the full form

$$\begin{bmatrix} \beta_a - \beta & \kappa_{ab} \\ \kappa_{ba} & \beta_b - \beta \end{bmatrix} \begin{bmatrix} A \\ B \end{bmatrix} = 0 \tag{13.13}$$

For non-trivial solutions, a determinant must vanish:

$$\det = (\beta_a - \beta)(\beta_b - \beta) - \kappa_{ab} \cdot \kappa_{ba} = 0 \tag{13.14}$$

From the above, two eigenvalues are found as

$$\beta = \frac{1}{2}(\beta_a + \beta_b) \pm \gamma \equiv \begin{cases} \beta_+ \\ \beta_- \end{cases} \tag{13.15}$$

where

$$\gamma = \sqrt{\Delta^2 + \kappa_{ab} \cdot \kappa_{ba}}, \quad \Delta = \frac{1}{2}(\beta_b - \beta_a) \tag{13.16}$$

Eigenvectors are

$$\vec{V}_1 = \begin{bmatrix} \kappa_{ab} \\ \Delta + \gamma \end{bmatrix} \quad \text{or} \quad \begin{bmatrix} -\Delta + \gamma \\ \kappa_{ba} \end{bmatrix} \quad \text{for } \beta_+ \qquad (13.17)$$

and

$$\vec{V}_2 = \begin{bmatrix} \kappa_{ab} \\ \Delta - \gamma \end{bmatrix} \quad \text{or} \quad \begin{bmatrix} -\Delta - \gamma \\ \kappa_{ba} \end{bmatrix} \quad \text{for } \beta_- \qquad (13.18)$$

The general solution is therefore

$$\begin{bmatrix} a(z) \\ b(z) \end{bmatrix} = \overleftrightarrow{V} \begin{bmatrix} e^{i\beta_+ z} & 0 \\ 0 & e^{i\beta_- z} \end{bmatrix} \overleftrightarrow{V}^{-1} \begin{bmatrix} a(0) \\ b(0) \end{bmatrix} \qquad (13.19)$$

where matrix $\overleftrightarrow{V}$ is formed from eigenvectors as

$$\overleftrightarrow{V} = \begin{bmatrix} \vec{V}_1; \vec{V}_2 \end{bmatrix} \qquad (13.20)$$

After some algebra, one finds the final solution

$$\begin{bmatrix} a(z) \\ b(z) \end{bmatrix} = \overleftrightarrow{S}(z) \begin{bmatrix} a(0) \\ b(0) \end{bmatrix} \qquad (13.21)$$

with

$$\overleftrightarrow{S}(z) = \begin{bmatrix} \cos\gamma z - i\frac{\Delta}{\gamma}\sin\gamma z & i\frac{\kappa_{ba}}{\gamma}\sin\gamma z \\ i\frac{\kappa_{ba}}{\gamma}\sin\gamma z & \cos\gamma z + i\frac{\Delta}{\gamma}\sin\gamma z \end{bmatrix} \cdot e^{\frac{i}{2}(\beta_a + \beta_b)z} \qquad (13.22)$$

As a special case, consider a situation when at $z = 0$ the optical power is incident only in waveguide 1, $a(0) = 1$, $b(0) = 0$. One finds in this case

$$|b(z)|^2 = \left|\frac{\kappa_{ba}}{\gamma}\right|^2 \sin^2 \gamma z$$

At $\gamma z = \frac{\pi}{2}, 3\frac{\pi}{2}, \ldots, (2n+1)\frac{\pi}{2}$, the power transfer from guide 'a' to guide 'b' is maximum. Since

$$\left|\frac{\kappa_{ba}}{\gamma}\right|^2 = \frac{|\kappa_{ba}|^2}{\left[\frac{1}{2}(\beta_b - \beta_a)\right]^2 + |\kappa_{ba}|^2} < 1$$

for $\beta_a \neq \beta_b$ the power transfer between waveguides is never complete. For more discussion consult Chuang [7].

13.2.5 Optical isolators

An optical isolator is a passive device which allows the transmission of an optical signal in only one direction. At the same time the reflections in the opposite direction will be eliminated.

Two key parameters of an isolator are: the insertion loss which is the loss in the forward direction, and its isolation which is the loss in the reverse direction. Typical insertion loss is about 1 dB, whereas the isolation loss is 40–50 dB.

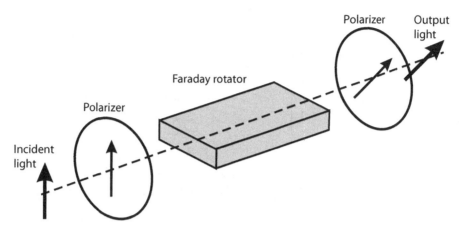

Fig. 13.8 Illustration of the principle of Faraday rotation.

An optical isolator based on the Faraday effect is shown in Fig. 13.8. The device consists of two linear polarizers and a 45° Faraday rotator. A light beam entering the device from the left passes through a linear polarizer which polarizes it vertically. Then it passes through a Faraday rotator which rotates it (here, say by 45°) with respect to vertical direction. After passing through a rotator, it travels through another linear polarizer aligned at 45° which allows the beam to pass through. Therefore, a vertically polarized beam can be transmitted through the device. However, all beams polarized at other angles will be blocked.

The angle of rotation α in a Faraday rotator is given by

$$\alpha = VBL \qquad (13.23)$$

where V is the Verdet constant, B is magnetic induction and L is the interaction length. For glass (crown) at a temperature of 18°C, the Verdet constant is $V = 2.68 \times 10^{-5}$ deg/Gauss· mm.

13.3 Applications of BPM to photonic devices

The development of new types of components for WDM applications requires accurate modelling before any fabrication attempts. Over the years many approaches and numerical methods have been developed. They can generally be divided into two main groups [8]: time-harmonic (monochromatic CW operation) and general time-dependent (pulse operation). The important role in numerical analysis of those components is played by BPM.

There are many specific applications of the BPM to modelling different aspects of photonic devices or circuits, such as passive waveguiding devices [9], channel-dropping filters [10], multimode waveguide devices [11], polarization splitters [12], multimode interference devices [13] and many more.

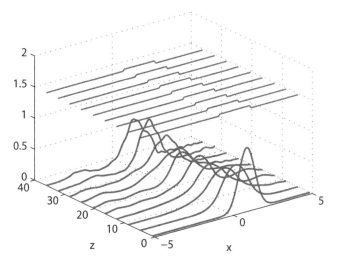

Fig. 13.9 Propagation of Gaussian pulse in a waveguide by BPM.

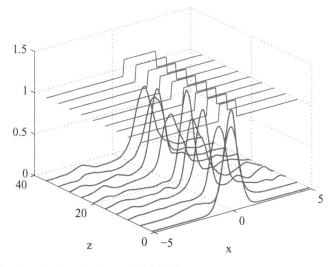

Fig. 13.10 Propagation of Gaussian pulse in a tapered waveguide by BPM.

For a recent summary on integrated optics consult the recent book by Hunsperger [14].

In this section we show the results of applications of the BPM developed earlier to the simplest two waveguiding structures which are the building blocks of WDM devices. Here, we consider a rib waveguide and a tapered waveguide. In Figs. 13.9 and 13.10 we show propagation of a Gaussian pulse and also profiles of the refractive indices. The MATLAB code used to generate those figures is provided in the Appendix, Listings 13A.1 and 13A.2.

To avoid reflections at the boundaries, transparent boundary conditions were implemented.

13.4 Projects

1. Function *y_junction.m* provided in the Appendix can be used to create a Y-junction. Use this function and other BPM routines to develop a program which will propagate a Gaussian pulse in the Y-junction. Analyse various configurations from the literature. Consult Refs. [15] and [16].
2. Implement the method of finding propagation constants using BPM. Consult Refs. [17] and [18].
3. Write MATLAB code to analyse propagation of Gaussian pulses in a coupled rib waveguide. Determine coupling length. Consult Refs. [19] and [20].
4. Analyse a fibre Bragg grating filter using BPM.

Appendix 13A: MATLAB listings

In Table 13.1 we provide a list of MATLAB files created for Chapter 13 and a short description of each function.

Listing 13A.1 Program bpm_wg.m. Describes propagation of a Gaussian pulse in a waveguide using BPM.

```
% File name: bpm_wg.m
% Driver function which propagates Gaussian pulse
% in a strip waveguide using BPM with transparent boundary conditions
clear all
x_0 = 1;                    % center of Gaussian pulse
L_x=10.0;                   % transversal dimension (along x-axis)
w_0=1.0;                    % width of input Gaussian pulse
lambda = 1.6;               % wavelength
k_0=2*pi/lambda;            % wavenumber
```

Table 13.1 List of MATLAB functions for Chapter 13.

Listing	Function name	Description
13A.1	*bpm_wg.m*	Gaussian pulse in a waveguide with TBC
13A.1.1	*wg_struct.m*	Constructs waveguiding structure used by $bpm_wg.m$
13A.1.2	*step.m*	Propagates BPM solution along one step
13A.2	*bpm_taper.m*	Gaussian pulse in a tapered waveguide with TBC
13A.2.1	*taper_struct.m*	Constructs tapered structure used by $bpm_taper.m$
13A.3	*plot_y.m*	Plots Y-junction
13A.3.1	*y_junction.m*	Function defines Y-junction

```
N_x=128;                                % points on x axis
Delta_x=L_x/(N_x-1);                    % x axis spacing
h=5*Delta_x;                            % propagation step
N_z=100;                                % number of propagation steps
plot_index=zeros(N_x,N_z);              % storage for plotting ref. index
plot_field=zeros(N_x,N_z);              % storage for plotting field
x=linspace(-0.5*L_x,0.5*L_x,N_x);       % coordinates along x-axis
x = x';
E=exp(-((x - x_0)/w_0).^2);             % initial Gaussian field
ref_index = wg_struct(x);               % structure is uniform along z-axis
n_eff = 1.40;                           % assumed value of prop. constant
%
z = 0; z_plot = zeros(N_z); for r=1:N_z
    z = z + h;
    z_plot(r) = z + h;
    plot_index(:,r)=wg_struct(x);
    plot_field(:,r)=abs(E).^2;
    E = step(Delta_x,k_0,h,ref_index,n_eff,E);
end;
%
for k = 1:N_z/10:N_z                    % choosing 2D plots every 10-th step
    plot(x,plot_index(:,k),'LineWidth',1.2)
    plot(x,plot_field(:,k),'LineWidth',1.5)
    hold on
end pause close all
%
for k = 1:N_z/10:N_z                    % choosing 3D plots every 10-th step
    y = z_plot(k)*ones(size(x));        % spread out along y-axis
    plot3(x,y,plot_index(:,k),'LineWidth',1.2)
    plot3(x,y,plot_field(:,k),'LineWidth',1.5)
    hold on
end grid on xlabel('x')
ylabel('z')                             % along propagation direction
pause close all
```

Listing 13A.1.1 Program wg_struct.m. Constructs a waveguiding structure.

```
function ref_index = wg_struct(x)
% Construction of the waveguide structure used by BPM
width = 2.0;                            % film width
n_c = 1.48;                             % refractive index of cover
n_s = 1.49;                             % refractive index of substrate
n_f = 1.52;                             % refractive index of film
ref_index = n_s*(x<0)+n_f*((x>=0)&(x<=width))+n_c*(x>width);
```

Listing 13A.1.2 Function *step.m* is slightly modified from the function of the same name used in Chapter 12.

```
% File name: step.m
function E_new = step(Delta_x,k_0,h,ref_index,n_eff,E_old)
% Function propagates BPM solution along one step

N_x = size(E_old,1);                   % determine size of the system

last_term = k_0^2*Delta_x^2*(ref_index.^2 - n_eff^2);

prefactor = 1/(2*k_0*n_eff*Delta_x^2);

main = ones(N_x,1) - 0.5*last_term; above = ones(N_x-1,1); below =
above;
P = prefactor*(diag(above,-1)-2*diag(main,0)+diag(below,1)); % matrix P
%
L_plus = eye(N_x) + 0.5i*h*P;          % step forward
L_minus = eye(N_x)-0.5i*h*P;           % step backward
%
%---- Implementation of boundary conditions
%
pref = 0.5i*h/(2*k_0*Delta_x^2);
k=1i/Delta_x*log(E_old(2)/E_old(1)); if real(k)<0
    k=1i*imag(k);
end;
left = pref*exp(1i*k*Delta_x);         % left correction for next step
L_plus(1) = L_plus(1)+left; L_minus(1) = L_minus(1)-left;
%
k=-1i/Delta_x*log(E_old(N_x)/E_old(N_x-1)); if real(k)<0
    k=1i*imag(k);
end;
right = pref*exp(1i*k*Delta_x);        % right correction for nest step
L_plus(N_x) = L_plus(N_x) + right; L_minus(N_x) = L_minus(N_x) -
right;
%
E_new = L_minus\L_plus*E_old;          % determine new solution
```

Listing 13A.3 Function plots a Y-junction.

```
% File name: plot_y.m
% Plots y-junction
clear all
L_x=10.0;                   % computational window along x-axis
N_x=100;                    % number of steps along x-axis
```

```
L_z = 20; N_z = 100;
h = L_z/N_z;                            % step size
plotting=zeros(N_x,N_z);                % storage for plotting
z=0;
%
for i=1:N_z
    z = z + h;
    z_plot(i) = z + h;
    x = linspace(-L_x,L_x,N_x);
    plotting(:,i)=y_junction(z,x,N_x,N_z);
end
%
for k = 1:10:N_z                % choosing 3D plots every 10-th step
    y = z_plot(k)*ones(size(x));    % spread out along y-axis
    plot3(x,y,plotting(:,k),'.-','LineWidth',1.0)
    hold on
end xlabel('x'); grid on pause close all
```

Listing 13A.3.1 Function defines a Y-junction.

```
function index = y_junction(z,x,N_x,N_z)
% Construction of the Y-junction structure used by BPM
% z - coordinate along propagation direction
% x - perpendicular coordinate
% Structure is described in only 2D, so we do not need cover
%
w = 2.0;                                % film width
n_s = 1.0;                              % refractive index of substrate
n_f = 1.52;                             % refractive index of film
a = 10; b = 20;
% Construction of Y-junction
slope = 0.1;
%
%index = zeros(N_x,N_z);
upper_1 = w/2 + slope*z; upper_2 = slope*z; lower_1 = - slope*z;
lower_2 = -w/2- slope*z;
%
n_1 = n_s*(x<-w/2)+n_f*((x>=-w/2)&(x<=w/2))+n_s*(x>w/2);

    n_2 = n_s*(x<=lower_2)+n_f*((x>=lower_2)&(x<=lower_1))+...
    n_s*((x>=lower_1)&(x<=upper_2))+n_f*((x>=upper_2)&(x<=upper_1))+...
    n_s*(x>upper_1);

index = ((0<z)&(z<=a))*n_1+((a<z)&(z<=b))*n_2; end
```

References

[1] G. Keiser. *Optical Fiber Communications. Third Edition*. McGraw-Hill, Boston, 2000.

[2] E. S. Koteles. Integrated planar waveguide demultiplexers for high density WDM applications. In R. T. Chen and L. S. Lome, eds., *Wavelength Division Multiplexing*, pages 3–32. SPIE, 1999.

[3] B. Chomycz. *Planning Fiber Optic Networks*. McGraw-Hill, New York, 2009.

[4] G. P. Agrawal. *Fiber-Optic Communication Systems. Second Edition*. Wiley, New York, 1997.

[5] K. O. Hill, Y. Fujii, D. C. Johnson, and B. S. Kawasaki. Photosensitivity in optical fiber waveguides: applications to reflection filter fabrication. *Appl. Phys. Lett.*, **32**:647–9, 1978.

[6] R. Kashyap. *Fiber Bragg Gratings*. Academic Press, San Diego, 1999.

[7] S.-L. Chuang. *Physics of Optoelectronic Devices*. Wiley, New York, 1995.

[8] R. Scarmozzino, A. Gopinath, R. Pregla, and S. Helfert. Numerical techniques for modeling guided-wave photonic devices. *IEEE J. Select. Topics Quantum Electron.*, **6**:150–62, 2000.

[9] L. Eldada, M. N. Ruberto, R. Scarmozzino, M. Levy, and R. M. Osgood. Laser-fabricated low-loss single-mode waveguiding devices in GaAs. *J. Lightwave Technol.*, **10**:1610, 1992.

[10] M. Levy, L. Eldada, R. Scarmozzino, R. M. Osgood, P. S. D. Lin, and F. Tong. Fabrication of narrow-band channel-dropping filters. *IEEE Photon. Technol. Lett.*, 4:1378, 1992.

[11] I. Ilic, R. Scarmozzino, R. M. Osgood, J. T. Yardley, K. W. Beeson, and M. J. McFarland. Modeling multimode-input star couplers in polymers. *J. Lightwave Technol.*, 12:996–1003, 1994.

[12] M. Hu, J. Z. Huang, R. Scarmozzino, M. Levy, and R. M. Osgood. Tunable Mach-Zehnder polarization splitter using heigh-tapered Y-branches. *IEEE Photon. Technol. Lett.*, **9**:773–5, 1997.

[13] D. S. Levy, Y. M. Li, R. Scarmozzino, and R. M. Osgood. A multimode interference-based variable power splitter in GaAs-AlGaAs. *IEEE Photon. Technol. Lett.*, **9**:1373–5, 1977.

[14] R. G. Hunsperger. *Integrated Optics. Theory and Technology. Sixth Edition*. Springer, New York, 2009.

[15] M. H. Hu, J. Z. Huang, R. Scarmozzino, M. Levy, and R. M. Osgood. A low-loss and compact waveguide Y-branch using refractive-index tapering. *IEEE Photon. Technol. Lett.*, **9**:203–5, 1997.

[16] T. A. Ramadan, R. Scarmozzino, and R. M. Osgood. Adiabatic couplers: design rules and optimization. *J. Lightwave Technol.*, **16**:277–83, 1998.

[17] G. Lifante. *Integrated Photonics. Fundamentals*. Wiley, Chichester, 2003.

[18] H. J. W. M. Hoekstra. On beam propagation methods for modelling in integrated optics. *Optical and Quantum Electronics*, **29**:157–71, 1997.

[19] C. Xu. *Finite-Difference Techniques for Simulation of Vectorial Wave Propagation in Photonic Guided-Wave Devices*. PhD thesis, University of Waterloo, 1994.

[20] H.-P. Nolting, J. Haes, and S. Helfert. Benchmark tests and modelling tasks. In G. Guekos, ed., *Photonic Devices for Telecommunications*, pages 67–109. Springer, Berlin, 1999.

14 Optical link

In the present chapter we will combine some of the methods developed earlier to create the simple point-to-point optical simulator which represents the simplest photonic system. It involves the transmitter, optical fibre and receiver. Some issues of how to quantify the quality of transmission in such a system will also be reviewed. Performance evaluation and tradeoff analysis are the central issues in the design of any communication system. Using only analytical methods, it is practically impossible to evaluate realistic communication systems. One is therefore left with computer-aided techniques.

In the last 10–15 years the design of photonic systems has moved from the back-of-the-envelope calculations to the use of sophisticated commercial simulators, see for example, products advertised by Optiwave, like OptiSystem [1], by RSoft Design Group the Optical Communication Design Suite [2] and by VPI Photonics line of products [3], to just name a few important players. They contain sophisticated physical models and allow for rapid assessment of new component technologies in the system under design.

Around 1995 the design of optical communication systems (operating over medium distances) would involve only a balance of power losses and pulse spreading. Later on, the demand on billion-dollar systems required complex analysis during the design process. This in turn created the need for sophisticated simulators.

Computer simulations can quickly provide answers to several important questions essential to every engineer designing optical communication system, like: what repeater spacing is needed for a given bit rate, or what is the required power generated by a transmitter?

There exists extensive literature on the subject of modelling and simulation of fibre optic communication systems. Early extensive work on the description and calculation of the transmission properties of optical fibres for communications was published by Geckeler [4]. There were several books published by van Etten and van der Plaats [5], Einarsson [6], Liu [7], Lachs [8], Keiser [9] and Palais [10], to list a few.

Literature published in journals is also extensive. Some of the relevant journal publications are: [11], [12], [13], [14], [15], [16], [17], [18], [19], [20]; that includes statistical design of long optical fibre systems is [21]. A general review intended for wide audience was written by Lowery [22].

14.1 Optical communication system

A generic optical link consisting of a transmitter, optical fibre and receiver was shown in the Introduction, Fig. 1.1. In that figure we showed a transmitter which employs a laser

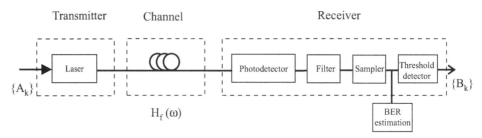

Fig. 14.1 More realistic optical link.

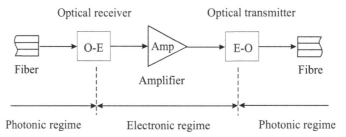

Fig. 14.2 Principle of regeneration of optical signals. O-E means optical to electronic; E-O means electronic to optical.

diode to generate light, a channel which is formed by an optical fibre and a receiver whose main element is a photodetector. Actual systems are more complicated.

In Fig. 14.1 we show a block diagram of a digital lightwave system (after Lima *et al.* [23]). Several detailed elements of the receiver are displayed. Some of them were already discussed in the earlier chapters. In the figure we also show notation for signals used to characterize operation of the particular elements.

Sequence A_k denotes the data values which are independent and identically distributed, $H_f(\omega)$ is the transfer function of the optical fibre and B_k is the output data sequence created by the receiver.

When light is transmitted over long distances, optical pulses degenerate due to various effects discussed previously. As a result, maximum transmission distance is limited. In order to increase that distance, repeaters are installed. Typical structure of the repeater is schematically shown in Fig. 14.2. An optical pulse is converted into electrical form, then amplified and regenerated and then converted back to light.

Such a process is costly and has also other limitations. Intensive research is conducted to have this process done entirely in the optical domain without conversion to an electronic regime.

Regenerators are devices which regenerate signals by removing noise and distortion after amplification. The process is normally possible only in digital systems. It serves 2R or 3R purposes. In general, the following terminology has been established:

- 1R (Re-amplification only) – signal is amplified only.
- 2R (Re-amplification and Re-shaping). Re-shaping corrects a pulse's shape.

Table 14.1 Typical values of losses for various components.

Wavelength [nm]	Losses [dB km^{-1}]
850	1.81
1300	0.35
1310	0.34
1380	0.40
1550	0.19

Table 14.2 Typical values of losses for various components.

Type of element	Losses [dB]
Connector	0.1–0.2
Splice losses	0.3

- 3R (Re-amplification, Re-shaping and Re-timing). Re-timing corrects the time drift in an optical pulse.

As a result of the regeneration process, clean digital pulses are created. The distance between repeaters varies, typically 50–70 km.

14.2 Design of optical link

There is a vast amount of literature on the design of optical communications links. Some of the useful papers and books are: Carnes *et al.* [24], Pepeljugoski and Kuchta [25], Iannone *et al.* [26], Binh [27], Mortazy *et al.* [28], Zheng *et al.* [29], Lowery *et al.* [30], Sabella *et al.* [31], de Melo *et al.* [32].

The design of the optical link involves several interrelated variables which describe operating characteristics of transmitter, fibre and receiver. The key practical requirements are: transmission distance, data rate and bit error rate.

The simplest approach to the design of optical link is based on the evaluation of power budget and rise time budget [33]. Power budget analysis is needed to ensure that the system will work over the proposed distance with given components. The rise times of the source, fibre and the detector will determine the bandwidth available for transmission.

In the tables we will summarize losses in typical optical components. Losses of a single-mode fibre from Corning Glass Works at various wavelengths are summarized in Table 14.1 [33]. Typical values of losses of other components are provided in Table 14.2.

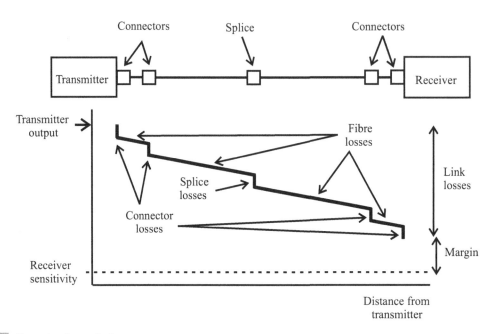

Fig. 14.3 Illustration of power budget.

14.2.1 Power budget analysis

In the analysis of power budget one determines the total allowed optical power loss between the transmitter and the receiver. Typical components which contribute to losses are: fibre, connector, splices, couplers and splitters. Typical values of losses in standard components are summarized in Table 14.2.

In Fig. 14.3 we show a typical power budget which illustrates loss of power over existing components as a function of distance from transmitter. In our example losses are created within connectors, splices and within an optical fibre.

In calculating power budget one considers passive and active components of the link. Passive loss is made up of fibre loss, connectors, splices, couplers and splitters, among others. An example of a system power budget is illustrated in Table 14.3.

In this example, one has an excess power margin of 29 dB − 28 dB = 1 dB (total 7 dB with the included system margin). As a general rule, the link loss margin should be greater than 3 dB to allow for link degradation over time, ageing of transmitters, etc. If during operation cables are accidentally cut, excess margin is needed to accommodate splices for restoration.

14.2.2 Rise time budget

The rise time budget is needed to determine whether the link will operate at the required bit rate. The bit rate is mostly limited by the dispersion. As explained earlier, dispersion

Table 14.3 Illustration of power budget determination.

Transmitter		
	Output power	−13 dBm
Receiver		
	Rx sensitivity	−42 dBm
	Margin	29 dBm
System loss		
	Fibre (3.5 × 5 km)	17.5 dB
	Connector (1 dB × 2)	2 dB
	Splicing (0.5 dB × 5)	2.5 dB
	System margin	6 dB
	Total	28 dB

is responsible for pulse broadening as it travels through the fibre. Those effects introduce rise time due to the material dispersion (τ_{mat}) and modal dispersion (τ_{mod}). Additionally, the response of the transmitter and the receiver must also be included. Therefore, total link rise time of the system is determined by the rms sum of all the rise times [33]:

$$\tau_{sys} = \sqrt{\sum_{i=1}^{N} \tau_i^2} = \sqrt{\tau_{tx}^2 + \tau_{mat}^2 + \tau_{mod}^2 + \tau_{rx}^2} \qquad (14.1)$$

For two main data coding schemes, namely RZ and NRZ, a general rule is that the system rise time should be determined as

$$\tau_{sys,RZ} \leq \frac{0.35}{B}, \quad \text{RZ} \qquad (14.2)$$

$$\tau_{sys,NRZ} \leq \frac{0.70}{B}, \quad \text{NRZ} \qquad (14.3)$$

where B is the bit rate. The rise times of the optical transmitter and receiver are known from the manufacturer. The rise time for multimode step index fibre is determined as [33] (subscript *im* stands for intermodal)

$$\tau_{im} = \frac{n_1 \Delta}{c} L \qquad (14.4)$$

For material dispersion

$$\begin{aligned}
\tau_{mat} &\simeq 85 L \Delta \lambda \, ps & (\lambda_0 \sim 850 \text{ nm}) \\
& 0.5 L \Delta \lambda \, ps & (\lambda_0 \sim 1300 \text{ nm}) \\
&\simeq 20 L \Delta \lambda \, ps & (\lambda_0 \sim 1500 \text{ nm})
\end{aligned} \qquad (14.5)$$

where L is in km and $\Delta\lambda$ in nm.

Table 14.4 Illustration of rise-time budget.

Component		Rise time (ns)	
System budget (NRZ)			1.75
Light source		1.0	
Fibre		0.23	
Photodetector			
	Transit time	0.5	
	Circuit	1.3	
	Total	1.4	
System rise time		1.75	1.75

For single-mode fibres

$$\begin{aligned}\tau_f &\simeq D \cdot L \cdot \Delta\lambda \\ &\simeq 2 \cdot L \cdot \Delta\lambda \quad (\lambda_0 \sim 1300 \text{ nm}, \lambda_z \sim 1300 \text{ nm}) \\ &\simeq 16 \cdot L \cdot \Delta\lambda \quad (\lambda_0 \sim 1550 \text{ nm}, \lambda_z \sim 1300 \text{ nm}) \\ &\simeq 2 \cdot L \cdot \Delta\lambda \quad (\lambda_0 \sim 1550 \text{ nm}, \lambda_z \sim 1550 \text{ nm})\end{aligned} \quad (14.6)$$

with $D \simeq 2$ ps/km·nm at 1300 nm and ~ 16 ps/km·nm at 1550 nm with fibres with $\lambda_z \sim 1300$ nm. Also $D \simeq 2$ ps/km·nm at 1550 nm for fibres with $\lambda_z \sim 1550$ nm. Here $\Delta\lambda$ is the source spectral width, L is fibre length, λ_0 is the operating wavelength and λ_z is the zero dispersion wavelength.

Typical rise time budget calculations are summarized in Table 14.4 [10].

14.3 Measures of link performance

There are various requirements which determine the performance of optical networks. One of them is network security, which will not be discussed here.

The simplest optical network is formed by a point-to-point optical link. Performance of such a link can be determined by the amount of information received at the acceptable level of errors [34]. Loss of information can be caused by a number of factors, of which some typical ones are (a) decrease in the amplitude of the received signal and (b) loss of timing.

Signal quality is determined by analysing the so-called eye diagram and also signal-to-noise ratio (SNR). The main parameter determining the quality of a digital link is the bit error rate (BER). The smaller the bit error rate, the better the link transmission quality. Typical error rates for optical fibre telecommunication systems range from 10^{-9} to 10^{-12} [35]. They depend on the SNR at the receiver. BER was discussed in Chapter 10 dealing with receivers. Here, we only discuss the eye diagram.

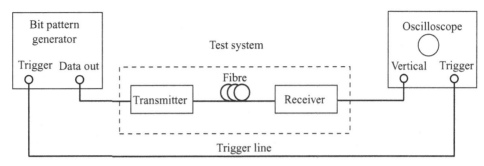

Fig. 14.4 Schematic of an experimental equipment setup for making eye diagram measurements.

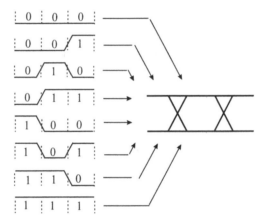

Fig. 14.5 Construction of an eye diagram for 3-bits.

14.3.1 Eye diagram

The eye diagram technique is a simple and convenient method of diagnosing problems with data systems. The eye diagram is generated in the time domain using an oscilloscope connected to the demodulated data, see Fig. 14.4. Pseudorandom data are generated and applied to the test system. After being transmitted through the test system, the data are applied to the vertical input of an oscilloscope. The oscilloscope is triggered at every symbol period (here using data from a separate line) or fixed multiple of symbol periods. As a result, the signal is retraced and superimposed on the same plot. Construction of an eye diagram for a 3-bit long sequence of data is shown in Fig. 14.5.

The result is an overlap of consecutive received symbols which form an eye pattern on the oscilloscope. The eye diagram provides a lot of information about the system's performance or about performance of its elements.

A created sequence of Gaussian pulses with random component for a predefined logical combination is shown in Fig. 14.6. For this sequence, the eye diagram is shown in Fig. 14.7. MATLAB code used in the generation of those results is shown in Appendix, Listing 14A.1.

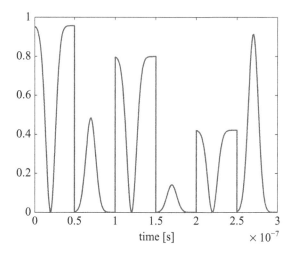

Fig. 14.6 Sequence of six Gaussian pulses with a random term. The logical data sequence is: 010101.

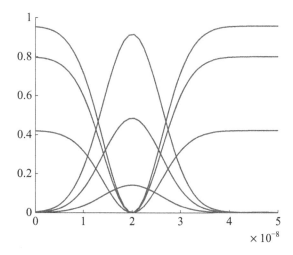

Fig. 14.7 Generated eye diagram for the sequence of Gaussian pulses shown in Fig. 14.6. Eye opening is located in the middle of all pulses.

14.4 Optical fibre as a linear system

In this section we consider the optical fibre modelled as a linear system and illustrate propagation of signals in such a system.

When an optical pulse propagates through optical fibre, it exhibits attenuation, delay and distortion. Delay can be handled through appropriate synchronization. Attenuation of a particular fibre determines the ultimate distance between transmitter and receiver. Distortion causes pulse broadening known as dispersion; it can produce situations when

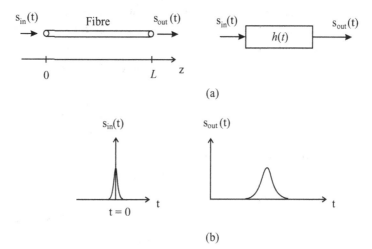

Fig. 14.8 Illustration of pulse transmission in an optical fibre represented as a linear system described by the impulse response function $h(t)$. (a) Schematic of optical fibre and its representation as a linear system, (b) pulse broadening during propagation.

one cannot distinguish between pulses representing different logical information due to their overlap.

Attenuation and dispersion determine practical limits on the information transmission performance of optical communication systems. Additional distortions, including noise, are produced in the receiver.

Propagation of optical signal of frequency ω through fibre of length L having attenuation coefficient α is schematically illustrated in Fig. 14.8. In a first approximation, the optical fibre is considered as a linear system with an impulse response $h(t)$. Its Fourier transform $H(\omega)$ known as a transfer function is

$$H(\omega) = \int_{-\infty}^{+\infty} h(t)\, e^{-i\omega t}\, dt \qquad (14.7)$$

The fibre is not an ideal system. Therefore during the passage, the properties of the transmitted pulse described by its time function $s(t)$ or its spectral function $S(\omega)$ are distorted, see Fig. 14.8. (Spectral function $S(\omega)$ and the time function $s(t)$ are related by the FT relation.)

In the frequency domain, the pulse propagation is described by the transfer function $H(\omega)$ as

$$H(\omega) = \frac{S_{out}(\omega)}{S_{in}(\omega)} = e^{-\alpha(\omega) - i\phi(\omega)} \qquad (14.8)$$

where $\alpha(\omega)$ is the system's attenuation and $\phi(\omega)$ its phase. $S_{out}(\omega)$ and $S_{in}(\omega)$ are FT of output and input time functions of a propagating pulse.

Some of the examples of spectral functions $H(\omega)$ which are used to model pulse spectra and filter functions will now be summarized. These functions are easy to program. All are dimensionless, normalized low-pass functions approaching unity at $\nu = 0$ and zero for

$\nu \longrightarrow \infty$ (we use the relation $\omega = 2\pi\nu$, where ω is angular frequency and ν is frequency). They are real functions and symmetric with respect to $\nu = 0$. The functions are [36]:

a) Gaussian pulse

$$H_G(\nu) = \exp\left(-\pi\nu^2\tau^2\right) \tag{14.9}$$

b) Rectangular pulse of duration τ

$$H_{RP}(\nu) = \frac{\sin(\pi\nu\tau)}{\pi\nu\tau} \tag{14.10}$$

c) Raised cosine spectrum

$$H_{RC}(\nu) = \cos^2\left(\frac{\pi\nu\tau}{2}\right) = \frac{1}{2}\left(1 + \cos\pi\nu\tau\right) \tag{14.11}$$

d) Spectrum of a trapezoidal pulse of duration τ and rise time t_r

$$H_{TP}(\nu) = \frac{\sin(\pi\nu\tau)}{\pi\nu\tau} \cdot \frac{1}{2}\left(1 + \cos\pi\nu t_r\right) \tag{14.12}$$

e) Spectrum with Nyquist slope

$$H_N(\nu) = \frac{1}{2}\left[1 - \text{sgn}(1 - 2\nu\tau)\left((1 - |2\nu\tau|)^N - 1\right)\right] \tag{14.13}$$

where N is the parameter which controls roll-off of this function.

f) Receiver filter that equalizes

$$H_E(\nu) = \frac{1 + \cos\pi\nu\tau}{2\exp\left(-\pi\nu^2\tau^2 Q\right)} \tag{14.14}$$

We will now summarize a description of the three main elements of an optical link: transmitter, fibre and receiver.

14.5 Model of optical link based on filter functions

In a typical practical situation the optical system transmits pulses. As they propagate from the transmitter to the receiver, they broaden and also their slopes get distorted. A presence of noise in all of a system's components also adds to the modification of propagating pulse.

We start with a simple analysis where the whole system is modelled by a single function. Here, we only consider a rectangular pulse.

14.5.1 Test analysis for a rectangular pulse

In this section we present the analysis of propagation of a rectangular pulse. Although not a very realistic situation, it serves as a test.

In Fig. 14.9 on the left we show the input rectangular pulse. It propagates through the system modelled as a rectangular filter (shown on right). The output after propagation is shown in Fig. 14.10. MATLAB code in shown in the Appendix, Listing 14A.2.

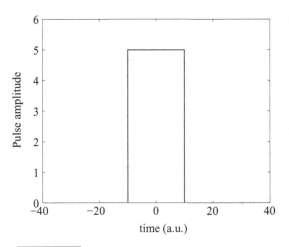
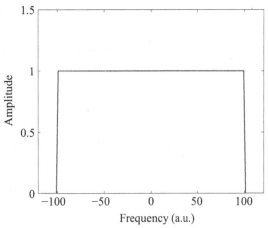

Fig. 14.9 Rectangular pulse (left) and rectangular filter.

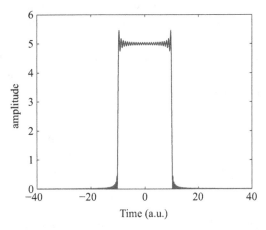

Fig. 14.10 Rectangular pulse after regeneration.

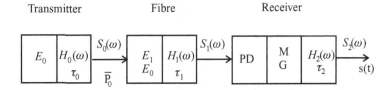

Fig. 14.11 Main elements of the transmission system.

Next, we build a more realistic model of the optical link, which is done in the following section. We summarize simple models for all the main components that constitute optical fibre system and then develop a program which allows for the analysis of pulse propagation in such a system.

Our approach is based on the work by Geckeler [36]. Main parts of transmission system are shown in Fig. 14.11. Here, E_0 is the energy of the optical pulse produced by the

Table 14.5 Parameters of typical generated pulses.

Parameter	Symbol	Value
Bit rate	R_0	50 Mbits/s^{-1}
Power	P_0	0.5 mW
Normalized pulse width	T_0	0.7

transmitter, τ_0 is the width of that pulse and $H_0(\omega)$ is the spectral function of the produced optical pulse. Detailed discussion of all the three elements now follows.

14.5.2 Transmitter

It is usually a semiconductor laser which can be directly modulated by changing bias current or by employing an external modulator. The main characteristics of light source are: wavelength, output power, modulation speed, spectral width, noise. For a directly modulated semiconductor laser, the performance is determined by frequency chirping (discussed in Chapter 7).

Here we assume a particular optical pulse as produced by a transmitter. Optical pulses were discussed in Chapter 6. Of particular importance are rectangular and Gaussian pulses.

Pulse produced by the transmitter in the time domain is described by the function $p_0(t)$. Its Fourier spectrum is

$$S_0(\omega) = \int_{-\infty}^{+\infty} p_0(t) \, e^{-j\omega t} \, dt$$

Fourier spectrum is related to spectral function $H_0(\omega)$ as

$$S_0(\omega) = p_{0,\max} \cdot \tau_0 \cdot H_0(\omega)$$

Parameters of the generated pulse are provided in Table 14.5. They refer to a system with a bit rate $R_0 = 50$ Mbits/s^{-1} and an LED transmitter operating at 1.3 μm wavelength with a maximum launched power of 0.5 mW.

14.5.3 Fibre

The frequency spectrum of propagating pulse after it leaves fibre is obtained by multiplying pulse spectral function and fibre spectrum.

Pulse propagation in the fibre is described by the product of two functions: D_1 which describes losses and its transfer function $H_1(\omega)$. Losses are represented by

$$D_1 = \frac{E_1}{E_0} = 10^{-\frac{\alpha L_1}{10 dB}} \quad (14.15)$$

The function $H_1(\omega)$ is a normalized filter function approximately expressed as a Gaussian low-pass filter:

$$H_1(\omega) = e^{-\frac{1}{4\pi}\pi \omega^2 \tau_1^2} \quad (14.16)$$

Table 14.6 Parameters of typical fibre.

Parameter	Symbol	Value
Losses	α	3 dB
Length	L_1	15 km
Mode coupling length	L_c	10 km
Bandwidth-length	B_L	500 MHz· km
Fibre parameter	τ_1	Eq. (14.17)

The parameter τ_1 can be determined from the measured bandwidth B_1 of a fibre. Fibre parameters are listed in Table 14.6.

The fibre parameter τ_1 is related to the bandwidth as

$$\tau_1 = \frac{1}{2B_1} \qquad (14.17)$$

Bandwidth is modelled after Geckeler [36] as

$$B_1 = B_L \left(\frac{1}{L_1} + \frac{1}{3L_c} \right) \qquad (14.18)$$

where L_c is the mode coupling length and B_L is the bandwidth-length product. The approximation (14.18) is valid for $0 < L_1 < 3L_c$. The parameters B_L and L_c are empirical parameters. Their values are summarized in Table 14.6.

14.5.4 Receiver

An optical receiver is described by the low-pass filter transfer function

$$H_2(\omega) = \frac{1}{2}\left[1 + \cos\left(\frac{1}{2}\omega\tau_2\right)\right] \quad \text{at } |\omega| \le \frac{2\pi}{\tau_2} \qquad (14.19)$$

with the parameter τ_2.

The signal spectrum at the receiver output is determined as [4]

$$S_2(\omega) = i_T \cdot T \cdot M \cdot R \cdot G \cdot H_0(\omega) \cdot H_1(\omega) \cdot H_2(\omega) \qquad (14.20)$$

where M is the multiplication factor of an avalanche photodiode (if used), G is its amplification factor and

$$i_T = P_{0,\text{max}} \cdot \frac{\tau_0}{T} \cdot D_1 \frac{\eta e}{h\nu} \qquad (14.21)$$

is the mean photocurrent of a pulse in a time slot of duration T.

The parameter of the receiver is provided in Table 14.7.

14.5.5 Implementation of link model

The above approach has been implemented following work by Geckeler [36] and [4]. MATLAB code is shown in Appendix, Listing 14A.3. Typical results are shown in Fig. 14.12.

Table 14.7 Parameter of a typical receiver.		
Parameter	Symbol	Value
Receiver parameter	τ_2	0.7

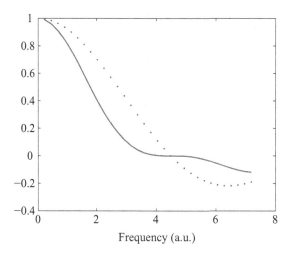

Fig. 14.12 Spectrum of rectangular pulse before (dotted line) and after propagation (solid line).

The signal spectrum generated by the transmitter is $sin(\pi \nu \tau_0)/(\pi \nu \tau_0)$, which represents rectangular pulse of duration τ_0. Input pulse is shown as a dotted line, the output signal after the receiver as a solid line. More details were provided by Geckeler in Refs. [36] and [4].

14.6 Problems

1. Analyse spectral functions summarized in Section 14.4. Plot the time dependence of each function. Evaluate Fourier transform and plot frequency spectra.
2. Analyse Gaussian pulse using the filter approach.
3. Analyse transmission of a Gaussian pulse by a single-mode fibre system. Determine the output pulse width as a function of the input pulse width.

14.7 Projects

1. Write MATLAB code to construct an eye diagram for the program *link.m*.
2. Add noise component to program *link.m*. Analyse the effect of noise on a system's performance.

Table 14.8 List of MATLAB functions for Chapter 14.

Listing	Function name	Description
14A.1	*bits_gen_eye.m*	Generates pattern of Gaussian pulses and eye diagram
14A.2	*rectangular.m*	Rectangular pulse before and after filtering
14A.3	*link.m*	Simulation of optical link (rectangular pulse)

Appendix 14A: MATLAB listings

In Table 14.8 we provide a list of MATLAB files created for Chapter 14 and a short description of each function.

Listing 14A.1 Function bits_gen_eye.m Function generates a 6-bits long pattern of Gaussian pulses with random terms and eye diagram.

```
% File name: bits_gen_eye.m
% Purpose:
% Generates 6-bits long pattern of Gaussian pulses with random terms
% Based on generated sequence, eye diagram is created
%
clear all
bits = [0 1 0 1 0 1];              % def. of bit's sequence
T_period = 50d-9;                  % pulse period [s]
t_zero = 20d-9;                    % center of incident pulse [s]
width = 6d-9;                      % width of the incident pulse [s]
% Generation of current pattern corresponding to bit pattern
I_p = 0;
N_div = 50;                        % number of divisions within each bit interval
t = linspace(0, T_period, N_div);  % the same time interval is
                                   % generated for each bit
g = exp(-0.5*((t_zero - t)/width).^2); % def. of a single Gaussian pulse
%
if bits(1)==0,    I_p_1 = rand(1)*(1- g); % generates first bit
    elseif bits(1)==1, I_p_1 = rand(1)*g;
end
temp = I_p_1;
number_of_bits = length(bits);
for k = 2:number_of_bits           % generation of remaining bits
    if bits(k)==1
        A = rand(1)*g;
        elseif bits(k)==0, A = rand(1)*(1- g);
    end
    A = [temp,A];
```

```
        temp = A;
end
temp_t = t;
for k = 2:number_of_bits
    t = linspace(0, T_period, N_div);
    t = [temp_t,(k-1)*T_period+t];
    temp_t = t;
end
%
h = plot(t,A,'LineWidth',1.5);
xlabel('time [s]','FontSize',14);   % size of x label
set(gca,'FontSize',14);    % size of tick marks on both axes
pause
close all
%
%=============== Eye diagram ============================
% put all bit plots on the first one
t_eye = linspace(0, T_period, N_div);
hold on
for m = 1:number_of_bits
    A_temp = A((1+(m-1)*N_div):(m*N_div));
    e = plot(t_eye,A_temp);
    set(e,'LineWidth',1.5);  % thickness of plotting lines
    set(gca,'FontSize',14);    % size of tick marks on both axes
end
pause
close all
```

Listing 14A.2 Function rectangular.m. Function which shows a rectangular pulse before and after filtering.

```
% File name: rectangular.m
% Analysis of rectangular pulse after filtering
%
clear all
hwidth = 10;                        % half-width of a pulse
time_step = 0.001;                  % time step
range = 40;
t = -range:time_step:range;         % time range
%------ Creation of rectangular pulse -------
y = (5/2)*(sign(t+hwidth)-sign(t-hwidth));
plot(t,y,'LineWidth',1.5);
xlabel('time (a.u','FontSize',14);
ylabel('Pulse amplitude','FontSize',14);
set(gca,'FontSize',14);              % size of tick marks on both axes
axis([-range range 0 6]);
```

```
pause
% %----- Fourier transform of a rectangular pulse -------
y_shift = fftshift(fft(y));
N=length(y_shift);
n=-(N-1)/2:(N-1)/2;
%------ Definitions of different filters ------------------
filter = (1/2)*(sign(n+100)-sign(n-100));
%filter = exp(-pi*n*0.0001);
%filter = 0.1*cos(pi*n*0.005).^2;
%filter = (1 + cos(0.05*n))/2;
plot(n,filter,'LineWidth',1.5);
axis([-3*range 3*range 0 1.5]);
xlabel('Frequency (a.u.)','FontSize',14);
ylabel('Amplitude','FontSize',14);
set(gca,'FontSize',14);              % size of tick marks on both axes
pause
%--------- Regeneration of signal using filtered output ----------
y_reg=filter.*y_shift;               % Filter original frequency spectrum
y_reg_inv=ifft(y_reg);               % Inverse Fourier transform
plot (t,abs(y_reg_inv),'LineWidth',1.3);
axis([-range range 0 6]);
xlabel('Time(a.u.)','FontSize',14);
ylabel('amplitude','FontSize',14);
set(gca,'FontSize',14);              % size of tick marks on both axes
pause
close all
```

Listing 14A.3 Function link.m. Function models propagation of a rectangular pulse in an optical link.

```
% File name: link.m
% Optical link simulations
% Follow code on p.133 of S. Geckeler (1983) paper
% No model of a transmitter
% No noise
%
clear all
%----------- Pulse produced by transmitter ----------------------
R_0 = 50d5;              % Bit rate of binary digital signal, 50 Mbit/s
P_0 = 5d-4;              % Max power of optical pulse, 0.5 mW
T_0 = 0.7;               % Normalized pulse duration = tau_0/T
%------------ Fibre parameters -------------------------
L_1 = 15;                % Fibre length, 15 km
A_L = 3.0;               % Attenuation coefficient, 3 dB/km
D_1 = 10^(-A_L*L_1/10);  % Power ratio
P_1 = P_0*T_0*D_1/2;     % Mean optical power at the end of fibre
```

```
B_L = 5d8;                      % Bandwidth-length product, 500 MHz km
L_C = 10;                       % Coupling length of a multimode fibre, 10 km
B_1 = B_L/L_1 + B_L/(3*L_C);    % Fibre bandwidth
T_1 = R_0/(2*B_1);
%-------------- Receiver parameters -------------------------
T_2 = 0.7;
%============== System evaluation ========================================
% Output signal spectrum
D_F = pi/(20*T_2);
N = 32;
f = zeros(1,N);
f_A = zeros(1,N);
f_B = zeros(1,N);
f_C = zeros(1,N);
H_F = zeros(1,N);
H_F(1) = 1.0;
for j = 1:N
    f(j) = j*D_F;
    f_A(j) = sin(f(j)*T_0)/(f(j)*T_0);    % transmitter
    f_B(j) = exp(-f(j)^2*T_1*T_1/pi);     % fibre
    f_C(j) = (1/2)*(1 + cos(f(j)*T_2));   % receiver
    H_F(j) = f_A(j)*f_B(j)*f_C(j);
end
%
plot(f,f_A,'.',f,H_F,'LineWidth',1.5)    % Plot signal after propagation
xlabel('Frequency (a.u.)','FontSize',14);
set(gca,'FontSize',14);         % size of tick marks on both axes
pause
s = ifft(H_F);
s = s/max(s);
ff = linspace(0,2/3,N);
%
plot(fftshift(abs(s)),'LineWidth',1.5);
xlabel('time','FontSize',14);
ylabel('time output function','FontSize',14);
set(gca,'FontSize',14);         % size of tick marks on both axes
grid
pause
close all
```

References

[1] Optiwave, www.optiwave.com/, 3 July 2012.

[2] RSoft, www.rsoftdesign.com/, 3 July 2012.

[3] VPI photonics, www.vpiphotonics.com/, 3 July 2012.

[4] S. Geckeler. *Optical Fiber Transmission System*. Artech House, Inc., Norwood, MA, 1987.

[5] W. van Etten and J. van der Plaats. *Fundamentals of Optical Fiber Communications*. Prentice Hall, New York, 1991.

[6] G. Einarsson. *Principles of Lightwave Communications*. Wiley, Chichester, 1996.

[7] M. M.-K. Liu. *Principles and Applications of Optical Communications*. Irwin, Chicago, 1996.

[8] G. Lachs. *Fiber Optic Communications. Systems, Analysis, and Enhancements*. McGraw-Hill, New York, 1998.

[9] G. Keiser. *Optical Fiber Communications. Third Edition*. McGraw-Hill, Boston, 2000.

[10] J. C. Palais. *Fiber Optic Communications. Fifth Edition*. Prentice Hall, Upper Saddle River, NJ, 2005.

[11] D. G. Duff. Computer-aided design of digital lightwave systems. *IEEE J. Select. Areas Commun.*, **2**:171–85, 1984.

[12] P. J. Corvini and T. L. Koch. Computer simulation of high-bit-rate optical fiber transmission using single-frequency lasers. *J. Lightwave Technol.*, **5**:1591–5, 1987.

[13] G. P. Shen, and T.-M. Agrawal. Computer simulation and noise analysis of the system performance at 1.5 μm single-frequency semiconductor lasers. *J. Lightwave Technol.*, **5**:653–9, 1987.

[14] A. F. Elrefaie, J. K. Townsend, M. B. Romeiser, and K. S. Shanmugan. Computer simulation of digital lightwave links. *IEEE J. Select. Areas Commun.*, **6**:94–104, 1988.

[15] J. C. Cartledge and G. S. Burley. The effect of laser chirping on lightwave system performance. *J. Lightwave Technol.*, **7**:568–73, 1989.

[16] K. Hinton and T. Stephens. Modeling high-speed optical transmission systems. *IEEE J. Select. Areas Commun.*, **11**:380–92, 1993.

[17] K. B. Letaief. Performance analysis of digital lightwave systems using efficient computer simulation techniques. *IEEE Trans. on Communications*, **43**:240–51, 1995.

[18] A. Lowery, O. Lenzmann, I. Koltchanov, R. Moosburger, R. Freund, A. Richter, S. Georgi, D. Breuer, and H. Hamster. Multiple signal representation simulation of photonic devices, systems, and networks. *IEEE J. Select. Topics Quantum Electron.*, **6**:282–96, 2000.

[19] S. H. S. Al-Bazzaz. Simulation of single mode fiber optics and optical communication components using VC++. *IJCSNS International Journal of Computer Science and Network Security*, **8**:300–8, 2008.

[20] L. N. Binh. MATLAB simulink simulation platform for photonic transmission systems. *I.J. Communications, Network and System Sciences*, **2**:97–117, 2009.

[21] T. J. Batten, A. J. Gibbs, and G. Nicholson. Statistical design of long optical fiber systems. *J. Lightwave Technol.*, **7**:209–17, 1989.

[22] A. J. Lowery. Computer-aided photonics design. *Spectrum*, **4**:26–31, 1997.

[23] R. A. A. Lima, M. C. R. Carvalho, and L. F. M. Conrado. On the simulation of digital optical links with EDFAs: an accurate method for estimating BER through Gaussian approximation. *IEEE J. Select. Topics Quantum Electron.*, **3**:1037–44, 1997.

[24] H. Carnes, R. Kearns, and E. Basch. Digital optical system design. In E. E. B. Basch, ed., *Optical-Fiber Transmission*, pages 461–86. Howard W. Sams and Co., Indianapolis, 1987.

[25] P. K. Pepeljugoski and D. M. Kuchta. Design of optical communications data links. *IBM J. Res. and Dev.*, **47**:223–37, 2003.

[26] E. Iannone, F. Matera, A. Mocozzi, and M. Settembre. *Nonlinear Optical Communication Networks*. Wiley, New York, 1998.

[27] L. N. Binh. *Optical Fiber Communications Systems*. CRC Press, Boca Raton, 2010.

[28] E. Mortazy and M. K. Moravvej-Farshi. A new model for optical communication systems. *Optical Fiber Technology*, **11**:69–80, 2005.

[29] H. Zheng, S. Xie, Z. Zhou, and B. Zhou. Simulation of optical fiber transmission systems using SPW and an HP workstation. *SIMULATION*, **71**:312–15, 1998.

[30] A. J. Lowery and P. C. R. Gurney. Two simulators for photonic computer-aided design. *Applied Optics*, **37**:6066–77, 1998.

[31] R. Sabella and P. Lugli. *High Speed Optical Communications*. Kluwer Academic Publishers, Dordrecht, 1999.

[32] C. F. de Melo Jr, C. A. Lima, L. D. S. Alcantara, R. O. dos Santos, and J. C. W. A. Costa. A Simulink toolbox for simulation and analysis of optical fiber links. In J. Javier Sanchez-Mondragon, ed., *Sixth International Conference on Education and Training in Optics and Photonics*, volume **3831**, pages 240–51. SPIE, 2000.

[33] G. Ghatak and K. Thyagarajan. *Introduction to Fiber Optics*. Cambridge University Press, Cambridge, 1998.

[34] L. D. Green. *Fiber Optic Communications*. CRC Press, Boca Raton, 1993.

[35] G. Keiser. *Optical Fiber Communications. Fourth Edition*. McGraw-Hill, Boston, 2011.

[36] S. Geckeler. Modelling of fiber-optic transmission systems on a desk-top computer. *Siemens Res. and Dev. Reports*, **12**:127–34, 1983.

15 Optical solitons

In this chapter we concentrate on optical solitons and their propagation in optical fibre. They are pulses of a special shape and owe their existence due to the presence of dispersion and nonlinearity in optical fibre. They are able to propagate ultra-long distances while maintaining their shape.

Here, we provide some basic knowledge for understanding the underlying physical principles of soliton creation and propagation. We generalize the theory of linear pulses developed in Chapter 5 and include the nonlinear part of polarization. We will then derive a nonlinear equation which describes propagation of solitons in optical fibre.

Soliton propagation in an optical fibre was first demonstrated at Bell Laboratories by L. F. Mollenauer, R. S. Stolen and J. P. Gordon. We refer to the book by Mollenauer and Gordon [1] for the description of basic principles and a brief history of solitons.

Some other relevant books on optical solitons and also on applications in optical communications are [2], [3], [4], [5], [6].

15.1 Nonlinear optical susceptibility

Solitons exist due to nonlinearity and dispersion. Dispersion was discussed earlier in the book. Here, we will concentrate on nonlinear effects. Optical responses including nonlinear effects are described as [7]

$$P(t) = \epsilon_0 \left\{ \chi^{(1)} E(t) + \chi^{(2)} E^2(t) + \chi^{(3)} E^3(t) + \cdots \right\} \quad (15.1)$$
$$\equiv P^{(1)}(t) + P^{(2)}(t) + P^{(3)}(t) + \cdots$$

where we have expressed polarization $P(t)$ as a power series in the field strength $E(t)$. The quantities $\chi^{(1)}$, $\chi^{(2)}$, $\chi^{(3)}$ are known as susceptibilities; $\chi^{(1)}$ is a linear susceptibility and $\chi^{(2)}$, $\chi^{(3)}$ are known as the second order- and third-order nonlinear susceptibilities.

For a typical solid-state system $\chi^{(1)}$ is of the order of unity whereas $\chi^{(2)}$ is of the order of $1/E_{at}$ and $\chi^{(3)}$ of the order of $1/E_{at}^2$, where $E_{at} = e/(4\pi\epsilon_0 a_0^2)$ is the characteristic atomic electric field strength and $a_0 = 4\pi\epsilon_0 \hbar^2/me^2$ is the Bohr radius of the hydrogen atom. Explicitly [7]

$$\chi^{(2)} \simeq 1.94 \times 10^{-12} m/V$$
$$\chi^{(3)} \simeq 3.78 \times 10^{-24} m^2/V^2$$

Formal expression for the third-order susceptibility is [7]

$$P_i(\omega_0 + \omega_n + \omega_m) = \epsilon_0 D \sum_{jkl} \chi^{(3)}_{ijkl}(\omega_0 + \omega_n + \omega_m, \omega_0, \omega_n, \omega_m)$$
$$\times E_j(\omega_0) E_k(\omega_n) E_l(\omega_m)$$

where i, j, k, l refer to the Cartesian components of the fields and the degeneracy factor D represents the number of distinct permutations of the frequencies $\omega_0, \omega_n, \omega_m$.

$\chi^{(j)}$ ($j = 1, 2, \ldots$) is the j-th order susceptibility. The linear susceptibility $\chi^{(1)}$ contributes to the linear refractive index $\bar{n}_0$ (real and imaginary parts; the imaginary part being responsible for attenuation). The second-order susceptibility $\chi^{(2)}$ is responsible for the second harmonic generation. For SiO_2 the second-order nonlinear effect is negligible since SiO_2 has the inversion symmetry. Therefore optical fibres normally do not show the second-order nonlinear effects.

The third-order susceptibility $\chi^{(3)}$ is responsible for the lowest order nonlinear effects in optical fibres. Generally, it manifests itself as the change in the refractive index with optical power or as a scattering phenomenon. It is linked with the optical Kerr effect, four-wave mixing, third-harmonic generation, stimulated Raman scattering, etc. An excellent discussion of third-order optical susceptibilities has been reported by Hellwarth [8].

Assuming linear polarization of propagating light and neglecting tensorial character of $\chi^{(3)}_{ijkl}$, one finds the following relation for the nonlinear polarization:

$$P^{NL}(\omega) = 3\epsilon_0 \chi^{(3)}(\omega = \omega + \omega - \omega) |E(\omega)|^2 E(\omega)$$

Total polarization, which consists of linear and nonlinear parts, is written as

$$P(\omega) = \epsilon_0 \chi^{(1)} E(\omega) + 3\epsilon_0 \chi^{(3)} |E(\omega)|^2 E(\omega)$$
$$= \epsilon_0 \chi_{eff} E(\omega)$$

The effective susceptibility is field dependent as

$$\chi_{eff} = \chi^{(1)} + 3\chi^{(3)} |E(\omega)|^2$$

and it is linked to refractive index as

$$\bar{n} = 1 + \chi^{(3)} \equiv \bar{n}_0 + \bar{n}_2 I \qquad (15.2)$$

Here I denotes the time-averaged intensity of the optical field. We start with the discussion of the main features of nonlinear effects.

15.2 Main nonlinear effects

15.2.1 Kerr effect

It was discovered by J. Kerr in 1875. He found that a transparent liquid becomes doubly refracting (birefringent) when placed in a strong electric field. Generally, Kerr effect

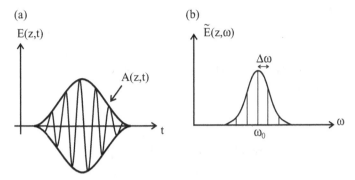

Fig. 15.1 Illustration of the propagating modulated pulse (a) and its spectrum (b).

describes situations where refractive index depends on electric field as

$$\bar{n}\left(\omega, |E|^2\right) = \bar{n}_0(\omega) + \bar{n}_2(\omega) |E|^2$$

Here, $\bar{n}_2$ is known as Kerr coefficient and it is related to susceptibility as

$$\bar{n}_2(\omega) = \frac{3}{4\pi} \chi^{(3)}_{xxx}$$

for a linearly polarized wave in the x direction. For silica its value is approximately 1.3×10^{-22} m²/V². Kerr effect originates from the non-harmonic motion of electrons bound in molecules. Consequently, it is fast effect, the response time of the order of 10^{-15} s.

15.2.2 Stimulated Raman scattering

Scattering phenomena are responsible for Raman and Brillouin effects. During those scatterings, the energy of the optical field is transferred to local phonons: in Raman scattering optical phonons are generated whereas in the Brillouin scattering the acoustic phonons.

15.3 Derivation of the nonlinear Schrödinger equation

Solitons in optical fibres are described by the so-called nonlinear Schrödinger (NSE) equation, which will be now derived. In the derivation we use the concept of the Fourier spectrum of the propagating pulse, see Fig. 15.1.

A medium where solitons propagate exhibits Kerr nonlinearity. In such a medium, refractive index depends on intensity of electric field $I(t)$ given by Eq. (15.2), again reproduced here:

$$\bar{n}(t) = \bar{n}_0 + \bar{n}_2 \cdot I(t) \tag{15.3}$$

where [7]

$$I(t) = 2\bar{n}_0 \varepsilon_0 c |A(z,t)|^2 \tag{15.4}$$

Here $A(z,t)$ is the slowly varying envelope connected to the optical pulse described by the optical pulse $E(z,t)$ as, see Fig. 15.1:

$$E(z,t) = A(z,t)\, e^{i(\omega_0 t - \beta_0 z)} \tag{15.5}$$

Fourier transform of the optical field is [3], [9]

$$E(z,t) = \int_{-\infty}^{+\infty} d\omega \, \tilde{E}(z,\omega)\, e^{i(\omega t - \beta z)} \tag{15.6}$$

where $\tilde{E}(z,\omega)$ is the Fourier spectrum of the pulse, β propagation constant and ω_0 the frequency at which the pulse spectrum is centered (also known as carrier frequency), see Fig. 15.1.

For quasi-monochromatic pulses with $\Delta\omega \equiv \omega - \omega_0 \ll \omega_0$, it is useful to expand propagation constant $\beta(\omega)$ in a Taylor series:

$$\beta(\omega) = \beta_0 + \beta_1 \cdot (\omega - \omega_0) + \frac{1}{2}\beta_2 \cdot (\omega - \omega_0)^2 + \Delta\beta_{NL} \tag{15.7}$$

where we have neglected higher order derivatives. Here $\Delta\beta_{NL} = \bar{n}_2 k_0 I$ is the nonlinear contribution to the propagation constant.

Substitute expansion (15.7) into Eq. (15.6):

$$\begin{aligned}
E(z,t) &= e^{-i\beta_0 z} \int_{-\infty}^{+\infty} d(\Delta\omega)\, \tilde{E}(z,\omega)\, \exp\left(i\omega t - i\beta_1 z \Delta\omega - \frac{1}{2}i\beta_2 z \Delta\omega^2 - i \cdot z \cdot \Delta\beta_{NL}\right) \\
&= e^{i(\omega_0 t - \beta_0 z)} \int_{-\infty}^{+\infty} d(\Delta\omega)\, \tilde{E}(z,\omega_0 + \Delta\omega) \\
&\quad \times \exp\left(it\Delta\omega - i\beta_1 z \Delta\omega - \frac{1}{2}i\beta_2 z \Delta\omega^2 - i \cdot z \cdot \Delta\beta_{NL}\right) \\
&\equiv e^{i(\omega_0 t - \beta_0 z)}\, A(z,t)
\end{aligned}$$

where we have introduced

$$\begin{aligned}
A(z,t) &= \int_{-\infty}^{+\infty} d(\Delta\omega)\, \tilde{E}(z,\omega_0 + \Delta\omega) \\
&\quad \times \exp\left(it\Delta\omega - i\beta_1 z \Delta\omega - \frac{1}{2}i\beta_2 z \Delta\omega^2 - i \cdot z \cdot \Delta\beta_{NL}\right) \\
&\equiv \int_{-\infty}^{+\infty} d(\Delta\omega)\, \tilde{E}(z,\omega_0 + \Delta\omega)\, e^{ig(z,t)} \tag{15.8}
\end{aligned}$$

Our next step is to obtain a differential equation describing evolution of the amplitude $A(z,t)$ from Eq. (15.8) which is in the integral form. To do this, one needs to take partial derivatives of Eq. (15.8). One obtains

$$\frac{\partial A(z,t)}{\partial t} = \int_{-\infty}^{+\infty} d(\Delta\omega)\, \widetilde{E}(z, \omega_0 + \Delta\omega)\, i\Delta\omega\, e^{ig(z,t)}$$

$$\frac{\partial^2 A(z,t)}{\partial t^2} = \int_{-\infty}^{+\infty} d(\Delta\omega)\, \widetilde{E}(z, \omega_0 + \Delta\omega)\, (i\Delta\omega)^2\, e^{ig(z,t)}$$

$$\frac{\partial A(z,t)}{\partial z} = \int_{-\infty}^{+\infty} d(\Delta\omega)\, \widetilde{E}(z, \omega_0 + \Delta\omega) \left(-i\beta_1 \Delta\omega - \frac{1}{2} i\beta_2 \Delta\omega^2 - i\cdot \Delta\beta_{NL} \right) e^{ig(z,t)}$$

When evaluating time derivatives, we have assumed in the above that $I(t)$ does not depend on time. Addition of the above combination of derivatives produces

$$\frac{\partial A(z,t)}{\partial z} + \beta_1 \frac{\partial A(z,t)}{\partial t} - i\frac{1}{2}\beta_2 \frac{\partial^2 A(z,t)}{\partial t^2}$$
$$= \int_{-\infty}^{+\infty} d(\Delta\omega)\, \widetilde{E}(z, \omega_0 + \Delta\omega)$$
$$\times \left[\left(-i\beta_1 \Delta\omega - \frac{1}{2} i\beta_2 \Delta\omega^2 - i\cdot \Delta\beta_{NL} \right) + \beta_1 i\Delta\omega - i\frac{1}{2}\beta_2 (i\Delta\omega)^2 \right] \times e^{ig(z,t)}$$
(15.9)

The term in the bracket is, $[\cdots] = -i \cdot \Delta\beta_{NL} = -i\overline{n}_2 k_0 I$. Eq. (15.9) thus gives

$$\frac{\partial A(z,t)}{\partial z} + \beta_1 \frac{\partial A(z,t)}{\partial t} - i\frac{1}{2}\beta_2 \frac{\partial^2 A(z,t)}{\partial t^2}$$
$$= \int_{-\infty}^{+\infty} d(\Delta\omega)\, \widetilde{E}(z, \omega_0 + \Delta\omega)\, (-i\overline{n}_2 k_0 I)\, e^{ig(z,t)}$$
$$= -i\overline{n}_2 k_0 I \int_{-\infty}^{+\infty} d(\Delta\omega) \widetilde{E}(z, \omega) e^{ig(z,t)}$$
$$\equiv -i\overline{n}_2 k_0 I\, A(z,t)$$

when using (15.8). The final equation describing solitons is therefore [3], [10]

$$\frac{\partial A(z,t)}{\partial z} + \beta_1 \frac{\partial A(z,t)}{\partial t} + i\frac{1}{2}\beta_2 \frac{\partial^2 A(z,t)}{\partial t^2} = i\gamma\, |A(z,t)|^2\, A(z,t) - \frac{\alpha}{2} A(z,t) \qquad (15.10)$$

where we have defined nonlinear coefficient γ (after [3]) as

$$\gamma = \frac{2\pi \overline{n}_2}{\lambda A_{eff}} \qquad (15.11)$$

Here A_{eff} is the effective core area.

Our interests here lie in the pulse evolution during propagation and not in the time of pulse arrival. We can therefore simplify the above equation by transforming it to a coordinate system which moves with group v_g. In this moving frame, new time T and new coordinate Z are

$$Z = z \qquad (15.12)$$
$$T = t - \beta_1 z$$

To obtain the transformed equation, we must evaluate derivatives with respect to new variables as follows:

$$\frac{\partial A}{\partial t} = \frac{\partial A}{\partial T}\frac{\partial T}{\partial t} + \frac{\partial A}{\partial |Z}\frac{\partial Z}{\partial t} = \frac{\partial A}{\partial T}$$

since $\frac{\partial T}{\partial t} = 1$ and $\frac{\partial Z}{\partial t} = 0$. From the above one finds

$$\frac{\partial^2 A}{\partial t^2} = \frac{\partial^2 A}{\partial T^2}$$

Using the above results, one has

$$\frac{\partial A}{\partial z} = \frac{\partial A}{\partial T}\frac{\partial T}{\partial z} + \frac{\partial A}{\partial |Z}\frac{\partial Z}{\partial z} = -\beta_1 \frac{\partial A}{\partial T} + \frac{\partial A}{\partial |Z}$$

The last result is used in Eq. (15.10) to replace $\frac{\partial A}{\partial t}$. The transformed equation is

$$\frac{\partial A}{\partial z} + i\frac{1}{2}\beta_2 \frac{\partial^2 A}{\partial T^2} - i\gamma |A|^2 A + \frac{1}{2}\alpha A = 0 \qquad (15.13)$$

where in the final step we replaced Z by z. It is known as a nonlinear Schrödinger equation (NSE).

To further analyse NSE, we will introduce two characteristic lengths describing dispersion (L_D) and nonlinearity (L_{NL}). Those are defined as

$$L_D = \frac{T_0^2}{|\beta_2|} = \frac{T_0^2\, 2\pi c}{|D|\lambda^2} \qquad (15.14)$$

and

$$L_{NL} = \frac{1}{\gamma P_0} \qquad (15.15)$$

where P_0 is the peak power of the slowly varying envelope $A(z, T)$ and T_0 is a temporal characteristic value of the initial pulse, which is often defined as full width half maximum (the pulse 3 dB width). Those two lengths characterize how far a pulse must propagate to show the respective effect. Physically, L_D is the propagation length at which a Gaussian pulse broadens by a factor of $\sqrt{(2)}$ due to group velocity dispersion (GVD).

GVD dominates pulse propagation in fibres whose length L is $L \ll L_{NL}$ and $L \geq L_D$. In such situation. the nonlinearity in NLSE can be ignored and the equation can be solved analytically. Nonlinear effects dominate in fibre where $L \ll L_D$ and $L \geq L_{NL}$. In this limit the dispersion term can be ignored.

Typical values of parameters used in the simulations are summarized in Table 15.1.

For numerical analysis normalize variables

$$U = \frac{1}{\sqrt{P_0}} A \quad \text{and} \quad \tau = \frac{T}{T_0} \qquad (15.16)$$

The width parameter T_0 is related to the full-width at half-maximum (FWHM) intensity of the input pulse. Specifically

$$T_s = 2T_0 ln(1 + \sqrt{(2)}) \approx 1.763 T_0 \qquad (15.17)$$

Table 15.1 Typical parameters used in simulation of solitons.

Parameter	Symbol	Value	Unit
wavelength	λ	1.55	μm
nonlinear coeff.	γ	1.3	1/ kmW
GVD	β_2	15×10^{-24}	s^2/km
width parameter	T_0	100	ps
peak power	P_0	0.15	mW
losses	α	0.01	dB/km^{-1}
chirp parameter	c	1.2	dimensionless
soliton period	z_0	1047.2	km

After simple algebra, Eq. (15.13) takes the form

$$\frac{\partial U}{\partial z} - i\frac{sign(\beta_2)}{2L_D}\frac{\partial^2 U}{\partial \tau^2} + i\frac{1}{L_{NL}}|U|^2 U + \frac{1}{2}\alpha U = 0 \quad (15.18)$$

Another normalized form of the Schrödinger equation exists in the literature. We obtain it in the lossless case, i.e. with $\alpha = 0$. To derive it, normalize the z coordinate as follows:

$$\xi = \frac{z}{L_D} \quad (15.19)$$

After a few algebraic steps, one obtains

$$\frac{\partial U}{\partial \xi} - i\frac{sign(\beta_2)}{2}\frac{\partial^2 U}{\partial \tau^2} + iN^2|U|^2 U = 0 \quad (15.20)$$

where N is known as soliton order and is defined as

$$N^2 = \frac{L_D}{L_{NL}} = \frac{\gamma P_0 T_0^2}{|\beta_2|} \quad (15.21)$$

The last popular form of the NLSE is found by introducing u as

$$u = NU = \left(\frac{\gamma P_0 T_0^2}{|\beta_2|}\right)^{1/2} A \quad (15.22)$$

Eq. (15.20) then takes the form

$$\frac{\partial u}{\partial \xi} - i\frac{sign(\beta_2)}{2}\frac{\partial^2 u}{\partial \tau^2} - i|u|^2 u = 0 \quad (15.23)$$

15.4 Split-step Fourier method

In this section we will discuss the numerical solution of the nonlinear Schrödinger equation (NSE) which describes propagation of optical solitons using the so-called split-step Fourier method (SSFM).

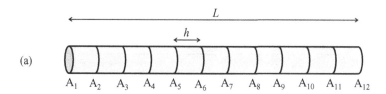

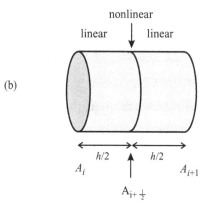

Fig. 15.2 Illustration of split-step Fourier method. (a) Division of optical fibre into N regions (here $N = 11$) of equal lengths. (b) Illustration of operation of linear and nonlinear operations at arbitrary segments.

The SSFM is a numerical technique used to solve nonlinear partial differential equations like the NSE. The method relies on computing the solution in small steps and on taking into account the linear and nonlinear steps separately. The linear step (dispersion) can be made in either frequency or time domain, while the nonlinear step is made in the time domain. The method is widely used for studying nonlinear pulse propagation in optical fibres. For more detail see [11]

A nonlinear Schrödinger equation, Eq. (15.13) contains dispersive and nonlinear terms. To introduce SSFTM, write the NLSE equation in the following form:

$$\frac{\partial A(z, T)}{\partial z} = \left(\widehat{L} + \widehat{N}\right) A(z, T) \tag{15.24}$$

where

$$\widehat{L}A = -\frac{\alpha}{2} - \frac{i}{2}\beta_2 \frac{\partial^2 A}{\partial T^2} \tag{15.25}$$

contains losses and dispersion in the linear medium and nonlinear term

$$\widehat{N}A = i\gamma |A|^2 A \tag{15.26}$$

accounts for the nonlinear effects in the medium.

The basis of the SSFM is to split a propagation from z to $z + h$ (h is a small step) into two operations (assuming that they act independently): during first step nonlinear effects are included and in the second step one accounts for linear effects, see Fig. 15.2.

Formal solution of Eq. (15.24) over a small step h is thus

$$A(z+h,t) = e^{h(\widehat{L}+\widehat{N})}A(z,t) \qquad (15.27)$$

In the first-order approximation, the above formula can be written as

$$A(z+h,t) = e^{h\widehat{L}} e^{h\widehat{N}} A(z,t) + O(h^2) \qquad (15.28)$$

The basis of this approximation is established by Baker-Hausdorf lemma [12], which is

$$e^{\widehat{A}} e^{\widehat{B}} = e^{\widehat{A}+\widehat{B}} e^{\frac{1}{2}[\widehat{A},\widehat{B}]} \qquad (15.29)$$

given that operators $\widehat{A}$ and $\widehat{B}$ commute with $[\widehat{A}, \widehat{B}]$.

The basis of the method is suggested by Eq. (15.28). It tells us that $A(z+h,t)$ can be determined by applying the two operators independently. The propagation from z to $z+h$ is split into two operations: first the nonlinear step and then the linear step assuming that they act independently. If h is sufficiently small, Eq. (15.28) gives good results.

The value of step h can be determined by assuming that the maximum phase shift $\phi_{\max} = \gamma |A_p|^2 h$, where A_p is the peak value of $A(z,t)$ due to the nonlinear operator is smaller than the predefined value. Iannone et al. [2] reported that $\phi_{\max} \leq 0.05$ rad.

For a practical implementation of the SSFM, we need to establish practical expressions for dispersive and nonlinear terms. In the following we will therefore analyse the effect of both terms independently neglecting losses.

Let us analyse the effect of the dispersive term alone. For that we temporally switch off the nonlinear term. After Fourier transform, the 'linear equation' becomes

$$\frac{\partial \widetilde{A}(z,\omega)}{\partial z} = -\frac{i}{2}\omega^2 \beta_2 \widetilde{A}(z,\omega)$$

which has the solution

$$\widetilde{A}(z,\omega) = \widetilde{A}(0,\omega) e^{-i\omega^2 \beta_2 z/2}$$

The action of the nonlinear term alone is described by the equation

$$\frac{\partial A(z,t)}{\partial z} = i\gamma |A(z,t)|^2 A(z,t)$$

The 'natural' solution is in the time domain. It produces

$$A(z,t) = A(0,t) e^{i\gamma |A|^2 A}$$

15.4.1 Split-step Fourier transform method

The propagation medium (say, cylindrical optical fibre) is divided into small segments, each of length h, see Fig. 15.2. Further, each individual segment of length h is subdivided into two of equal lengths. The linear operator operates over each subsegment in the frequency domain, whereas the nonlinear operator operates only locally at the central point.

Operation of the linear operator $\widehat{L}$, Eq. (15.25) over first subsegment is done as follows:

$$e^{h\widehat{L}/2}A(z,t) = F^{-1}\left\{e^{h\widehat{L}/2}\,F\,\{A(z,t)\}\right\} \tag{15.30}$$

i.e. one must Fourier transform original amplitude from time domain into frequency domain, apply linear operator $\widehat{L}$ and then apply inverse Fourier transform to get the amplitude back to time domain.

The operation of nonlinear operator defined by Eq. (15.26) is as follows:

$$A_{i+1/2,L}(z,t) = A_{i+1/2,R}(z,t)\,e^{h\widehat{N}} \tag{15.31}$$

where $A_{i+1/2,L}$ is the value of field amplitude at an infinitesimal point left from $i+\frac{1}{2}$. Finally, the operation of linear operator over second subsegment of length $\frac{h}{2}$ is done exactly the same way as over the first segment.

To summarize, the method over each segment of length h consists of three steps:

$$\text{step 1}\quad \begin{cases}\widetilde{A}_i(z,\omega) = F\{A_i(z,t)\}\\ \widetilde{A}_{i-}(z,\omega) = \widetilde{A}_i(z,\omega)\cdot\exp\left(-i\tfrac{1}{2}\omega^2\beta_2 h\right)\\ A_{i-}(z,t) = F^{-1}\{\widetilde{A}_{i-}(z,\omega)\}\end{cases}$$

$$\text{step 2}\quad A_{i+}(z,t) = A_{i-}(z,t)\cdot\exp\left(i\gamma|A|^2 Ah\right)$$

$$\text{step 3}\quad \begin{cases}\widetilde{A}_{i+1}(z,\omega) = F\{A_{i+1}(z,t)\}\\ \widetilde{A}_{i+1}(z,\omega) = \widetilde{A}_{i+1}(z,\omega)\cdot\exp\left(-i\tfrac{1}{2}\omega^2\beta_2 h\right)\\ A_{i+1}(z,t) = F^{-1}\{\widetilde{A}_{i+}(z,\omega)\}\end{cases}$$

where F indicates Fourier transform (FT) and F^{-1} inverse FT.

15.4.2 Symmetrized split-step Fourier transform method (SSSFM)

The simulation time of Eq. (15.28) essentially depends on the size of step h. To reduce simulation time, a new method was invented which allows us to use larger steps, h.

Mathematically, one uses the following second-order approximation:

$$A(z+h,t) = e^{\frac{1}{2}h\widehat{L}}\exp\left\{\int_z^{z+h}\widehat{N}(z')dz'\right\}e^{\frac{1}{2}h\widehat{L}}A(z,t) + O(h^3) \tag{15.32}$$

In this approach one assumes that the nonlinearities are distributed over h, which is more realistic. For small h, one can approximately evaluate

$$\int_z^{z+h}\widehat{N}(z')dz' \approx \frac{h}{2}\left[\widehat{N}(z)+\widehat{N}(z+h)\right]$$

In the SSSFM algorithm Eq. (15.31) is replaced by

$$A_{i+1/2,L}(z,t) = A_{i+1/2,R}(z,t)\,e^{h[\widehat{N}(z)+\widehat{N}(z+h)]} \tag{15.33}$$

The above scheme requires iteration to find A_{i+1} since it is not known at $z+\frac{1}{2}h$. Initially $\widehat{N}(z+h)$ will be assumed to be the same as $\widehat{N}(z)$.

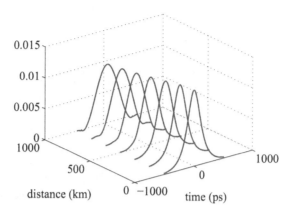

Fig. 15.3 Evolution in time over one soliton period for $N = 1$ soliton.

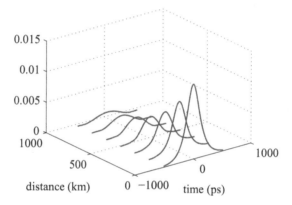

Fig. 15.4 Evolution in time over one soliton period for $N = 1$ soliton with damping.

15.5 Numerical results

Based on the developed software, in this section we will report on numerical analysis of soliton propagation in optical fibre. The SSFM has been implemented and MATLAB code is provided in Appendix, Listing 15A.1. We start with the single solitons.

15.5.1 Single solitons

In our analysis we used fibre and pulse parameters summarized in Table 15.1. The input soliton was considered to be in the form

$$A_{in}(\xi = 0, T) = \sqrt{A_0}\, sech\left(\frac{T}{T_0}\right) \tag{15.34}$$

where N is the soliton order given by Eq. (15.21). In Fig. 15.3 we illustrated the propagation of $N = 1$ soliton without damping and in Fig. 15.4 the evolution of the same soliton with damping ($\alpha = 0.01$ dB km^{-1}).

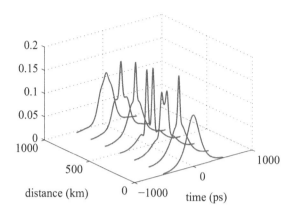

Fig. 15.5 Evolution in time over one soliton period for $N = 3$ soliton. Note soliton splitting near $z_0 = 0.5$ and its recovery beyond that point.

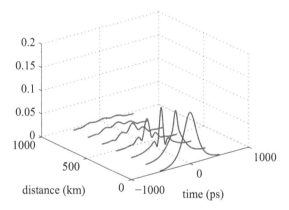

Fig. 15.6 Evolution in time over one soliton period for $N = 3$ soliton with damping.

A higher-order soliton with $N = 3$ is defined as

$$A_{in}(\xi = 0, T) = N^3 \sqrt{A_0}\, sech\left(\frac{T}{T_0}\right) \tag{15.35}$$

Propagation of higher-order soliton with $N = 3$ is shown in Fig. 15.5 (no damping) and in Fig. 15.6 (with damping). One can observe periodic evolution of the undamped soliton.

15.5.2 Chirped solitons

Here we analyse how chirp affects single soliton propagation. We assume the following input:

$$A_{in}(\xi = 0, T) = \sqrt{A_0}\, sech\left(\frac{T}{T_0}\right)\, \exp\left[-iC/2\left(\frac{T}{T_0}\right)^2\right] \tag{15.36}$$

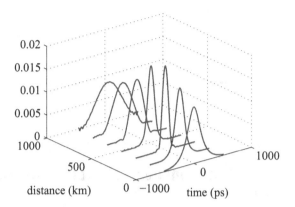

Fig. 15.7 Evolution in time of a chirped soliton for $N = 1$ without damping.

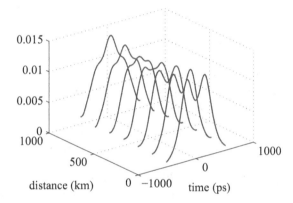

Fig. 15.8 Evolution in time of two interacting solitons without damping.

where C is the chirped parameter. Evolution of such soliton in the case of $N = 1$ and $C = 1.6$ is shown in Fig. 15.7. The pulse is initially compressed and then broadens.

15.5.3 Two interacting solitons

The effect of nonlinearity produces mutual interaction between soliton pulses if they are launched close together [13]. This interaction is important from a practical point of view and also from a fundamental perspective related to soliton propagation. For example, it was shown [13] that nonlinear interaction between solitons can result in a bandwidth reduction by a factor of 10. See also the review in [14].

First, we consider an interaction between two solitons of the same strength and $N = 1$. We assume the initial pulses of the form [15]

$$A_{in}(\xi = 0, T) = \sqrt{A_0}\,sech\left(\frac{T}{T_0} - 1\right) + \sqrt{A_0}\,sech\left(\frac{T}{T_0} + 1\right) \qquad (15.37)$$

The result of two interacting solitons is shown in Fig. 15.8.

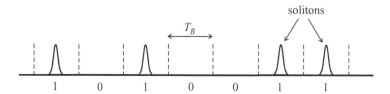

Fig. 15.9 Stream of solitons.

15.6 A few comments about soliton-based communications

As mentioned at the beginning of this chapter, an important application of solitons is in the transmission of information in optical fibre systems. Soliton pulses are stable when they propagate over long distances. Losses in fibres are an important limiting factor, so it becomes necessary to compensate periodically for fibre loss. This can be done by using EDFA.

In the transmission of information, solitons are considered like linear pulses. In a bit stream each soliton has its own bit slot and it represents logical one or zero, see Fig. 15.9. Here T_B is the duration of the bit slot and it is related to bit rate B as

$$T_B = \frac{1}{B} = 2aT_0 \qquad (15.38)$$

where T_0 is the soliton width and $2a$ specifies the distance between neighbouring solitons. To prevent their interactions, the neighbouring solitons should be well separated.

Solitons are very well suited for long-haul communication because of their high information carrying capacity and the possibility of periodic amplification. However, soliton systems are still waiting for the full field deployment. The reader might consult recent publications on soliton-based optical communication systems [16], [17], [18], [19].

15.7 Problems

1. Implement a symmetrized SSFM based on Eq. (15.33).
2. Generalize the implementation by including the third-order dispersion effect where $\beta_3 \neq 0$. Analyse the influence of β_3 on propagation of $N = 1$ and $N = 3$ solitons.
3. Implement SSFM for a normalized form of NLSE given by Eq. (15.23). Using your implementation, conduct the analysis for $N = 1$ and $N = 3$ solitons.
4. Analyse the propagation of Gaussian pulses in fibres with nonlinearity using SSFM. Consider regular Gaussian and also chirped Gaussian pulses.

Listing	Function name	Description
	Table 15.2 List of MATLAB functions for Chapter 15.	
15A.1	*ssftm_full.m*	Analysis of various types of solitons

Appendix 15A: MATLAB listings

In Table 15.2 we provide a list of MATLAB files created for Chapter 15 and a short description of each function.

Listing 15B.1 Program ssftw_full.m. Here we conduct analysis of various solitons using the split-step Fourier transform method.

```
% File name: ssftm_full.m
% Split-step Fourier transfer method for the analysis of solitons
% Contains the following initial pulses
% N=1 soliton
% N=3 solitons
% Chirp N=1 soliton
% Two interacting solitons
% For the analysis of the appropriate case it should be uncommented
%
clear all
tic;
%===== Material parameters ==================
%alpha=0.01;                % Fibre loss (dB/km)
alpha = 0.0;
alpha=alpha/(4.343);        % Fibre loss (1/km)
gamma = 1.3;                % Fibre nonlinearity [1/W/km]
beta_2=15d-24;              % 2nd order disp. (s^2/km)
%====== Time parameters ===========
T =- 512e-12:1e-12: 511e-12;
delta_t=1e-12;
%======= Input pulse and its parameters ========
T_0 = 100d-12;              % Initial pulse width (s)
P_0 = 0.15d-3;              % Input power (Watts)
A_0 = sqrt(P_0);
L_D =((T_0^2)/(abs(beta_2)));  % Dispersion length in [km]
L_NL = (1/(gamma*P_0));
h = 2d0;                    % step size [km]
z_plot = (0:h:((pi/2)*L_D));   % z-values to plot [in km]
n = 1;                      % controls stepping
z_0 = (pi/2)*L_D;           % Soliton period [km]
N = 3;                      % Soliton number
```

```
%A = N^2*A_0*sech(T/T_0);         % N=3 soliton
%A = A_0*sech(T/T_0);             % N=1 soliton
C = 1.6;                          % chirp parameter
%A = A_0*sech(T/T_0).*exp((-1i*C/2).*(T/T_0).^2);  % N=1 soliton with chirp
A = A_0*sech(T/T_0-1.6)+A_0*sech(T/T_0+1.6);       % interacting solitons
%
L = max(size(A));
Delta_omega=1/L/delta_t*2*pi;
omega = (-L/2:1:L/2-1)*Delta_omega;
A_f = fftshift(fft(A));           % Pulse in frequency domain
for jj=h:h:z_0
    A_f = A_f.*exp(-alpha*(h/2)-1i*beta_2/2*omega.^2*(h/2));
    A_t =ifft(A_f);               % Pulse in time domain
    A_t = A_t.*exp(1i*gamma*((abs(A_t)).^2)*(h));
    A_f = fft(A_t);
    A_f = A_f.*exp(-alpha*(h/2)-1i*beta_2/2*omega.^2*(h/2));
    A_t = ifft(A_f);
    plotting(:,n) = abs(A_t);
    n = n+1;
end
toc;
cputime=toc;
%
z_plot = h:h:z_0;                 % creates distance for plotting
length(z_plot)
for k = 1:80:length(z_plot)       % Choosing 3D plots every 10-th step
    y = z_plot(k)*ones(size(T));% Spread out along y-axis
    plot3(T*1d12,y,plotting(:,k),'LineWidth',1.5)
    hold on
end
xlabel('time (ps)','FontSize',14);
ylabel('distance (km)','FontSize',14);
set(gca,'FontSize',14);           % size of tick marks on both axes
grid on
pause
close all
```

References

[1] L. F. Mollenauer and J. P. Gordon. *Solitons in Optical Fibers. Fundamentals and Applications*. Elsevier, Amsterdam, 2006.

[2] E. Iannone, F. Matera, A. Mocozzi, and M. Settembre. *Nonlinear Optical Communication Networks*. Wiley, New York, 1998.

[3] G. P. Agrawal. *Fiber-Optic Communication Systems. Second Edition*. Wiley, New York, 1997.

[4] Y. S. Kivshar and G. P. Agrawal. *Optical Solitons. From Fibers to Photonic Crystals.* Academic Press, Amsterdam, 2003.

[5] A. Hasegawa and M. Matsumoto. *Optical Solitons in Fibers.* Springer, Berlin, 2003.

[6] A. Hasegawa and Y. Kodama. *Solitons in Optical Communications.* Clarendon Press, Oxford, 1995.

[7] R. W. Boyd. *Nonlinear Optics. Third Edition.* Academic Press, Amsterdam, 2008.

[8] R. W. Hellwarth. Third-order optical susceptibilites of liquids and solids. *Prog. Quant. Electr.*, **5**:1–68, 1977.

[9] A. Yariv and P. Yeh. *Photonics. Optical Electronics in Modern Communications. Sixth Edition.* Oxford University Press, New York, Oxford, 2007.

[10] G. P. Agrawal. *Nonlinear Fiber Optics.* Academic Press, Boston, 1989.

[11] G. P. Agrawal. *Nonlinear Fiber Optics, Fourth Edition.* Elsevier Science & Technology Books, Boston, 2006.

[12] R. L. Liboff. *Introductory Quantum Mechanics.* Addison-Wesley, Reading, MA, 1992.

[13] P. L. Chu and C. Desem. Mutual interaction between solitons of unequal amplitudes in optical fibre. *Elect. Lett.*, **21**:1133–4, 1985.

[14] C. Desem and P. L. Chu. Soliton-soliton interactions. In J. R. Taylor, ed., *Optical Solitons – Theory and Experiment*, pages 107–51, Cambridge University Press, Cambridge, 1992.

[15] Z. B. Wang, H. Y. Yang, and Z. Q. Li. The numerical analysis of soliton propagation with split-step Fourier transform method. *Journal of Physics: Conference Series*, **48**:878–82, 2006.

[16] A. Hasegawa. Soliton-based optical communications: An overview. *IEEE J. Select. Topics Quantum Electron.*, **6**:1161–72, 2000.

[17] M. F. Ferreira, M. V. Facao, S. V. Latas, and M. H. Sousa. Optical solitons in fibers for communication systems. *Fiber and Integrated Optics*, **24**:287–313, 2005.

[18] R. Gangwar, S. P. Singh, and N. Singh. Soliton based optical communication. *Progress in Electromagnetics Research, PIER*, **74**:157–66, 2007.

[19] N. Vijayakumar and N. S. Nair. Solitons in the context of optical fibre communications: a review. *Journal of Optoelectronics and Advanced Materials*, **9**:3702–14, 2007.

16 Solar cells

In this chapter we provide the fundamentals of solar cell operations. We outline basic physical principles, summarize a model based on equivalent circuit and discuss ways to increase efficiency of single-junction solar cell by employing multijunction structures or by creating intermediate bands.

General introductory books describing solar cells are [1], [2] and [3]. Anderson [1] and Rabl [2] provide a practical background in solar-energy conversion including techniques for estimating the solar radiation incident upon a collector, fundamentals of optics for solar collectors, discussion of types of concentrators and economical analysis, among others. Moeller [3] concentrates on fundamental, physical and material aspects of semiconductors for photovoltaic energy conversion.

16.1 Introduction

World consumption of electric energy circa 2009 was around 12–13 TW [4]. This energy was created by several methods. Let us look briefly at two of them: nuclear power and solar energy.

Assuming that a single nuclear plant produces 1 GW of power, creation of 10 TW of power would require 10 000 nuclear power plants. This huge number will certainly create social problems. Also, the uranium needed for those plants could be diminished in less than 20 years.

As far as the solar energy is concerned, the Sun supplies more solar energy on Earth in one hour than we use globally in one year. To make another comparison, a football field covered with silicon solar cells with an efficiency around 17% at 1-Sun illumination produces approximately 500 kW of power. The USA uses about 3 TW, so to produce that amount one will need about 6 million football fields. However, by increasing efficiency of solar cells, for example by using multijunction devices and increasing concentration of solar energy, the area of semiconductor solar cells can be significantly reduced [5].

Those rough estimates indicate that solar energy can be a practical alternative as a source of electric energy, and also that increasing the efficiency of solar cells is important. One should, however, remember that the incidence of solar energy is not uniform.

Solar irradiation spectrum (spectrum of solar energy) is shown in Fig. 16.1. This graph will be important later when we discuss in detail multijunction cells. In the figure we show spectral intensity, that is intensity of solar energy per unit wavelength. Above the Earth's atmosphere it is denoted as AM0 (air-mass zero). The integrated (over the whole

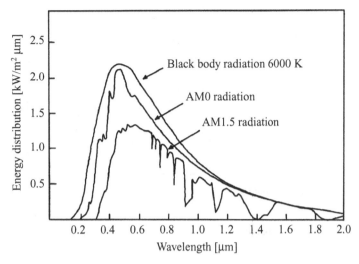

Fig. 16.1 Solar irradiation spectrum. Reprinted and adapted with permission from C. H. Henry, *J. Appl. Phys.* **51**, 4494 (1980). Copyright 1980, American Institute of Physics.

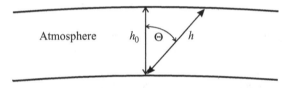

Fig. 16.2 Definition of air-mass.

spectrum) intensity above the Earth's atmosphere is constant and its approximate value is 1.353 kW m^2. It gives the total power flow through a unit area perpendicular to the direction of the Sun.

The Earth's atmosphere significantly absorbs radiation from the Sun. In Fig. 16.1 the radiation at the Earth's surface is labelled as AM1.5. In the figure several absorption bands and lines due to various effects are shown. For comparison, the spectrum of black body radiation at the temperature of 6000 K is also shown. One can notice that the solar spectrum above the Earth's atmosphere (AM0) resembles black body radiation.

The air mass coefficient defines the direct optical path length through the Earth's atmosphere, expressed as a ratio relative to the path length vertically upwards, i.e. at the zenith.

Let us finish this section with an explanation of AM (air-mass) convention. Generally air mass m (AMm) is defined (see Kasap [6]) as the ratio of the actual radiation path h to the shortest path h_0, $m = h/h_0 = \cos \Theta$, see Fig. 16.2.

Changing the light intensity incident on a solar cell changes all solar cell parameters, including the short-circuit current, the open-circuit voltage, the efficiency and the impact of series and shunt resistances. The light intensity on a solar cell is measured in the number of 'suns', where 1 sun corresponds to standard illumination at AM1.5, or 1 kW m^2. For

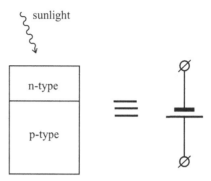

Fig. 16.3 A p-n junction operating as a solar cell.

example, a system with 10 kW m² incident on the solar cell would be operating at 10 suns, or at 10X.

16.2 Principles of photovoltaics

The simplest solar cell (or photovoltaic (PV) cell) is a single p-i-n junction operating under forward bias. It is formed by using, e.g. properly doped silicon, see Fig. 16.3. It can produce potential difference, thus acting as a battery. It contains a very narrow and heavily doped n-region. It is covered with thin antireflecting coating to reduce sun reflection which penetrates p-n junction from n-side, see Fig. 16.4 for more details.

Short wavelengths penetrate only the n-region, longer wavelengths penetrate depletion region of width w, and finally the longest wavelengths reach the p-region, see Fig. 16.5.

Photons of frequency ν having energies $h\nu \geq E_g$, where E_g is the bandgap energy, generate electron-hole (e-h) pairs. Due to a built-in electric field E_0 in the depletion region, electrons move towards metal contact in the n-region whereas holes travel in opposite direction towards metallic contact in the p-region. Since the n-region is very narrow, most of the photons are absorbed in the depletion region. Each generated electron increases charge in the n-region by $-e$; similarly each hole makes the p-region more positive by $+e$. Thus potential difference is created between metallic electrodes.

A semiconductor can only efficiently convert photons into current with energies equal to the bandgap. Photons with energies smaller than the bandgap are not absorbed, and photons with higher energies reduce their energies to the bandgap energy by thermalization of the photogenerated carriers; a process which involves losses.

The basic relation which provides the link between photons' energy and their wavelength is

$$\lambda[\mu m] = \frac{1.24}{E[eV]} \quad (16.1)$$

If both terminals of solar cell are shorted, the excess electrons on the n-side will start flowing through an external wire contributing to electrical current known as photocurrent,

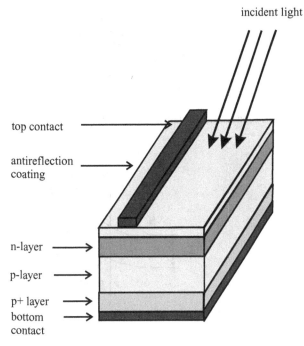

Fig. 16.4 Solar cell perspective.

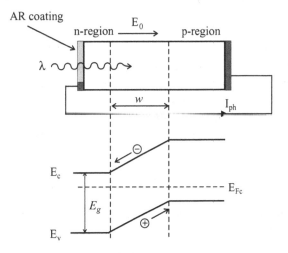

Fig. 16.5 The principle of photocurrent generation in a solar cell in short circuit.

I_{ph}, see Fig. 16.6. The short circuit conventional current I_{sc} flows opposite to photocurrent. If I_{light} is the light intensity, one can write

$$I_{sc} = -I_{ph} = -K \cdot I_{light}$$

where K is some (device dependent) constant.

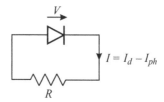

Fig. 16.6 Solar cell driving an external load resistance R.

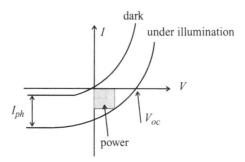

Fig. 16.7 Finding an operating point of a solar cell.

When the illuminated solar cell is loaded with resistance R, the conventional current I is $I = I_d - I_{ph}$, see Fig. 16.6, where I_d is a forward diode current:

$$I_d = I_0 \left[\exp\left(\frac{eV}{\eta k_B T}\right) - 1 \right] \tag{16.2}$$

and where I_0 is a constant, V is the voltage across diode and η is known as diode fidelity factor. It is equal to 1 for diffusion controlled and 2 for space charge layer recombination controlled characteristics.

The I-V characteristic of solar cell is shown in Fig. 16.7. It is obtained from a dark diode by shifting its I-V curve downward by I_{ph}. The analytical expression for conventional current is

$$I = -I_{ph} + I_0 \left[\exp\left(\frac{eV}{\eta k_B T}\right) - 1 \right] \tag{16.3}$$

With the load resistance R connected to the solar cell, see Fig. 16.6, the current-voltage relation is

$$I = -\frac{V}{R} \tag{16.4}$$

Operating point of solar cell with load resistance R is obtained by solving Eqs. (16.3) and (16.4). The graphical solution is illustrated in Fig. 16.8., where we have plotted Eqs. (16.3) and (16.4). Crossing of both lines determines operating point (V_1, I_1) of a solar cell.

The point at which a characteristic intersects the vertical current axis is known as a short circuit condition and the corresponding current as short-circuit current I_{sc}. Similarly, the point at which I-V characteristic intersects the horizontal voltage axis is known as the open

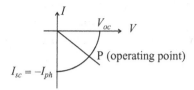

Fig. 16.8 Finding an operating point of a solar cell. V_{oc} is an open-circuit voltage.

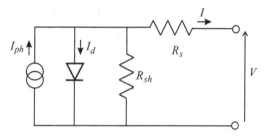

Fig. 16.9 Equivalent circuit of a solar cell.

circuit condition. It defines the open-circuit voltage V_{oc}, which is the maximum voltage which can be drawn from the solar cell. It corresponds to zero current.

Power delivered to the load R is $P_{load} = I_1 \cdot V_1$, which is the area of the rectangle, see Fig. 16.7. The goal of the solar cell design is to maximize that power.

16.3 Equivalent circuit of solar cells

Over the years several methods of modelling solar cells have been introduced. The simplest approach is based on equivalent circuit models. Some recently published works with an emphasis on using MATLAB are [7], [8], [9].

16.3.1 Basic model

The equivalent circuit of solar cell which includes parasitic effects is shown in Fig. 16.9. It consists of current source which generates photoelectric current I_{ph}, ideal diode, shunt resistor and series resistor. Based on this model, the goal is to obtain current-voltage relation. From Kirchoff rule one obtains (from now on we reverse direction of flow of conventional current, which is a common practice)

$$I = I_{ph} - I_d \tag{16.5}$$

where I is the cell total current and I_d is the diode current. Combining the above and relation (16.3), one finds

$$I = I_{ph} - I_0 \left[\exp\left(\frac{eV_d}{\eta k_B T} \right) - 1 \right] \tag{16.6}$$

This equation describes an ideal situation without parasitic effects. In a real solar cell there exists leakage current flowing through shunt resistance R_{sh}. Potential drop across the device is represented by series resistance R_s. The inclusion of shunt resistance modifies solar current I as

$$I = I_{ph} - I_d - \frac{V_d}{R_{sh}} \tag{16.7}$$

The impact of series resistance is included as

$$V_d = V + I \cdot R_s \tag{16.8}$$

With the parasitic effects included, the solar current I takes the form

$$I = I_{ph} - I_0 \left[\exp\left(\frac{eV_d}{\eta k_B T}\right) - 1 \right] - \frac{V + I \cdot R_s}{R_{sh}} \tag{16.9}$$

The photocurrent I_{ph} depends on the solar insolation and cell's temperature and it is given by [10], [11]

$$I_{ph} = \left[I_{sc} + K_I \left(T_c - T_{ref} \right) \right] S_{in} \tag{16.10}$$

where I_{sc} is the cell's short-circuit current at a 25° C and 1 kW m^2, K_I is the cell's short-circuit current temperature coefficient, T_{ref} is the cell's reference temperature and S_{in} is the solar insolation in kW m^2.

The expression for I_0 is [10], [11]

$$I_0 = I_{RS} \left(\frac{T_c}{T_{ref}}\right)^{3/n} \exp\left[\frac{E_{gap}}{Ak_B \left(\frac{1}{T_{ref}} - \frac{1}{T_c}\right)} \right] \tag{16.11}$$

where I_{RS} is cell's reverse saturation at a reference temperature and a solar radiation, E_{gap} is the bandgap energy of the semiconductor used in the cell and A is an ideal factor which depends on PV technology. Some typical values are: $A = 1.2$ for Si-mono, $A = 1.3$ for Si-poly and $A = 1.5$ for CdTe (after [10]).

16.3.2 Other models

There exist other models of solar cells which are based on a basic model. Here, we will only mention two such models: the double exponential model and ideal PV cell model.

Double exponential model [12] is shown in Fig. 16.10. This model contains a light generated current source, two diodes D_1 and D_2 and a series and parallel resistances. The model is derived from the physics of the p-n junction and can describe cells constructed from polycrystalline silicon.

Simple models. The shunt resistance R_{sh} is inversely related to shunt leakage current to the ground. PV efficiency shows little sensitivity to the variation of R_{sh} and one can assume that $R_{sh} = \infty$. The resulting model is similar to the basic one, shown in Fig. 16.9, without R_{sh}. The current-voltage expression for this model is

$$I = I_{ph} - I_0 \left[\exp\left(\frac{e(V + I \cdot R_s)}{\eta k_B T}\right) - 1 \right] \tag{16.12}$$

Equivalent circuit of solar cells

Table 16.1 Parameters for equivalent circuit model of Si solar cell.

Description of parameter	Symbol	Value
Illumination	I_{ph}	10 mA
eta	η	1.5
Current	I_0	3×10^{-6} mA
Series resistance	R_s	0 Ω, 20 Ω, 50 Ω
Shunt resistance	R_{sh}	415 Ω

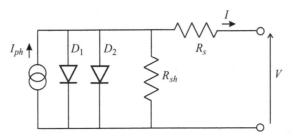

Fig. 16.10 Double exponential model of a solar cell.

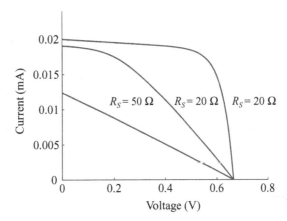

Fig. 16.11 Current-voltage characteristics of an ideal solar cell based on Si for several values of series resistance.

An ideal PV cell model assumes no series loss and no leakage to ground; i.e. $R_s = 0$ and $R_{sh} = \infty$. The equivalent circuit consists of only current source and one diode. Current is expressed as

$$I = I_{ph} - I_0 \left[\exp\left(\frac{eV}{\eta k_B T}\right) - 1 \right] \tag{16.13}$$

We conducted analysis using typical parameters, which are summarized in Table 16.1. Current-voltage characteristics for several values of series resistance are shown in Fig. 16.11 and in Fig. 16.12 we presented current-voltage characteristics at three different values

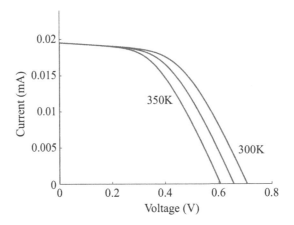

Fig. 16.12 Current-voltage characteristics of an ideal solar cell based on Si for several values of temperature.

of temperatures. MATLAB code used to generate those characteristics is shown in the Appendix, Listings 16A.1, 16A.2 and 16A.3. These are typical results as reported in the literature.

16.4 Multijunctions

As determined by Shockley and Queisser [13], the maximum theoretical limit for a single junction solar cell is about 31%. In order to increase theoretical efficiency one needs to increase number of p-n junctions in the cell. Henry [14] found that maximum theoretical efficiency at a concentration of 1 sun is 31%. At a concentration of 1000 suns with the cell at 300 K, the maximum efficiencies are 37%, 50%, 56% and 72% for cells with 1, 2, 3 and 36 p-n junctions (with different, properly chosen energy gaps), respectively.

Increase in efficiencies can be obtained by splitting solar spectrum into several parts and using different materials for conversion in various parts. This principle is illustrated in Fig. 16.13 for a triple solar cell built from $Ga_{0.49}In_{0.51}P(1.9\,eV)$, $Ga_{0.99}In_{0.01}As(1.4\,eV)$ and $Ge(0.7\,eV)$ (adopted from Dimroth [15]). Germanium is used mainly due to its robustness and its low cost of production. Those different semiconductor materials are grown on one substrate. Tunnel junctions are used to form an electrical series connection of the subcells.

Suitable materials must be chosen for each cell so that photons of appropriate wavelengths are absorbed. The bandgap of each material is determined using Eq. (16.1). Energy bandgap E_g must decrease from top cell to bottom cell in order not to absorb all photons by the top cell. Also, all layers of semiconductors must be lattice matched, so that there is no built-in strain.

The anti-reflective (AR) coating typically consists of several layers. Design of AR layers can be performed using methods discussed in Chapter 3.

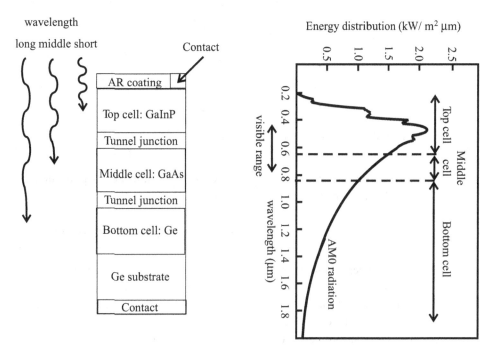

Fig. 16.13 Triple-junction solar cell. (From F. Dimroth, High-efficiency solar cells from III-V compound semiconductors, *Phys. Stat. Sol.* (c) **3**, 373–9 (2006). Copyright Wiley-VCH Verlag GmbH and Co. KGaA. Reproduced with permission.)

The presence of tunnel junctions is to provide low electrical resistance between subcells. Modelling of those layers specifically for the purpose of MJ solar cells has been described recently by Baudrit and Algora [16].

16.4.1 Quantum dots in multijunctions

The triple-junction subcells shown are connected in series and their voltages add up, but the current flowing through the structure is determined by the smallest current produced by subcells. Total power generated by the structure is greater than that of the single junction cell.

The thickness of each subcell is determined by its absorption and the light power available in the spectral range where the subcell operates. Referring to Fig. 16.13, the bottom cell (formed by Ge, including substrate) is much thicker than two upper subcells. Due to this and also because it covers much larger spectral range, the Ge subcell produces much larger current compared to two upper subcells.

One of the possible solutions of reducing current of the Ge subcell and at the same time increasing current of the GaAs subcell, is to introduce InAs-based quantum dots into the GaAs subcell [17]. These quantum dots are designed to have an effective bandgap slightly smaller than that of GaAs. Therefore they can 'steal' current from bottom Ga subcell roughly by about 15%. Increasing also the thickness of top GaInP subcell to match the increased current of GaAs results in an overall current increase by about 15%. With

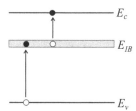

Fig. 16.14 Illustration of a concept of intermediate-band solar cell.

the minimal change in voltage over GaAs subcell, this concept resulted in an increase of triple-junction conversion efficiency by about 13%.

16.4.2 Intermediate band solar cells (IBSC)

The concept of intermediate band solar cells (IBSC) was first described by Luque and Marti [18] and by Wolf [19]. The main idea was to create an additional band between conduction and valence bands, see Fig. 16.14. The additional intermediate band can be created using several approaches, like, for example, minibands.

The standard cell is only able to absorb photons with energy equal to or greater than $E_{gap} = E_c - E_v$. Within this new scheme it is possible to have additional absorption via the IB.

The electronic states in the IB should be accessible via direct transitions. Therefore photon of energy E_{IV} can transfer an electron from the valence band into IB whereas another photon with energy E_{CI} will be able to pump electrons from IB to conduction band. Thus the performance of solar cell is increased.

Solar cells based on this design are attractive because of their predicted photon conversion efficiency of up to about 60% [20]. The authors of Ref. [20] show that the ordered three-dimensional arrays of quantum dots, i.e., quantum dot supracrystals, can be used to implement the intermediate-band solar cell with the efficiency significantly exceeding the Shockley-Queisser limit for a single junction cell. The increase is due to the utilization of photogenerated hot carriers which can produce higher voltages and higher photocurrents.

Modelling of such structures has been conducted using the CFD Research Corporation 3D device simulator, NanoTCAD [21]. They performed accurate simulations of quantum dot solar cell performance and degradation due to effects of space radiation.

A recent summary of some aspects of intermediate-band solar cells was published in *Nature Photonics* [22].

16.4.3 Role of simulations

Numerical simulations and modelling of solar cells play an important role. One particular example has been mentioned in the preceding section. Optimization of modern structures, especially for multijunction or quantum dots by experimental trial-and-error methods, is definitely too costly. This opens the room for simulations.

The optical and electrical interactions between the numerous layers in a multi-junction solar cell are very complex. Over the years several groups have developed simulation models of various levels of sophistication. The simplest approach based on equivalent circuits model was summarized in an earlier section. There were also many successful attempts to develop microscopic models of various complexity. Full discussion of microscopic models, including those based on drift-diffusion approach, is beyond the scope of this book.

Also, a few commercial products are found on the market: Sentaurus from Synopsys, Atlas from Silvaco or Solar Cell utility from RSoft. Some recent examples of using commercial software to simulate and design solar cells are described in [23], [24] and [25]. The authors of Ref. [25] describe details of using Sentaurus to simulate quantum well solar cells.

More sophisticated approaches to designing of quantum well solar cells using quantum transport are also under development [26], [27]. Such approaches combine elements from semiconductor optics and quantum transport in nanostructures. A very fundamental approach based on non-equilibrium Green's functions (NEGF) is discussed extensively in [26]. The developed approach treats absorption, transport and relaxation on equal footing and within a framework based on non-equilibrium quantum statistical mechanics.

We finish this discussion by mentioning the role played by plasmonics in improving properties of photovoltaic devices, see a recent summary [28]. Plasmonics describe guiding and localizing light at the nanoscale smaller than the wavelength of light in free space. Using plasmonics allows for new designs of solar cells in which light is fully absorbed in a single quantum well. Proposed solutions can result in improved absorption in photovoltaic devices.

Appendix 16A: MATLAB listings

In Table 16.2 we provide a list of MATLAB files created for Chapter 16 and a short description of each function.

Listing 16A.1 Program solar_Si.m. MATLAB program which plots I-V characteristics of an ideal solar cell for several values of series resistance based on equivalent circuit model.

```
% File name: solar_Si.m
% Simulates ideal solar cell based on Si
```

Table 16.2 List of MATLAB functions for Chapter 16.

Listing	Function name	Description
16A.1	solar_Si.m	Generates I-V characteristics for several values of series resistance
16A.2	current_V.m	Function used by solar_Si.m and solar_Si_T.m
16A.3	solar_Si_T.m	Generates I-V characteristics for several values of temperature

```
clear all
eta = 1.5;
I_ph = 0.02;                        % illumination current, in mA
T = 330;
V_init = 0;
V_final = 1.0;
V_step = 0.01;
V=0;
hold on
for R_s = [0.01 20 50]              % series resistance, in ohms
curr_total = fzero(@(curr) current_V(curr,V,T,eta,I_ph,R_s),0.2);
for V = V_init:V_step:V_final
    curr = fzero(@(curr) current_V(curr,V,T,eta,I_ph,R_s),0.2);
    curr_total = [curr_total, curr];
end
V_plot = V_init:V_step:V_final;     % creation voltage values for plot
V_plot = [0, V_plot];
plot(V_plot, curr_total,'LineWidth',1.5)
set(gca,'FontSize',14);             % size of tick marks on both axes
axis([0 V_final*0.8 0 I_ph*1.2]);
ylabel('Current (mA)','Fontsize',14)
xlabel('Voltage (V)','Fontsize',14)
text(0.65, 0.01, 'R_s = 0 \Omega','Fontsize',14)
text(0.45, 0.01, 'R_s = 20 \Omega','Fontsize',14)
text(0.2, 0.01, 'R_s = 50 \Omega','Fontsize',14)
end
pause
close all
```

Listing 16A.2 Function current_V.m. MATLAB function which creates an expression for current in the equivalent circuit model.

```
function fun = current_V(I,V,T,eta,I_ph,R_s)
% Expression for current used in solar cell model based on equivalent
% circuit
q = 1.6d-19;            % charge of electron
k_B = 1.38d-23;         % Boltzmann constant
R_sh = 415;             % shunt resistance in ohms
I_0 = 3d-9;             % reverse saturation current of the diode
VV = V+I*R_s;
fun = I_ph - I - I_0*(exp(q*VV/(eta*k_B*T))-1) - VV/R_sh;
```

Listing 16A.3 Function solar_Si_T.m. MATLAB program which plots current-voltage characteristics of an ideal solar cell for three temperatures using the equivalent circuit model.

```
% File name: solar_Si_T.m
% Simulates effect of temperature for ideal solar cell based on Si
clear all
eta = 1.5;
I_ph = 0.02;                            % illumination current, in mA
R_s = 10;                               % series resistance, in ohms
V_init = 0;
V_final = 1.0;
V_step = 0.01;
V=0;
hold on
for T = [300 325 350]                   % temperature
curr_total = fzero(@(curr) current_V(curr,V,T,eta,I_ph,R_s),0.2);
for V = V_init:V_step:V_final
    curr = fzero(@(curr) current_V(curr,V,T,eta,I_ph,R_s),0.2);
    curr_total = [curr_total, curr];
end
V_plot = V_init:V_step:V_final;         % creation voltage values for plot
V_plot = [0, V_plot];
plot(V_plot, curr_total,'LineWidth',1.5)
set(gca,'FontSize',14);                 % size of tick marks on both axes
axis([0 V_final*0.8 0 I_ph*1.2]);
ylabel('Current (mA)','Fontsize',14)
xlabel('Voltage (V)','Fontsize',14)
text(0.65, 0.005, '300K','Fontsize',14)
text(0.38, 0.01, '350K','Fontsize',14)
end
pause
close all
```

References

[1] E. E. Anderson. *Fundamentals of Solar Energy Conversion*. Addison-Wesley, Reading, MA, 1983.

[2] A. Rabl. *Active Solar Collectors and Their Applications*. Oxford University Press, New York, Oxford, 1985.

[3] H. J. Moeller. *Semiconductors for Solar Cells*. Artech House, Boston, 1993.

[4] K. Tanabe. A review of ultrahigh efficiency III-V semiconductor compound solar cells: multijunction tandem, lower dimensional, photonic uo/down conversion and plasmonic nanometallic structures. *Energies*, 2:504–30, 2009.

[5] G. S. Kinsey, R. A. Sherif, R. R. King, C. M. Fetzer, H. L. Cotal, and N. H. Karam. Concentrator multijunction solar cells for utility-scale energy production, 2005. unpublished.

[6] S. O. Kasap. *Optoelectronics and Photonics: Principles and Practices*. Prentice Hall, Upper Saddle River, NJ, 2001.

[7] F. M. Gonzalez-Longatt. Model of photovoltaic module in Matlab. *2do Congreso Iberoamericano de Estudiantes de Ingenieria Electrica, Electronica y Computacion (II Cibelec, 2005)*, pages 1–5, 2005.

[8] R. K. Nema, S. Nema, and G. Agnihotri. Computer simulation based study of photovoltaic cells/modules and their experimental verification. *International Journal of Recent Trends in Engineering*, **1**:151–6, 2009.

[9] M. Azab. Improved circuit model of photovoltaic array. *International Journal of Electrical Power and Energy Systems Engineering*, **2**:185–8, 2009.

[10] H.-L. Tsai, C.-S. Tu, and Y.-J. Su. Development of generalized photovoltaic model using Matlab/Simulink. *Proceedings of the World Congress on Engineering and Computer Science*, WCECS 2008, 2008.

[11] R. Hernanz, C. Martin, Z. Belver, L. Lesaka, Z. Guerrero, and E. P. Perez. Modelling of photovoltaic module. *International Conference on Renewable Energies and Power Quality*, ICREPQ'10, 2010.

[12] J. A. Gow and C. D. Manning. Development of a photovoltaic array model for use in power-electronics simulation studies. *IEE Proc.-Electr. Power Appl.*, **146**:193–200, 1999.

[13] W. Shockley and H. J. Queisser. Detailed balance limit of efficiency of p-n junction solar cells. *J. Appl. Phys.*, **32**:510–19, 1961.

[14] C. H. Henry. Limiting efficiencies of ideal single and multiple energy gap terrestrial solar cells. *J. Appl. Phys.*, **51**:4494–500, 1980.

[15] F. Dimroth. High-efficiency solar cells from III-V compound semiconductors. *Phys. Stat. Sol.(C)*, **3**:373–9, 2006.

[16] M. Baudrit and C. Algora. Tunnel diode modeling, including nonlocal trap-assisted tunneling: a focus on III-V multijunction solar cell simulation. *IEEE Trans. Electron Devices*, **57**:2564–71, 2010.

[17] B. Riel. Quantum dots: higher CPV efficiencies, same production cost. *Future Photovoltaics*, May, 2010.

[18] A. Luque and A. Marti. Increasing the efficiency of ideal solar cells by photon induced transitions at intermediate levels. *Phys. Rev. Lett.*, **78**:5014–17, 1997.

[19] W. Wolf. Limitations and possibilities for improvement of photovoltaic solar energy converters. *Proc. IRE*, **48**:1246–63, 1960.

[20] Q. Shao, A. A. Balandin, A. I. Fedoseyev, and M. Turowski. Intermediate-band solar cells based on quantum dot supracrystals. *Appl. Phys. Lett.*, **91**:163503, 2007.

[21] A. I. Fedoseyev, M. Turowski, A. Raman, E. W. Taylor, S. Hubbard, S. Polly, S. Shao, and A. A. Balandin. Space radiation effects modeling and analysis of quantum dot based photovoltaic cells. In E. W. Taylor and D. A. Cardimona, eds., *Proc. of SPIE. Nanophotonics and Macrophotonics for Space Environments III*, volume **7467**, page 746705. 2009.

[22] A. Luque and A. Marti. Towards the intermediate band. *Nature Photonics*, **5**:137–8, 2011.

[23] M. Hermle, G. Letay, S. P. Philipps, and A. W. Bett. Numerical simulation of tunnel diodes for multi-junction solar cells. *Prog. Photovolt: Res. Appl.*, **16**:409–18, 2008.

[24] S. P. Philipps, M. Hermle, G. Letay, F. Dimroth, B. M. George, and A. W. Bett. Calibrated numerical model of a GaInP-GaAs dual-junction solar cell. *Phys. Stat. Sol. (RRL)*, **2**:166–8, 2008.

[25] P. Kailuweit, R. Kellenbenz, S. P. Philipps, W. Guter, A. W. Bett, and F. Dimroth. Numerical simulation and modeling of GaAs quantum-well solar cells. *J. Appl. Phys.*, **107**:064317, 2010.

[26] U. Aeberhard. *A microscopic theory of quantum well photovoltaics*. PhD thesis, Swiss Federal Institute of Technology, Zurich, 2008.

[27] S. Agarwal. *Design guidelines for high efficiency photovoltaics and low power transistors using quantum transport*. PhD thesis, Purdue University, 2010.

[28] H. A. Atwater and A. Polman. Plasmonics for improved photovoltaic devices. *Nature Materials*, **9**:205–13, 2010.

17 Metamaterials

In this chapter we review the basic concept of metamaterials as those possessing simultaneously negative permittivity and permeability over the same frequency range. Theoretical principles and basic experimental results are reviewed. Possible applications including cloaking, slow light and optical black holes are described.

17.1 Introduction

Metamaterials are artificially created structures with predefined electromagnetic properties. They are fabricated from identical elements (atoms) which form one-, two- or three-dimensional structures. They resemble natural solid state structures. Metamaterials typically form a periodic arrangement of artificial elements designed to achieve new properties usually not seen in Nature [1]. In a sense, they are composed of elements in the same way as matter consists of atoms.

Metamaterials are characterized and defined by their response to electromagnetic wave. Optical properties of such materials are determined by an effective permittivity ε_{eff} and permeability μ_{eff} valid on a length scale greater than the size of the constituent units. In order to introduce such a description, one requires that the size of artificial inclusions characterized by d be much smaller than wavelength λ, i.e. $d \ll \lambda$.

The name meta originates from Greek, $\mu\epsilon\tau\alpha$ and means 'beyond'. Main characteristics of metamaterials (MM) are:

- man-made,
- have properties not found in Nature,
- have rationally designed properties,
- are constructed by placing inclusions at desired locations.

With modern fabrication techniques it is possible to create structures which are much smaller than the wavelength of visible light. An example of the unusual properties that such structures could have is the negative refractive index. Although the problem has a long history (for example, Mandelstam [2] in 1945 discussed negative refraction and negative group velocity), it was only in 1967 when Victor Veselago [3] indicated the possibility of materials having simultaneously negative μ and ε. He demonstrated the technological potential that materials with a negative refractive index could have for applications in imaging. It recently made its way to optics, mostly due to rapid progress in nanofabrication.

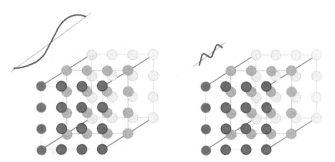

Fig. 17.1 Schematics of the elementary cells and the wavelength of external electromagnetic wave in two extreme cases.

The most well-known property of metamaterials is the negative index of refraction (NIM – negative index materials). Another popular name used is left-handed material. This terminology will soon be explained.

The situation is illustrated in Fig. 17.1. For a longer wavelength, one cannot sense the properties of individual constituent atoms, shown on the left, whereas the shorter wavelength, on the order of the distance between 'atoms', can effectively be used to determine some of the atom's properties (like their locations). In metamaterials we deal with the situation on the left.

Metamaterials can have controlled magnetic and electric responses over a broad range of frequencies. Those responses depend on the properties of individual elements (atoms). In the long-wavelength limit, where $a \ll \lambda$, where a is the characteristic dimension and λ is the wavelength of electromagnetic wave, one should perform some sort of averaging procedure to determine effective parameters of MM. Details of those procedures are discussed in the literature [4]. In the end, it is possible to achieve the condition where $\varepsilon_{\mathit{eff}} < 0$ and $\mu_{\mathit{eff}} < 0$.

A diagram which illustrates the classification of metamaterials is shown in Fig. 17.2. In general negative index materials do not exist in Nature; the rare exception is bismuth, which when placed in a waveguide shows a negative refractive index at a wavelength of $\lambda = 60 \,\mu$m [5]. There are no known naturally occurring NIMs in the optical range. However, artificially designed materials (metamaterials) can act as NIM. One should, however, notice that the occurrence of NIM needs both negative $\varepsilon_{\mathit{eff}} < 0$ and $\mu_{\mathit{eff}} < 0$ over the same frequency range.

Metamaterials can open new avenues to achieve unprecedented physical properties and functionality unattainable with naturally existing materials. Optical NIMs promise to create entirely new prospects for controlling and manipulating light, optical sensing and nanoscale imaging and photolithography.

17.1.1 Short history of MM

The earliest discussion of the concept of negative refraction goes probably to Schuster [6]. In the book entitled *An Introduction to the Theory of Optics* he noted that negative dispersion of the refractive index with respect to wavelength, i.e. $dn/d\lambda < 0$, can produce negative refraction of light entering such a medium from a vacuum. Some other early works

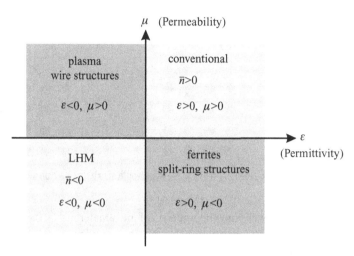

Fig. 17.2 Classification of materials according to the sign of electric permittivity and magnetic permeability.

on negative refraction include Lamb [7] published in 1904 and Pocklington [8] published in 1905. Historical aspects were summarized by Simovski and Tretyakov [9] and by Moroz [10].

In recent years the field of metamaterials has received remarkable attention with the number of published papers growing exponentially. This is due to unusual properties of such systems (see [4] for a recent review) and also important practical applications like perfect lenses [11], invisibility cloaking [12], [13], slow light [14] and enhanced optical nonlinearities [15]. Parallel to theoretical developments, there has been noted spectacular experimental progress in the field [16].

The subject of metamaterials is attracting enormous attention from researchers. We searched ISI Web of Knowledge, and as of October, 2011 there were 4545 published papers on the subject of 'metamaterials'. The plot of the number of published papers is shown in Fig. 17.3. As can be seen, the number of published papers grows exponentially with some saturation observed recently.

The modern research progress started in 1999 with the pioneering work by Pendry and collaborators [17] who discussed artificial electromagnetic structures, such as split-ring resonators (SRR) (in general, two concentric split rings), for which they predicted the existence of negative magnetic permeability. In those structures the incident electromagnetic wave excites circulating currents in the loops. These currents in turn create oscillating magnetic dipole moments. With proper design, those currents are resonantly enhanced, leading to a negative magnetic permeability.

An SRR meta-atom can, in the long-wavelength (quasi-static) regime, be modelled as a resistor/inductor/capacitor (RLC) resonant circuit [18]. Stacks of such structures exhibit negative permeability.

The creation of a negative electric permittivity is relatively easy. For instance, it is well-known that metals at optical frequencies exhibit a plasmon-like permittivity that can assume

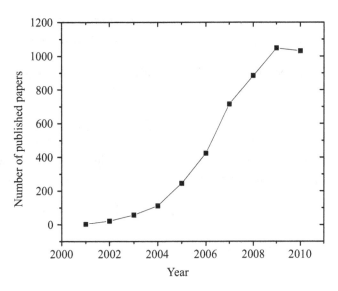

Fig. 17.3 Number of papers published on the subject of 'metamaterials'.

negative values. At smaller (e.g. microwave) frequencies, it was shown by Pendry *et al.* in 1996 [19] that an array of long wires acts as a diluted Drude metal when the electric field is oriented along the wire axis. Such a system allows for the existence of a negative permittivity below the effective plasma frequency. A combination of both magnetic and electric elements in a suitable design leads to negative index of refraction over a specified frequency band.

A report on the first negative-index material (NIM) designed and fabricated along the above principles was published in 2000 by Smith *et al.* [20]. It operated at microwave frequencies. The individual SRR had its resonance frequency at 4.845 GHz. The diameter of the wires was about 0.8 mm, which resulted in a plasma frequency of $\omega_p = 13 \times 10^9 \text{s}^{-1}$. There, the authors performed transmission measurements on a NIM structure having a range of frequencies over which the refractive index was negative for one direction of propagation. In a subsequent paper, Shelby *et al.* [21] (again in a transmission experiment) demonstrated a negative refraction of microwaves incident on the interface between air and NIM.

In 2003, a group from Boeing Phantom Works reported [22] a very accurate measurements of a Snell's law using a NIM wedge at frequencies from 12.6 to 13.2 GHz.

Following those pioneering demonstrations, in the next few years the operation frequency has been increased by more than four orders of magnitude. In particular, Yen *et al.* [23] demonstrated a NIM response at 1 THz ($\lambda = 300$ μm) by making use of SRRs placed on a double-sided polished quartz substrate. Further design and fabrication improvements have led to the possibility of creating a three-dimensional double-negative medium with subwavelength meta-atoms in the optical regime [24], [25]. Indeed, in 2008, a three-dimensional photonic NIM at optical frequencies has been demonstrated in Ref. [26].

17.2 Veselago approach

In this section we summarize an early approach as developed by Veselago [3].

17.2.1 Wave equation

In order to proceed with a quantitative description, at this stage we will introduce Maxwell's equations, which in the absence of free charges and currents are

$$\nabla \times \mathbf{E} = -\mu \frac{\partial \mathbf{H}}{\partial t} \quad \text{and} \quad \nabla \times \mathbf{H} = \varepsilon \frac{\partial \mathbf{E}}{\partial t} \quad (17.1)$$

Take $\nabla \times$ (rotation) operation on the first equation and then use second equation

$$\nabla \times \nabla \times \mathbf{E} = -\varepsilon \frac{\partial}{\partial t}(\nabla \times \mathbf{H}) = -\mu \varepsilon \frac{\partial^2 \mathbf{E}}{\partial t^2}$$

Use general relation valid for arbitrary vector $\mathbf{A}$

$$\nabla \times \nabla \times \mathbf{A} = \nabla(\nabla \cdot \mathbf{A}) - \nabla^2 \mathbf{A} = -\nabla^2 \mathbf{A}$$

In the last step, we used $\nabla \cdot \mathbf{E} = 0$. One obtains the wave equation

$$\nabla^2 \mathbf{E} = -\mu \varepsilon \frac{\partial^2 \mathbf{E}}{\partial t^2} = 0 \quad (17.2)$$

If we disregard losses and consider ε and μ as real numbers, then one can observe that wave equation is unchanged when we *simultaneously* change signs of ε and μ.

17.2.2 Left-handed materials

Start with the basic Maxwell's equations (17.1). Assume time-harmonic fields

$$\mathbf{E}(x, y, z, t) = \mathbf{E}(x, y, z)e^{i\omega t} + c.c.$$

and introduce wave vector $\mathbf{k}$ by assuming plane-wave dependence

$$\mathbf{E}, \mathbf{H} \sim e^{i\mathbf{k}\cdot\mathbf{r}}$$

Maxwell's equations take the form

$$\mathbf{k} \times \mathbf{E} = -\mu \omega \mathbf{H} \quad (17.3)$$
$$\mathbf{k} \times \mathbf{H} = \varepsilon \omega \mathbf{E} \quad (17.4)$$

From Eqs. (17.3) and (17.4) and definition of cross product, one can immediately see that for $\varepsilon > 0$ and $\mu > 0$ vectors $\mathbf{E}$, $\mathbf{H}$ and $\mathbf{k}$ form a right-handed triplet of vectors, and if $\varepsilon < 0$ and $\mu < 0$ they form a left-handed system, see Fig. 17.4.

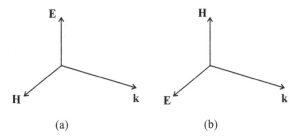

Fig. 17.4 (a) Right-hand orientation of vectors **E**, **H**, **k** for the case when $\varepsilon > 0$, $\mu > 0$. (b) Left-hand orientation of vectors **E**, **H**, **k** for the case when $\varepsilon < 0$, $\mu < 0$. The figure is taken, with permission of the Canadian Association of Physicists (CAP), from an article by Wartak *et al.* [27].

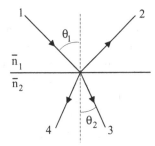

Fig. 17.5 Reflection and refraction in negative index material. The figure is taken, with permission of the Canadian Association of Physicists (CAP), from an article by Wartak *et al.* [27].

17.2.3 The refraction of a ray

Consider propagation of a ray through the boundary between left-handed and right-handed media, in Fig. 17.5. Light crossing the interface at non-normal incidence undergoes refraction, that is a change in its direction of propagation. The angle of refraction depends on the absolute value of the refractive index of the medium. Here, 1 is the incident ray, 2 is the reflected ray, 3 is the refracted ray assuming medium '2' is right-handed, and 4 is the refracted ray when assuming that medium '2' is left-handed. In a metamaterial the refraction of light would be on the same side of the normal as the incident beam, see Fig. 17.5. The relation between angles is determined by Snell's law.

17.3 How to create metamaterial?

17.3.1 Metamaterials with negative effective permittivity in the microwave regime

At optical frequencies metals are characterized by an electric permittivity that varies with frequency according to Drude relation

$$\varepsilon(\omega) = \varepsilon_0 \left[1 - \frac{\omega_p^2}{\omega(\omega + i\gamma)} \right] \qquad (17.5)$$

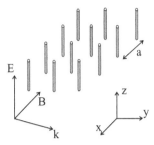

Fig. 17.6 Schematic illustration of a periodic arrangement of infinitely long thin wires along z direction used in the creation of an effective plasma medium at microwave frequencies. The figure is taken, with permission of the Canadian Association of Physicists (CAP), from an article by Wartak *et al.* [27].

Here $\omega_p^2 = \frac{Ne^2}{m\varepsilon_0}$ is the plasma frequency, i.e. the frequency with which the plasma consisting of free electrons oscillates in the presence of an external electric field. Typical values for ω_p are in the ultraviolet regime. The other symbols are: N is the electron density; e is charge of an electron and m its mass. The parameter γ describes damping and its value, for example for copper is $\gamma \approx 4 \times 10^{13}$ rad s^{-1}.

In the limit when $\gamma = 0$, from Eq. (17.5) it follows that $\varepsilon < 0$ for $\omega < \omega_p$; i.e. the medium is characterized by a negative permittivity. Considering typical values of ω_p, the resulting range of negative values of ε is in the ultraviolet regime. Unfortunately, in this frequency range $\omega \ll \gamma$, and as a result losses dominate the behaviour of ε. Thus, using metals to achieve negative ε over this frequency range will be impractical (high losses) and the propagation of light will be mainly evanescent.

To achieve negative ε at microwave frequencies, Pendry *et al.* [28] proposed to use periodic structure consisting of long thin metallic wires of radius r arranged on a horizontal plane (xy), see Fig. 17.6. The unit cell of this periodic structure is a square whose sides have length equal to a.

When electric field $\mathbf{E} = E_0 e^{-i(\omega t - kz)} \mathbf{z}$ is incident on this structure, it forces free electrons to move inside the wires in the direction of the incident field. Effective electron density N_{eff} of such structures which participate in plasma oscillations is $N_{eff} = N\frac{\pi r^2}{a^2}$ (with N being the electron density inside each wire, r the radius of a wire and a distance between wires), which is significantly smaller compared to N, thus reducing the effective plasma frequency. For example, for a wire with radius $r = 1$ μm and wire spacing $a = 5$ mm, one finds that $N_{eff} \approx 1.3 \times 10^{-7} N$; i.e. the effective electron density of the new medium is reduced by seven orders of magnitude compared to that of the free electron gas inside an isolated wire.

Additionally, in such engineered structures the effective mass m_{eff} of the electrons is significantly larger compared to that of a free electron. To determine effective mass of an electron in this wired medium, we use the classical equation of motion of a moving electron:

$$d(m\mathbf{v})/dt = e[\mathbf{E} + \mathbf{v} \times \mathbf{B}] \rightarrow \frac{d}{dt}[m\mathbf{v} + e\mathbf{A}] = -e\nabla(\varphi - \mathbf{v} \cdot \mathbf{A}) \qquad (17.6)$$

where e is the charge of an electron and $\mathbf{v}$ the velocity of an electron. One also has $\mathbf{B} = \nabla \times \mathbf{A}$ and $\mathbf{E} = -\nabla\varphi - \partial\mathbf{A}/\partial t$. We divide the xy plane into circles of radius R_c, centred at each wire and having area equal to that of the square unit cell; i.e. $R_c = a/\sqrt{\pi}$. Furthermore, we assume that the wires are sufficiently apart from each other so that the magnetic field inside each circle arises only from the current I that flows perpendicularly to the centre of the circle, and that the field at the circumference of each circle vanishes; i.e. $H(R_c) = 0$. Magnetic field intensity at a distance R from each wire is given by

$$H = \frac{I}{2\pi R}\left(1 - \frac{R^2}{R_c^2}\right) \quad (17.7)$$

Magnetic field **H** associated with a vector potential **A** according to the relation $\mathbf{H} = \mu_0^{-1}\nabla \times \mathbf{A}$ gives the following expression for the vector potential **A**:

$$\mathbf{A} = \frac{\mu_0 I}{2\pi}\left(ln(R_c/R) + \frac{R^2 - R_c^2}{2R_c^2}\right)\mathbf{z} \quad (17.8)$$

where **z** is a unit vector along the z-direction. It has been assumed that $\mathbf{A}(R \geq R_c) = 0$. For distances very close to the wires, i.e. for $R \to r << R_c$, Eq. (17.8) gives

$$\mathbf{A} \approx \frac{\mu_0 I}{2\pi}ln(a/r)\mathbf{z} \quad (17.9)$$

Since both **v** and **A** point along **z**, the right hand side of the second part of Eq. (17.6) will vanish (for a given R); i.e. the so-called 'conjugate momentum' $(m\mathbf{v} + e\mathbf{A})$ of an electron will be *conserved along the z-direction*. As a result, an electron will be moving in the above engineered medium with an effective mass, $m_{eff} = eA(r)/v$. Realizing that the current I can be simply re-expressed as $I = -(\pi r^2)(Nev)$, we finally obtain using Eq. (17.9) the effective mass of a moving electron inside our effective medium $m_{eff} = 0.5 \times \mu_0 Ne^2 r^2 ln(a/r)$. Thus, for copper wires of radius $r = 1$ μm, being separated by $a = 5$ mm, we obtain: $m_{eff} \approx 1.3 \times 10^4$ m; i.e. the effective mass of an electron in our engineered medium is increased by more than four orders of magnitude. This, combined with the fact that the effective electron density is reduced by approximately seven orders of magnitude, leads to an effective plasma frequency that is in the microwave regime:

$$\omega_p^2 = \frac{N_{eff}e^2}{m_{eff}\varepsilon_0} = 5.1 \times 10^{10}[\text{rad s}^{-1}]^2$$
$$\to v_p = \omega_p/2\pi = 8.2 \ GHz \quad (17.10)$$

It should be noted that, based on Eq. (17.10), the calculated wavelength, $\lambda_p = c/v_p$, which corresponds to our medium's effective plasma frequency, turns out to be considerably larger compared to the periodicity of the structure ($\lambda_p \approx 7a$), justifying the description of the periodic structure as an effective medium. Therefore we are able to construct an engineered medium that can exhibit a *negative* electric permittivity in the *microwave* regime with reasonably low losses.

17.3.2 Magnetic properties: split-ring resonators

In the previous section, we examined how to construct an artificial structure possessing negative effective permittivity ε. Here, we will show how to create negative effective permeability μ in the microwave regime using the so-called split-ring resonators (SRR).

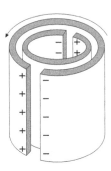

Fig. 17.7 Perspective view of split-ring resonator.

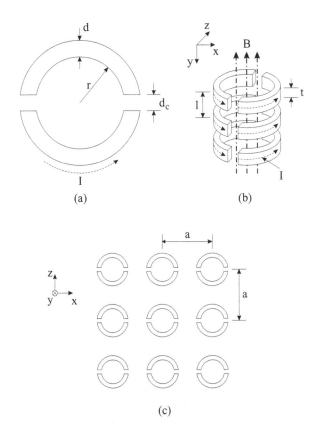

Fig. 17.8 Arrangement of SRR in space. Reprinted with permission from H. Chen *et al.*, *J. Appl. Phys.*, **100**, 024915 (2006). Copyright 2006, American Institute of Physics.

The perspective view of SRR is shown in Fig. 17.7. The structure consists of two cut cylinders. The model used to represent SRR is illustrated in Fig. 17.8. Here we will describe description of split-ring resonators (SRR) based on equivalent circuit model approach [18]. The structure under consideration is shown in Fig. 17.8.

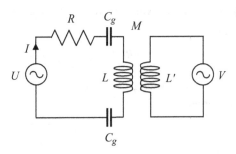

Fig. 17.9 Equivalent circuit of SRR.

We will establish the equivalent circuit model of this structure and show that it can produce negative effective permittivity μ. The unit cell shown above can be modelled using equivalent circuit model shown in Fig. 17.9.

Apply a time-varying external magnetic field H_0 in the y-direction. It will induce a current I flowing in each SRR unit. From Faraday's law, the voltage V due to the external field H_0 is

$$V = i\omega\mu_0\pi r^2 H_0 \qquad (17.11)$$

where r is the radius of the ring. Along the y-direction, SRR loops form a column (stack) which behaves like a solenoid. In each loop there is a current I flowing. Neglect the fringing effects, i.e. spreading of the magnetic field lines. The magnetic flux in such a column is thus

$$\Phi = \pi r^2 \mu_0 \frac{I}{l} \qquad (17.12)$$

where l is the separation between loops. The inductance L appearing in the circuit model is

$$L = \mu_0 \pi \frac{r^2}{l} \qquad (17.13)$$

The inductance L is defined per ring of an infinite column of rings.

Let L' be the total inductance of the SRR units in the other column (excluding column represented by inductance L). Coupling between L and L' is represented by the mutual inductance M. For very long columns in y-direction, the mutual inductance M is

$$M = \frac{\Phi_L}{I} = \lim_{n\to\infty} \frac{\pi r^2}{na^2} \frac{\phi_d}{I} = \lim_{n\to\infty} \frac{\pi r^2}{na^2}(n-1)L$$
$$= \frac{\pi r^2}{a^2} L = F \cdot L \qquad (17.14)$$

Here Φ_L is the flux of the depolarization field located in the interior of L, $\Phi = L \cdot I$ is the depolarization field generated by one column of the rings, $\phi_d = (n-1)\Phi = (n-1)L \cdot I$ is the flux of the total depolarization field, and $F = \frac{\pi r^2}{a^2}$ is the fractional volume of the periodic unit cell in the xz plane occupied by the interior of the SRR.

Applying Kirchoff's voltage drop law for a loop in the equivalent circuit shown in Fig. 17.9 gives

$$V = R \cdot I + \frac{I}{-i\omega C} + (-i\omega L) I - (-i\omega M) I \tag{17.15}$$

where $C = \frac{1}{2}C_g$ is the total capacitance in the loop.

From the previous equations

$$V = i\omega \mu_0 \pi r^2 H_0 = i\omega L\, I\, H_0 \tag{17.16}$$

From Eq. (17.15) one has

$$V = \left(R - \frac{1}{i\omega C} - i\omega L + i\omega M\right) I$$

Solving for current

$$I = \frac{V}{R - \frac{1}{i\omega C} - i\omega L + i\omega M} = \frac{i\omega\, L\, I\, H_0}{R - \frac{1}{i\omega C} - i\omega L + i\omega F \cdot L} \tag{17.17}$$

$$= \frac{L\, H_0}{\frac{R}{i\omega L} + \frac{1}{\omega^2 LC} - 1 + F} = \frac{-L\, H_0}{(1-F) - \frac{1}{\omega^2 LC} + i\frac{R}{\omega L}}$$

Magnetic dipole moment per unit volume of the material is

$$M = \frac{\pi r^2}{la^2} I$$

with the current I inferred from the above equation to be

$$I = -\frac{H_0 l}{(1-F) - 1/(\omega^2 LC) + iR/(\omega L)} \tag{17.18}$$

As a result, the (relative) effective magnetic permeability associated with this medium will be (in the direction, x, that the incident magnetic field is polarized)

$$\mu_r = \frac{B/\mu_0}{B/\mu_0 - M_d} = 1 - \frac{F}{1 - 1/(\omega^2 LC) + iR/(\omega L)} \tag{17.19}$$

From Eq. (17.19) we can see that μ assumes negative values in the range: $1/\sqrt{LC} < \omega_p < 1/\sqrt{LC(1-F)}$, where $\omega_{m0} = 1/\sqrt{LC}$ is the resonance frequency of the Lorentzian variation of the medium's magnetic permeability, and $\omega_{mp} = 1/\sqrt{LC(1-F)}$ is the corresponding plasma frequency (where $\text{Re}\{\mu\} = 0$). Crucially, we note that the resonant wavelength (λ_{m0}) of the structure depends entirely on the rings' effective inductance (L) and capacitance (C), and can therefore be made considerably larger than the periodicity (a) of the structure, thereby fully justifying its description as an effective medium.

Plot of real and imaginary parts of effective magnetic permeability is shown in Fig. 17.10 and the MATLAB code is in Appendix, Listing 17A.1.

The combination of long wires and SRR in a unit cell is shown in Fig. 17.11. The effect of both results in a negative effective permittivity and permeability over the same frequency band.

We finish this section by showing in Fig. 17.12 schematic plots of effective permittivity ($\varepsilon_{\mathit{eff}}$) and permeability ($\mu_{\mathit{eff}}$).

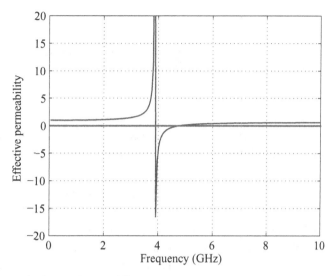

Fig. 17.10 Real and imaginary parts of magnetic permeability.

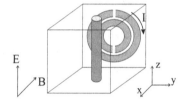

Fig. 17.11 Elementary cell of metamaterial formed by SRR and thin wire. The figure is taken, with permission of the Canadian Association of Physicists (CAP), from an article by Wartak *et al.* [27].

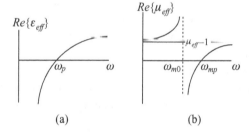

Fig. 17.12 (a) Permittivity of wire medium demonstrating plasma-like frequency dependent permittivity and (b) Frequency response of effective permeability. The figure is taken, with permission of the Canadian Association of Physicists (CAP), from an article by Wartak *et al.* [27].

17.4 Some applications of metamaterials

Materials with such unusual properties allow for unusual applications. We do not attempt to review all of them; we just concentrate on a few interesting possibilities. We start with the possibility of creating a so-called 'perfect lens'.

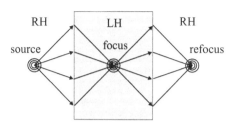

Fig. 17.13 Double focussing of a source with a planar double negative (left-handed) metamaterials slab surrounded by regular (right-handed) dielectrics.

17.4.1 Perfect lenses

Until recently, it was thought that the manipulation of light is limited by the fundamental law of diffraction to a relatively long wavelengths (say around 0.5λ). The sub-wavelength details are carried by evanescent harmonics which decay exponentially and are also subject to noise. A conventional lens only collects the propagating waves. The evanescent waves are lost due to their decay.

A planar slab consisting of NIM with sufficient thickness can act as a lens, known as Veselago lens. Such a slab lens with refractive index $n = -1$ placed in vacuum (see Fig. 17.13) can resolve details of an object with subwavelength precision [11]. Detailed mathematical analysis is discussed, for example, by Ramakrishna and Grzegorczyk [4] and Cai and Shalaev [29].

To understand the problem, consider a slab of thickness d and refractive index of, e.g. $n = -1$, surrounded by air. It will bring all rays emanating from a source to a double focus: first, at a point inside the NIM slab, at a distance $s = l < d$, where l is the distance of the source from the slab, and second at a point outside the slab, at a distance $d - l$. Hence, such a slab acts like a lens, and is able to bring the rays radiated by a source to a focus outside the slab, without reflections occurring at the media interfaces because the $n = -1$ slab is impedance-matched to free space.

Evanescent waves are associated with the high spatial frequencies of the electromagnetic waves created by source. They carry fine (subwavelength) features of the source. Therefore, a NIM slab can, in principle, enable us to obtain the image of an object with 'perfect' resolution, containing all the subwavelength features of an object and overcoming the usual diffraction limitations that characterize conventional lenses.

17.4.2 Stopped light in metamaterials

For decades scientists maintained that optical data cannot be stored statically and must be processed and switched on the fly. The reason for this conclusion was that stopping and storing an optical signal by dramatically reducing the speed of light itself was thought to be unfeasible.

At present, some of the most successful slow-light designs based on photonic-crystals (PhCs) [30] or coupled-resonator optical waveguides (CROWs) [31], can indeed slow

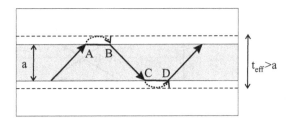

Fig. 17.14 Classical G-H shift in regular dielectrics. The figure is taken, with permission of the Canadian Association of Physicists (CAP), from an article by Wartak *et al.* [27].

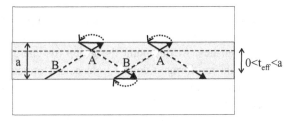

Fig. 17.15 G-H shift in NIM forming central layer. The figure is taken, with permission of the Canadian Association of Physicists (CAP), from an article by Wartak *et al.* [27].

down light efficiently by a factor of only 40; otherwise, large group-velocity-dispersion and attenuation-dispersion occur; i.e. the guided light pulses broaden and the attainable bandwidth is severely restricted.

Recently Tsakmakidis *et al.* [14] proposed a new method that can allow for a true stopping of light in NIM. The stopping of light in the proposed configuration is associated with a negative Goos-Hänchen (G-H) phase shift, a lateral displacement of light ray when it is totally reflected at the interface of two different dielectric media. Classical G-H shift between dielectrics having positive refractive indices is illustrated in Fig. 17.14.

In a structure where the central layer is formed by NIM, the G-H shift is reversed as it is illustrated in Fig. 17.15.

To more precisely understand the manner in which light is decelerated in this structure, let us imagine a ray of light propagating in a zig-zag fashion along a waveguide with a negative-index ('left-handed') core. The ray experiences negative Goos-Hänchen lateral displacements each time it strikes the interfaces of the core with the positive-index ('right-handed') claddings, see Fig. 17.15. Accordingly, the cross points of the incident and reflected rays will sit inside the left-handed core and the effective thickness of the guide will be smaller than its natural thickness. It is reasonable to expect that by gradually reducing the core physical thickness, the effective thickness of the guide will eventually vanish. Obviously, beyond that point the ray will not be able to propagate further down, and will effectively be trapped inside the negative index metamaterial (NIM) heterostructure, see Fig. 17.16.

The authors of [14] proposed stopping a light pulse by varying the thickness of the waveguide core to the point where the cycle-averaged power flow in the core and the

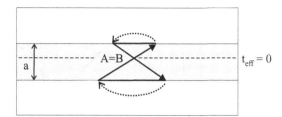

Fig. 17.16 G-H shift in NIM at critical thickness. The figure is taken, with permission of the Canadian Association of Physicists (CAP), from an article by Wartak *et al.* [27].

cladding become comparable. At the degeneracy point, where the magnitudes of these powers become equal, the total time-averaged power flow directed along the central axis of the core vanishes. At this point the group (or energy) velocity goes to zero and the path of the light ray forms a double light cone ('optical clepsydra') where the negative GH lateral shift experienced by the ray is equal to its positive lateral displacement as it travels across the core. Adiabatically reducing the thickness of the NIM core layer may thus, in principle, enable complete trapping of a range of light rays, each corresponding to a different frequency contained within a guided wavepacket.

This ability of metamaterial-based heterostructures to dramatically decelerate or even *completely stop* [32] light under realistic experimental conditions, has recently led to a series of experimental works [33], [34] that have reported an observation of so-called 'trapped rainbow' light-stopping in metamaterial waveguides.

17.4.3 Cloaking (invisibility)

Unusual light-bending properties constitute a rather generic feature of metamaterials. Part of the excitement surrounding these materials is that they could be engineered to 'cloak' objects from electromagnetic radiation such as light: that is, make them seem invisible at specific frequencies. A metamaterial 'invisibility' cloak can be designed such that it does not reflect waves back nor scatter them in other directions. Several methods were proposed to make extended bodies invisible, such as those based on cancellation of scattering [35] or on coordinate transformations. The method suggested by Pendry *et al.* [36] (see also Leonhardt [12]) relies on controlling the paths of electromagnetic waves. It was applied to a spherical volume and uses coordinate transformation that expels the paths of electromagnetic waves (rays) from a spherical volume, squeezing them into a spherical shell around the volume that is to be cloaked, thereby making it invisible to incident radiation, see Fig. 17.17. The light rays smoothly avoid the cloaked object and flow around it like a fluid. They appear to have properties of the free space when observed externally. The rays make a detour around the hidden part of the device. On the left side they must arrive at the same time as if they were propagating through empty space. Since in the cloaking region they travel longer distances, their phase velocity must exceed c. This is in principle possible [37] (for a specific frequency).

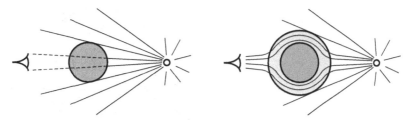

Fig. 17.17   The rays go around the inner object and then go to the eye. An observer at the left of the cloak would see the point source.

To introduce the relevant transformation, one starts with a mathematical point (placing it at the origin of a coordinate system), which is obviously invisibly small. To hide an extended object, say a sphere of radius R_1, we transform the mathematical point at the origin into a sphere of radius R_1 and the vicinity of the sphere into another sphere of radius $R_2 > R_1$. The transformation which does this is

$$r' = R_1 + \frac{R_2 - R_1}{R_2} r, \quad \theta' = \theta, \quad \phi' = \phi \qquad (17.20)$$

Maxwell's equations are form-invariant to coordinate transformations like the one above. Only the components of μ and ε are affected by the transformation. They become spatially varying and anisotropic tensors. Their forms were determined by Pendry *et al.* [36] and for $R_1 < r < R_2$ are

$$\begin{aligned}
\varepsilon'_{r'} &= \mu'_{r'} = \frac{R_2}{R_2 - R_1} \frac{(r' - R_1)^2}{r'} \\
\varepsilon'_{\theta'} &= \mu'_{\theta'} = \frac{R_2}{R_2 - R_1} \\
\varepsilon'_{\phi'} &= \mu'_{\phi'} = \frac{R_2}{R_2 - R_1}
\end{aligned} \qquad (17.21)$$

Inside the sphere with radius R_1 the permittivity and permeability can be arbitrary, whereas the space between the spheres is filled with a material having the permittivity and permeability tensors determined by Eqs. (17.21). As a result, any object inside the sphere with radius $r < R_1$ is concealed. These conclusions follow from exact manipulations of Maxwell's equations and are not restricted to a ray approximation.

17.4.4 Optical black holes

By extending the concept behind the invisibility cloak, one can speculate on the possibility of creating an optical analogue of a black hole. In this concept, by proper design of a new class of metamaterials, one can expect to concentrate and trap light waves, similarly to what can happen in a 'black hole', see Fig. 17.18. In such a system, light will be permanently trapped. Such a black hole design has recently been proposed by Narimanov and Kildishev [38]. Numerical simulations showed a highly efficient light absorption. The electromagnetic black hole was built recently [39], [40] and it operates at microwave

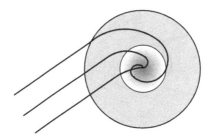

Fig. 17.18 Metamaterial black hole fabricated from properly crafted gradient-index material.

frequencies. The structure is composed of 60 concentric layers, and each layer is a thin printed circuit board etched with a numbers of subwavelength unit structures on one side and coated with 0.018 mm thick copper on the other side. The permittivity changes radially in the shell of the microwave black hole, and hence the unit cells are identical in each layer but have different sizes in adjacent layers. The structure can efficiently absorb electromagnetic waves coming from all directions owing to the local control of electromagnetic fields. It is expected that such a microwave black hole could find important applications in solar-light harvesting, thermal emitting, and cross-talk reduction in microwave circuits and devices. These specially designed analogues of black holes could also be used for controlling, slowing and trapping electromagnetic waves [40], as well as for investigating some of the more exotic physics associated with celestial mechanics.

17.5 Metamaterials with an active element

This last section is intended to briefly discuss the problem of losses in metamaterials and how to compensate for them.

As mentioned before, metamaterials show large losses which at present are orders of magnitude too large for practical applications and are considered as an important factor limiting practical applications of metamaterials. For example, detailed analytical studies show that losses limit the superresolution of a theoretical superlens [41]. There was some controversy whether loss elimination can be feasible [42], but as shown by Webb and Thylen [43] it is possible to completely eliminate losses in metamaterials.

Very recently, several computational [44], [45], [46] and experimental [47] works have demonstrated that optical losses can be fully overcome in realistic negative-refractive-index metamaterials. The specific loss-free design considered in [44] and [47] consisted of two metallic films perforated with small rectangular holes ('fishnets'), and with an active medium (laser dye) spacer between two films. Additionally, several reports [48], [49], [50] speculated about possible compensation for losses in metamaterials by introducing a gain element. For example, Wegener *et al.* [49] formulated a simple model where gain is represented by a fermionic two-level system which is coupled via a local-field to a single bosonic resonance representing the plasmonic resonance of the metamaterial. Also recently,

Fang *et al.* [50] described a model where the gain system is modelled by a generic four-level atomic system. They conducted numerical analysis using the FDTD technique. Gain material was introduced in the gap region of the split-ring resonators (SRR). The system had a magnetic resonance frequency at 100 THz. Some other reports of the design and analysis of active metamaterials are by Yuan *et al.* [51] and Sivan *et al.* [32].

17.6 Annotated bibliography

We finish this short review by highlighting some recently published textbooks where further useful information about metamaterials can be found. We start with the most recent:

- T. J. Cui, D. R. Smith, R. Liu, eds., *Metamaterials. Thory, Design, and Applications*, Springer, 2010.
  Concentrates on the recent progress in metamaterials, in particular the optical transformation theory, invisible cloaks, new type of antennas and optical metamaterials.
- W. Cai and V. Shalaev, *Optical Metamaterials. Fundamentals and Applications*, Springer, 2010.
  Details recent advances on optical metamaterials from fundamental aspects to current implementations.
- F. Capolino, ed., *Metamaterials Handbook. Vol. 1: Theory and Phenomena of Metamaterials, Vol. 2: Applications of Metamaterials*, CRC Press, Boca Raton, 2009.
  A very broad collection of topics covered by many authors.
- L. Solymar and E. Shamonina, *Waves in Metamaterials*, Oxford University Press, 2009.
  Offers a comprehensive treatment of all aspects of research in this field at a level that should appeal to final year undergraduates in physics or in electrical and electronic engineering.
- S. A. Ramakrishna and T. M. Grzegorczyk, *Physics and Applications of Negative Refractive Index Materials*, SPIE and CRC Press, 2009.
  Discussion of principles of negative refraction and comparison with other media that exhibit similar properties.

Appendix 17A: MATLAB listings

In Table 17.1 we provide a list of MATLAB files (in fact, only one file) created for Chapter 17 and a short description of its function.

Table 17.1 List of MATLAB functions for Chapter 17.

Listing	Function name	Description
17A.1	*ChenFig3.m*	Reproduces Fig. 3 from the paper by Chen *et al.* [18]

Listing 17A.1 Program ChenFig3.m.

```
% File name: ChenFig3.m
% Matlab code to plot Fig.3 of Chen et al
% using equivalent circuits approach
% Plot Fig3. of Chen et al, JAP,v.100, 024915 (2006)
clear all
% Fundamental constants
epsilon_zero=8.8542d-12;    % epsilon zero (F/m)
mu_zero = 4*pi*10^-7;       % mu zero (H/m)
%
% Geometrical dimensions, see Fig.1 of Chen
r = 2.0;                    % internal radius = 2 mmm
a = 5.0;                    % distance between elements = 5 mm
l_vert = 1.0;               % vertical separation = 1 mm
L_c = 1.0;                  % Dimension of capacitor = 1 mm
t = 0.8;                    % Dimension of capacitor = 0.8 mm
d_c = 0.2;                  % Dimension of capacitor = 0.2 mm
eps_r = 4.0;                % permittivity of the gap
sigma = 10^-6;              % resistance per unit length
%
% Intermediate parameters
F = pi*r^2/(a^2);           % fractional volume (dimensionless)
C_g = (epsilon_zero*eps_r*L_c*t/d_c)*10^-3; % gap capacitance unit (F)
L = (mu_zero*pi*r^2/l_vert)*10^-3;          % Eg.(2) unit (H)
% pause
R = 2*pi*r*sigma;                       % resistance of the ring
C = C_g/2.0;                            % total capacitance in the loop
%
% Determination of frequency range
N_max = 200;                            % number of points for plot
                                        % in the interval [0,10] GHz
freq_GHz = linspace(0,10,N_max);        % creation of frequency arguments
freq_Hz = freq_GHz*10^9;
omega_Hz = 2.0*pi*freq_Hz;
%
tt = 1./(omega_Hz.^2*L*C);
ttt = R./(omega_Hz.*L);
temp = 1./tt + 1i*ttt;
%
% effective permeability without coupling M, Eq.(9) of Chen
extra = F./(1. + F - tt + 1i*ttt);
%
mu_eff = 1. - F./(1. + F - tt + 1i*ttt);
result_re = real(mu_eff);
result_im = imag(mu_eff);
```

```
%
% plot(freq_GHz,result_im);
% pause
plot(freq_GHz,result_re,freq_GHz,result_im,'.','LineWidth',1.5);
axis([0 10 -20 20])
xlabel('Frequency (GHz)','FontSize',14);
ylabel('Effective permeability','FontSize',14);
set(gca,'FontSize',14);          % size of tick marks on both axes
grid
pause
close all
```

References

[1] M. Lapine and S. Tretyakov. Contemporary notes on metamaterials. *IET Microw. Antennas Propag.*, **1**:3–11, 2007.

[2] L. I. Mandelstam. Group velocity in crystal lattice. *Zh. Eksp. Teor. Fiz.*, **15**:475–8, 1945.

[3] V. G. Veselago. The electrodynamics of substances with simultaneously negative values of ϵ and μ. *Soviet Physics Uspekhi*, **10**:509–14, 1968.

[4] S. A. Ramakrishna and T. M. Grzegorczyk. *Physics and Applications of Negative Refractive Index Materials*. SPIE Press and CRC Press, Bellingham, WA, 2009.

[5] V. A. Podolskiy, L. Alekseev, and E. E. Narimanov. *J. Mod. Optics*, **52**:2343–9, 2005.

[6] A. Schuster. *An Introduction to the Theory of Optics*. Edward Arnold, London, 1904.

[7] H. Lamb. On group-velocity. *Proc. London Math. Soc*, **1**:473–9, 1904.

[8] H. C. Pocklington. *Nature*, **71**:607–8, 1905.

[9] C. R. Simovski and S. A. Tretyakov. Historical notes on metamaterials. In F. Capolino, ed., *Theory and Phenomena of Metamaterials*, pages **1**: 1–17. CRC Press and Taylor and Francis Group, Boca Raton and London, 2009.

[10] A. Moroz, www.wave-scattering.com/negative.html, 3 July 2012.

[11] J. B. Pendry. Negative refraction makes a perfect lens. *Phys. Rev. Lett.*, **85**:3966–9, 2000.

[12] U. Leonhardt. Optical conformal mapping. *Science*, **312**:1777–80, 2006.

[13] D. Schurig, J. J. Mock, B. J. Justice, S. A. Cummer, J. B. Pendry, A. F. Starr, and D. R. Smith. Metamaterial electromagnetic cloak at microwave frequencies. *Science*, **314**:977–80, 2006.

[14] K. L. Tsakmakidis, A. D. Boardman, and O. Hess. 'Trapped rainbow' storage of light in metamaterials. *Nature*, **450**:397–401, 2007.

[15] S. O'Brien, D. McPeake, S. A. Ramakrishna, and J. B. Pendry. Near-infrared photonic band gaps and nonlinear effects in negative magnetic metamaterials. *Phys. Rev. B*, **69**:241101(R), 2004.

[16] V. M. Shalaev. Optical negative-index metamaterials. *Nature Photonics*, **1**:41–8, 2007.

[17] J. B. Pendry, A. J. Holden, D. J. Robbins, and W. J. Stewart. Magnetism from conductors and enhanced nonlinear phenomena. *IEEE Trans. Microw. Theory Tech.*, **47**:2075–84, 1999.

[18] H. Chen, L. Ran, J. Huangfu, T. M. Grzegorczyk, and J. A. Kong. Equivalent circuit model for left-handed metamaterials. *J. Appl. Phys.*, **100**:024915, 2006.

[19] J. B. Pendry, A. J. Holden, W. J. Stewart, and I. Youngs. Extremely low frequency plasmons in metallic mesostructures. *Phys. Rev. Lett.*, **76**:4773–6, 1996.

[20] D. R. Smith, W. J. Padilla, D. C. Vier, S. C. Nemat-Nasser, and S. Schultz. Composite medium with simultaneously negative permeability and permittivity. *Phys. Rev. Lett.*, **84**:4184–7, 2000.

[21] R. A. Shelby, D. R. Smith, and S. Schultz. Experimental verification of a negative index of refraction. *Science*, **292**:77–9, 2001.

[22] C. G. Parazzoli, R. B. Greegor, K. Li, B. E. C. Koltenbah, and M. Tanielian. Experimental verification and simulation of negative index of refraction using Snell's law. *Phys. Rev. Lett.*, **90**:107401, 2003.

[23] T. J. Yen, W. J. Padilla, N. Fang, D. C. Vier, D. R. Smith, J. B. Pendry, D. N. Basov, and X. Zhang. Terahertz magnetic response from artificial materials. *Science*, **303**:1494–6, 2004.

[24] N. Liu, H. Guo, L. Fu, S. Kaiser, H. Schweizer, and H. Giessen. Three-dimensional photonic metamaterials at optical frequencies. *Nature Materials*, **7**:31–7, 2008.

[25] J. Yao, Z. Liu, Y. Liu, Y. Wang, C. Sun, G. Bartal, A. M. Stacy, and X. Zhang. Optical negative refraction in bulk metamaterials of nanowires. *Science*, **321**:930, 2008.

[26] J. Valentine, S. Zhang, T. Zentgraf, E. Ulin-Avila, D. A. Genov, G. Bartal, and X. Zhang. Three-dimensional optical metamaterial with a negative refractive index. *Nature*, **455**:376–80, 2008.

[27] M. S. Wartak, K. L. Tsakmakidis, and O. Hess. Introduction to metamaterials. *Physics in Canada*, **67**:30–4, 2011.

[28] J. B. Pendry, A. J. Holden, D. J. Robbins, and W. J. Stewart. Low frequency plasmons in thin-wire structures. *J. Phys. Condens. Matter*, **10**:4785–809, 1998.

[29] W. Cai and V. Shalaev, eds. *Optical Metamaterials. Fundamentals and Applications*. Springer, New York, 2010.

[30] B. Corcoran, C. Monat, C. Grillet, D. J. Moss, T. P. Eggleton, B. J. White, L. O. O'Faolain, and T. F. Krauss. Green light emission in silicon through slow-light enhanced third-harmonic generation in photonic crystal waveguides. *Nature Photonics*, **3**:206–10, 2009.

[31] F. Xia, L. Sekaric, and Y. Vlasov. Ultracompact optical buffers on a silicon chip. *Nature Photonics*, **1**:65–71, 2006.

[32] Y. Sivan, S. Xiao, U. K. Chettiar, A. V. Kildishev, and V. M. Shalaev. Frequency-domain simulations of a negative-index material with embedded gain. *Opt. Express*, **26**:24060–74, 2009.

[33] X. P. Zhao, W. Luo, J. X. Huang, Q. H. Fu, K. Song, C. Cheng, and C. R. Luo. Trapped rainbow effect in visible light left-handed heterostructures. *Appl. Phys. Lett.*, **95**:071111, 2009.

[34] V. N. Smolyaninova, I. I. Smolyaninov, A. V. Kildishev, and V. M. Shalayev. Experimental observation of the trapped rainbow. *Appl. Phys. Lett.*, **96**:211121, 2010.

[35] A. Alu and N. Engheta. Plasmonic materials in transparency and cloaking problems: mechanism, robustness, and physical insights. *Opt. Express*, **15**:3318–32, 2007.

[36] J. B. Pendry, D. Schuring, and D. R. Smith. Controlling electromagnetic fields. *Science*, **312**:1780, 2006.

[37] P. W. Milonni. *Fast Light, Slow Light and Left-Handed Light*. Institute of Physics Publishing, Bristol and Philadelphia, 2005.

[38] E. E. Narimanov and A. V. Kildishev. Optical black hole: Broadband omnidirectional light absorber. *Appl. Phys. Lett.*, **95**:041106, 2009.

[39] Q. Cheng and T. J. Cui. *An Electromagnetic Black Hole Made of Metamaterials*, 2009. arXiv:0910.2159v1.

[40] Q. Bai, J. Chen, N.-H. Shen, C. Cheng, and H.-T. Wang. Controllable optical black hole in left-handed materials. *Opt. Express*, **18**:2106–15, 2010.

[41] C. Hafner, C. Xudong, and R. Vahldieck. Resolution of negative-index slabs. *J. Opt. Soc. Am. A*, **23**:1768–78, 2006.

[42] M. I. Stockman. Criterion for negative refraction with low optical losses from a fundamental principle of causality. *Phys. Rev. Lett.*, **98**:177404, 2007.

[43] K. J. Webb and L. Thylen. Perfect-lens-material condition from adjacent absorptive and gain resonances. *Optics Letters*, **33**:747–9, 2008.

[44] S. Wuestner, A. Pusch, K. L. Tsakmakidis, J. M. Hamm, and O. Hess. Overcoming losses with gain in a negative refractive index metamaterial. *Phys. Rev. Lett.*, **105**:127401, 2010.

[45] K. L. Tsakmakidis, M. S. Wartak, J. J. H. Cook, J. M. Hamm, and O. Hess. Negative-permeability electromagnetically induced transparent and magnetically-active metamaterials. *Phys. Rev. B*, **81**:195128, 2010.

[46] A. Fang, Th. Koschny, and C. M. Soukoulis. Self-consistent calculations of loss-compensated fishnet metamaterial. *Phys. Rev. B*, **82**:121102, 2010.

[47] S. Xiao, V. P. Drachev, A. V. Kildishev, X. Ni, U. K. Chettiar, H. K. Yuan, and V. M. Shalaev. Loss-free and active optical negative-index metamaterials. *Nature*, **466**:735–40, 2010.

[48] A. Bratkovsky, E. Ponizovskaya, S.-Y. Wang, P. Holmstrom, L. Thylen, Y. Fu, and H. Agren. A metal-wire/quantum-dot composite metamaterial with negative ε and compensated optical loss. *Appl. Phys. Lett.*, **93**:193106, 2008.

[49] M. Wegener, J. L. Garcia-Pomar, C. M. Soukoulis, N. Meinzer, M. Ruther, and S. Linden. Toy model for plasmonic metamaterial resonances coupled to two-level system gain. *Optics Express*, **16**:19785–98, 2008.

[50] A. Fang, Th. Koschny, M. Wegener, and C. M. Soukoulis. Self-consistent calculation of metamaterials with gain. *Phys. Rev. B*, **79**:241104(R), 2009.

[51] Y. Yuan, B.-Iopa Popa, and S. A. Cummer. Zero loss magnetic metamaterials using powered active unit cells. *Optics Express*, **17**:16135–43, 2009.

Appendix A Basic MATLAB

In this Appendix we discuss basic elements of MATLAB, including code design and specific examples of programming in MATLAB. Some introductory books are: *MATLAB Programming for Engineers* by Chapman [1], *MATLAB. An Introduction with Applications* by Gilat [2] and *Mastering MATLAB 5. A Comprehensive Tutorial and Reference* by Hanselman and Littlefield [3] – a reference book with a broad range of examples.

MATLAB can operate in two modes:

1. from the command window (interactive mode), where one introduces commands at MATLAB prompt
2. from the m-file (script). All commands are written in a file (with *.m extension). Such files are activated (run) by typing their names (without an extension) at MATLAB prompt. Each *.m file can call several other *.m files.

We would always advocate using m-files. They can be reused, combined with other programs, etc. In short, all code can be used in both ways.

The Appendix is divided into several sections, each dealing with different aspects of MATLAB. In the code provided, we put many comments describing its workings. We suggest that in each case the reader runs the code and analyses its outputs.

In Table A.1 we provide a list of MATLAB files created for this Appendix A and a short description of each function.

Before we start a more detailed description of MATLAB and its rules, we will introduce and run sessions dealing with fundamental MATLAB issues.

Table A.1 List of MATLAB functions for Chapter A.

Listing	Function name	Description
A.1	*intro_session.m*	Preliminary use of MATLAB
A.2	*memory.m*	Compares execution times for various memory handling
A.3	*loops.m*	Compares execution times for various loops
A.4	*basic_2D_plot.m*	Basic 2D MATLAB plot
A.5	*sub_plots.m*	Introduces subplots
A.6	*pview.m*	3D plot
A.7	*file_write.m*	Writes to a file
A.8	*file_read.m*	Reads from file
A.9	*deriv.m*	Numerical differentiation using MATLAB function $diff()$

A.1 Working session with m-files

Here, we will illustrate preliminary use of MATLAB using the m-file. The full code named *intro$_s$ession.m* is shown below. We suggest run this code and observe the results. Try also to modify it and analyse the outputs. We put many comments inside the code to explain its operation.

```
% File name: intro_session.m
%---------------------- Initialization -------------------------
clear all
'its me'                % outputs text on screen
pause
d=6;                    % assigns value 6 to variable d
C=[1 2 3; 3 2 1; 4 5 6];% defines matrix C
C
who                     % lists defined matrices
pause
whos                    % lists the matrices and their sizes
pause
%----------------- Colon operator ------------------------------
x =C( : , 3)            % selects third column of C matrix
pause
format long             % shows variables on a screen in a long format
% format short
t = 0:0.1:2;
t                       % lists value of t
pause
t'                      % transpose operator, creates vertical vector
pause
%------------ Special values and special matrices --------------
pi                      % shows the value of pi
1i
Inf
clock                   % year, month, day, hour, minute, seconds
pause
date                    % date in string format
eps                     % the smallest number on my computer
pause
A = zeros(4)            % creates 4x4 matrix consisting of zeros
B = zeros(4,3)          % creates 3x4 matrix consisting of zeros
pause
A1 = ones(3)            % creates 3x3 matrix consisting of ones
B1 = ones(4,3)          % creates 3x4 matrix consisting of ones
```

```
G = ones(1,5)           % creates vector, length 5 consisting of ones
pause
D = eye(3)              % creates identity matrix

z=input('input   z');
z                       % shows inputed number
pause
%------------ Prints a short table of cos values ----------------
clc                     % clears screen
n = 21; x = linspace(0,1,n); y = cos(2*pi*x);
disp(' ')
disp('k      x(k)    cos(x(k))')
disp('------------------------')
for k=1:n
   degrees = (k-1)*360/(n-1);
   fprintf(' %2.0f   %5.0f %8.3f \n',k,degrees,y(k));
end
disp( ' ');
disp('x(k) is given in degrees.')
pause
%------------ Output Options ----------------------------------
a = 1.23456789;
format short
a                       % display a in short format
pause
format long
a                       % display a in long format
disp(a)                 % another display of a in long format
fprintf('Its me,\n %4.24f ', a)
pause
%--------------------- Data files --------------------------
save data_1 x y;        % as *.mat file
pause
load data_1;
z=x'
save data_5.dat z /ascii;
pause
%------------ Scalar and array operations ---------------------
clear all
2+5
A=[1 2 3; 3 2 1; 4 5 6]; B=[2 3 4; 4 5 6; 5 6 7];
%
C1=A+B                  % addition of two matrices
C2=A*B                  % matrix multiplication
```

```
C3=A.*B                    % element by element multiplication
C4=A./B                    % element by element division
C5=A.^B                    % element by element exponentiation
pause
```

A.2 Basic rules

Several of the special symbols used in MATLAB are listed in Table A.2. All variables in MATLAB are treated as matrices. Each operation is therefore considered as a matrix operation and as such has been optimized for vector and matrix operations. For the best use, one should vectorize algorithms and loops and also reserve in advance memory for all objects (matrices). We summarize basic rules and illustrate them with preliminary examples.

- vectorization of loops. This means replacement of *for* and *while* loops by proper vector or matrix notation.

Suppose that we need to generate 101 values of sin function in the interval from 0 to 1. It can be done in two ways:

1. using *for* loop (no vectorization)

```
t0 = clock; for n = 0:0.001:1
    x = sin(n)
end
time_no_vect = etime(clock,t0)
```

Run time on my computer is *time_no_vect* $= 0.2300$.

2. applying vectorization as shown below

```
t0 = clock;
n = 0:0.001:1;
```

Table A.2 Some of the special symbols used in MATLAB.

Symbol	Description
=	assigning value
[]	construction of an array
+	array addition
−	array subtraction
.*	array multiplication
*	matrix multiplication
./	array right division
'	transpose operator
%	beginning of a comment

```
x = sin(n);
time_vect = etime(clock,t0);
```

Now, run time on the same machine is *time_vect* = 0.0810. In order to determine time difference between start and end points, we used MATLAB function *etime*, which accurately computes time differences over many orders of magnitude.

- reserve memory.
- use colon (:) notation
- limit using *for* in loops
- use m-files with functions instead of scripts
- provide extensive comments
- create descriptive names for variables
- use dot (.) notation

Some of the above rules will be discussed now in more detail.

A.3 Some rules about good programming in MATLAB

Here we summarize several general rules of programming including those specific to MATLAB (but not solely).

1. Always comment your program, so another person (even yourself after several months) can understand it.
    - use meaningful variable names,
    - use units consistently in your formulas or (better) work with dimensionless (or scaled) variables.
2. All calculations in MATLAB can be either performed in the Command Window or m-files can be created containing the same source code as typed in the Command Window. m-files play the role of functions or subroutines.
3. A semicolon at the end of all MATLAB statements suppresses printing the assigned values on the screen, in the Command Window. This speeds up program execution.

A.3.1 Preallocate memory

Always preallocate memory for the generated matrix. It increases execution speed because MATLAB does not need to increase matrix size after determining each element. Run and experiment with program *memory_res.m* shown below. A significant difference in the execution time can be observed.

```
% File name: memoryMSW.m
% comparison of execution time with and without memory reservation
clear all
% regular loop without memory allocation
```

```
% we use functions tic and toc to measure elapsed time
N = 10000;
tic;      % start timer
n = 0;
for x = 1:N
    n = n + 1;
    y(n) = cos(2*pi*x);
end
toc       % measure elapsed time since last call to tic
%
% the same loop with preallocated memory
tic
n = 0;
y  =zeros(1,N);
for x = 1:N
    n = n + 1;
    y(n) = cos(2*pi*x);
end
toc
```

A.3.2 Vectorize loops

MATLAB has been designed for vectorial operations. They execute much faster than regular loops. The comparison of both methods is illustrated in the program *loops.m* shown below. We use MATLAB functions *tic* and *toc* to measure elapsed time.

```
% File name: loops.m
% comparison of execution time for regular loop and vector loop
clear all
% regular loop is executed first
% we use functions tic and toc to measure elapsed time
tic;         % start timer
n = 0;
for x = 1:0.1:1000
    n = n + 1;
    y(n) = cos(2*pi*x);
end
toc          % measure elapsed time since last call to tic
%
% the same loop is vectorized below
tic
x = 0:0.1:1000;
y = cos(2*pi*x);
toc
```

A.4 Basic graphics

Very good quality graphics is one of the reasons for MATLAB's popularity. Here we summarize the basic rules.

A.4.1 Basic 2D plot

A basic plot is produced in two steps:

1. Creating two arrays of equal lengths, one containing values of independent variable x and the other containing dependent variable $y = f(x)$.

Creation of an array of values of independent variable can be done in several ways. Below we created arrays of points ranging from 0 to 2π using different methods.

```
x1 = 0:pi/100:2*pi;      % 100 evenly spaced points
x2 = [0:0.25:2*pi];      % predefined spacing of 0.25
x3 = linspace(0,2*pi,100); % 100 evenly spaced points
```

2. Inputting created arrays x and y into the MATLAB *plot* function.

The steps are illustrated in the MATLAB program below.

```
clear all              % clears all variables in memory
x = 0:pi/100:2*pi;     % creates array containing independent variable
y = cos(2*x);          % creates array containing function values
plot(x,y)              % creates generic plot, without description
pause                  % pause, to view the plot
close all              % closes all figures
```

A.4.2 Two-dimensional plots

The main function used here is *plot*. It can be called up with different number of arguments

```
plot(y) plot(x,y plot(x,y,'line_type')
```

Here, we illustrate several ways to use it.

Listing: Basic 2D plot.

Here we show creation of the simplest two-dimensional plot by plotting function $y = 3x$, adding basic description and grid. The graph generated by *basic_2D_plot.m* file is shown in Fig. A.1.

MATLAB code is shown below. It explains creation of a basic two-dimensional graph.

```
% File name: basic_2D_plot.m
% Basic plotting of MATLAB are explained.
clear all
% Generate data for plot
n = 100;                     % number of plotting points
x = linspace(0,10,n);        % generates points on x-axis
```

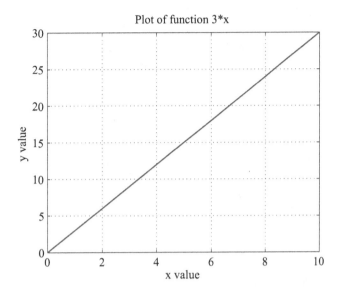

Fig. A.1 Basic two-dimensional plot.

```
y = 3*x;                              % function to be plotted
% Creation of basic plot and adding description
h = plot(x,y);                        % basic plot
pause                                 % stop to analyse plot
title('Plot of function 3*x','FontSize',14)   % adding title
xlabel('x value','FontSize',14);% adding text on x-axis and size of x label
ylabel('y value','FontSize',14);% adding text on y-axis and size of y label
%
pause(2)                              % stop for 2 seconds
grid                                  % adding grid
set(h,'LineWidth',1.5);               % new thickness of plotting lines
set(gca,'FontSize',14);               % new size of tick marks on both axes
%
pause                                 % final stop
close all                             % closing all windows
```

Listing: Creating subplots.

Function subplot (2,2,1) divides the Figure window into 2 vertical and 2 horizontal windows and places the current plot into window 1 (top, left). In the remaining code other types of plots are created and placed in the remaining subwindows.

The results of *sub_plots.m* are shown in Fig. A.2. MATLAB code is shown below. It explains the creation of subplots.

```
% File name: sub_plots.m
% Creation of subplots with different properties.
clear all
% Generate data for plot
```

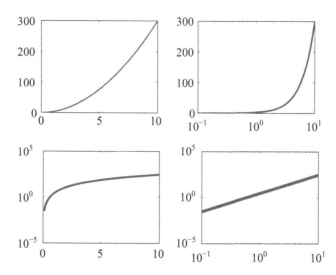

Fig. A.2 Creation of subplots with different properties.

```
n = 100;                          % number of plotting points
x = linspace(0,10,n);             % generates points on x-axis
y = 3*x.^2;                       % function to be plotted
subplot(2,2,1)                    % division of Figure window
h1 = plot(x,y);                   % basic plot shown in top-left window
set(gca,'FontSize',10);           % size of tick marks on both axes
pause                             % stop to contemplate effects
% The above function is now plotted on different log scales
% and plots are placed in the remaining sub-windows
subplot(2,2,2)
h2 = semilogx(x,y);               % log scale on x-axis
set(gca,'FontSize',14);           % size of tick marks on both axes
subplot(2,2,3)
h3 = semilogy(x,y);               % log scale on y-axis
set(gca,'FontSize',15); subplot(2,2,4)
h4 = loglog(x,y);                 % log scale on both axes
set(gca,'FontSize',16);           % size of tick marks on both axes
%
% Below we set new thicknesses of plotting lines
set(h1,'LineWidth',1.5); set(h2,'LineWidth',2);
set(h3,'LineWidth',2.5); set(h4,'LineWidth',3.5);
pause
close all
```

Types of lines (both colours and symbols) are summarized in Table A.3. Colours are automatically selected by MATLAB.

Table A.3 Colours and symbols used to create lines.

Symbol	Colour	Symbol	Type of line
y	yellow	.	point
m	magenta	o	circle
c	cyan	x	shown symbol
r	red	+	shown symbol
g	green	*	shown symbol
b	blue	-	continuous
w	white	:	shown symbol
k	black	-.	dash-dotted line

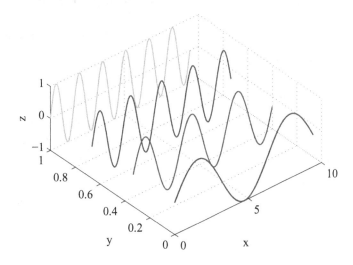

Fig. A.3 A 3D plot.

A.4.3 Some 3D plots

We show how to put several 2D plots on one graph and make it look like a 3D plot. The code is shown below and the generated graph is as Fig. A.3.

```
% File name: pview.m
% Allows multiple 2D plots to be stacked next to one another
% along one dimension; also provides 3D view of all plots
clear all
x = linspace(0,3*pi).';      % x-axis data
Z = [sin(x) sin(2*x) sin(3*x) sin(4*x)];
% Code below gives each curve different value on y-axis
Y = [zeros(size(x)) ones(size(x))/3 (2/3)*ones(size(x)) ones(size(x))];
plot3(x,Y,Z,'LineWidth',1.5)
grid on
xlabel('x','FontSize',14)
```

```
ylabel('y','FontSize',14)
zlabel('z','FontSize',14)
set(gca,'FontSize',14);        % size of tick marks on both axes
view(-40,60)
pause
close all
```

A.5 Basic input-output

A.5.1 Writing to a text file

To save the results of some calculations to a file in text format requires the following steps:

a) Open a new file, or overwrite an old file, keeping a 'handle' for the file.
b) Print the values of expressions to the file, using the file handle.
c) Close the file, using the file handle.

The file handle is just a variable which identifies the open file in your program. This allows you to have any number of files open at any one time.

```
% File name: file_write.m
% open file
fid = fopen('myfile.txt','wt');    % 'wt' means "write text"
if (fid < 0)
error('could not open file "myfile.txt"');
end;
for i=1:10                         % write to file
fprintf(fid,'Number = %3d Square = %6d\n',i,i*i);
end;
fclose(fid);                       % close the file
```

A.5.2 Reading from a text file

To read some results from a text file is straightforward if you just want to load the whole file into the memory. This requires the following steps:

a) Open an existing file, keeping a 'handle' for the file.
b) Read expressions from the file into a single array, using the file handle.
c) Close the file, using the file handle.

The fscanf() function is the inverse of fprintf(). However, it returns the values it reads as values in a matrix. You can control the 'shape' of the output matrix with a third argument.

```
% File name: file_read.m
% open file
```

```
fid = fopen('myfile.txt','rt'); % 'rt' means "read text"
if (fid < 0)
error('could not open file "myfile.txt"');
end;
% read from file into table with 2 rows and 1 column per line
out = fscanf(fid,'Number = %d Square = %d\n',[2,inf]);
fclose(fid);                    % close the file
xx = out';                      % convert to 2 columns and 1 row per line.
xx                              % output results to screen
```

A.6 Numerical differentiation

We will illustrate numerical differentiation using MATLAB function *diff(x)*. It computes the difference between elements in an array.

Given a vector $x = \{...x_{k-1}, x_k, x_{k+1}...\}$ containing points x_k, one can derive approximations to the first derivative $f'(x) = \frac{dy}{dx}$ of the function $y = f(x)$. Three basic possibilities are [4]

$$\text{Backward difference,} \qquad f'(x_k) \approx \frac{f(x_k) - f(x_{k-1})}{x_k - x_{k-1}} \tag{A.1}$$

$$\text{Forward difference,} \qquad f'(x_k) \approx \frac{f(x_{k+1}) - f(x_k)}{x_{k+1} - x_k} \tag{A.2}$$

$$\text{Central difference,} \qquad f'(x_k) \approx \frac{f(x_{k+1}) - f(x_{k-1})}{x_{k+1} - x_{k-1}} \tag{A.3}$$

To compute first derivative we use MATLAB function *diff(x)*, which determines the difference between adjacent values of the vector x. The derivative is then determined as

```
dy = diff(y)./diff(x);
```

The returned array of differences contains one less element than the original array. One should observe that trimming the last value of x produces a forward difference and trimming the first value gives a backward difference.

To obtain the second derivative, we apply the above algorithm a second time. Using backward difference, one has

$$f''(x_k) \approx \frac{f'(x_k) - f'(x_{k-1})}{x_k - x_{k-1}} \tag{A.4}$$

```
% File name: deriv.m
% Program evaluates first and second derivatives and plots results
clear all
font_size = 18;
N_max = 190;
x = linspace(0,2,N_max);
```

```
y1 = sin(pi.*x);
h1 = plot(x,y1);            % plot of original function
xlabel('x','FontSize',font_size);
ylabel('Original function','FontSize',font_size);
grid on
pause
%
temp1 = y1;
dy1 = diff(temp1)./diff(x);
xnew1 = x(1:length(x)-1);
h2 = plot(xnew1,dy1);       % plots first derivative
xlabel('x','FontSize',font_size);
ylabel('First derivative','FontSize',font_size);
grid on
pause
%
temp2 = dy1;
dy2 = diff(temp2)./diff(xnew1);
xnew2 = xnew1(1:length(xnew1)-1);
h3 = plot(xnew2,dy2);       % plots second derivative
xlabel('x','FontSize',font_size);
ylabel('Second derivative','FontSize',font_size);
grid on
pause
close all
```

A.7 Review questions

1. How are variables represented in MATLAB?
2. What is the role of the semicolon (;) operator in MATLAB?
3. What does MATLAB function *linspace*(0, 1, 10)?
4. What is the role of dot (.) operator in MATLAB?
5. What does MATLAB function *subplot*(3, 2, 1)?
6. Perform a polar plot.
7. Estimate π using the random method.

References

[1] S. J. Chapman. *MATLAB Programming for Engineers*. Brooks/Cole, Pacific Grove, CA, 2000.

[2] A. Gilat. *MATLAB. An Introduction with Applications*. John Wiley & Sons, Hoboken, NJ, 2008.

[3] D. Hanselman and B. Littlefield. *Mastering MATLAB 5. A Comprehensive Tutorial and Reference*. Prentice Hall, Upper Saddle River, NJ, 1998.

[4] S. E. Koonin. *Computational Physics*. The Benjamin/Cummings, Menlo Park, CA, 1986.

Appendix B Summary of basic numerical methods

In a book like this, the development of computer programs for various tasks and also execution of simulations for different processes and devices, plays an essential role. The fundamentals of many computer programs are supported by numerical methods. Therefore, in this Appendix we summarize main elements of numerical analysis with an emphasis on methods related to the development of programs used in this book, and also to understanding of operation of those programs.

There are many excellent textbooks devoted to numerical analysis. We found the books by Koonin [1], DeVries [2], Garcia [3], Gerald and Wheatley [4], Rao [5], Heath [6] and Recktenwald [7] of significant pedagogical value. The books by Press *et al.* [8] stand on their own as an excellent source of practical computer codes ready to use.

We concentrate on description and implementation of some practical numerical methods and not on the problems which those methods are typically used for. We start our discussion with a summary of methods of solving nonlinear equations.

There are many textbooks aimed to the introduction of numerical methods and their applications. Some of the most popular are: *Applied Numerical Analysis Using MATLAB* by Fausett [9], *Numerical Methods for Physics* by Garcia [3], *Introduction to Scientific Computing* by van Loan [10], *Advanced Engineering Mathematics with MATLAB* by Harman *et al.* [11], *A Friendly Introduction to Numerical Analysis* by Bradie [12].

They contain extensive code written in MATLAB (sometimes also in other languages) which should help with understanding and could be easily adopted to a particular problem.

Our aim here is to summarize some of the numerical methods and techniques which are directly relevant to the problems discussed in the main text.

In Table B.1 we provide a list of MATLAB files created for this Appendix B and a short description of each function.

B.1 One-variable Newton's method

The very efficient and popular method, also known as Newton-Raphson method, is used to find roots. Here we establish its principles by showing how to find root x^* for an arbitrary function of single variable $f(x)$ such that

$$f(x^*) = 0 \qquad (B.1)$$

One-variable Newton's method

Listing	Function name	Description
	Table B.1 List of MATLAB functions for Appendix B.	
B.1	*muller.m*	Function implements Muller's method
B.2	*dmuller.m*	Driver to test Muller's method
B.3	*fmuller.m*	Test functions for Muller's method
B.4	*ode_single.m*	Driver for RK method (single eq.)
B.5	*func_ode.m*	Function called by *ode_single.m*
B.6	*func_ode_sys.m*	Function called by *ode_sys.m*
B.7	*ode_sys.m*	Driver for RK method (system of eqs.)
B.8	*square_wave.m*	Implements Fourier series for square wave
B.9	*FT_example_1.m*	Preliminary Fourier transform
B.10	*FT_example_2.m*	Testing Fourier transform in MATLAB

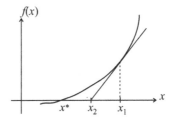

Fig. B.1 Graphical illustration of Newton's method.

The Newton method is based on a linear approximation of the function by employing a tangent to this function at a particular point. In Fig. B.1 we provide graphical illustration of Newton's method.

One starts from an initial guess point, say x_1, which should be not too far from root x^*. Evaluating tangent at point x_1 and its intersection with the x axis creates new point x_2 and the process is repeated. The relevant mathematical formulas are obtained by writing the Taylor expansion of $f(x)$ around initial guess x_1:

$$f(x^*) = f(x_1 - \delta x) = f(x_1) - \frac{df(x_1)}{dx}\delta x + O(\delta x^2) \tag{B.2}$$

In the previous, $f(x^*) = 0$. Solving for δx, one finds

$$\delta x = \frac{f(x_1)}{f'(x_1)}$$

From Fig. B.1 one observes that point x_2 can be obtained as

$$x_2 = x_1 - \delta x = x_1 - \frac{f(x_1)}{f'(x_1)} \tag{B.3}$$

The procedure can be generalized and for arbitrary step n, one has

$$x_{n+1} = x_n - \frac{f(x_n)}{f'(x_n)}, \quad n = 1, 2, \ldots \tag{B.4}$$

B.2 Muller's method

Muller's method is an efficient technique of finding both real and complex roots of a scalar-valued function $f(x)$ of a single variable x where there is no information about its derivatives. It uses quadratic interpolation involving three points [13]. The method is described in several textbooks, including Atkinson [14] and Fausett [9].

In order to illustrate how the method works, let us start with three initial guesses x_0, x_1, x_2. By applying Muller's method, a new improved approximate point x_3 is determined. In this method one uses quadratic function $y(x)$ which is used in fitting original function $f(x)$. Function $y(x)$ is of the form

$$y(x) = a\,(x - x_2)^2 + b\,(x - x_2) + c \tag{B.5}$$

It contains three constants a, b, c which can be determined by evaluating Eq. (B.5) at points x_0, x_1, x_2. One obtains

$$\text{for } x = x_2 \quad f(x_2) = c$$

$$\begin{aligned}\text{for } x = x_1 \quad & f(x_1) = a\,(x_1 - x_2)^2 + b\,(x_1 - x_2) + c \\ \text{or} \quad & f(x_1) - f(x_2) = a\,(x_1 - x_2)^2 + b\,(x_1 - x_2)\end{aligned}$$

$$\begin{aligned}\text{for } x = x_0 \quad & f(x_0) = a\,(x_0 - x_2)^2 + b\,(x_0 - x_2) + c \\ \text{or} \quad & f(x_0) - f(x_2) = a\,(x_0 - x_2)^2 + b\,(x_0 - x_2)\end{aligned}$$

Introduce notation $u_0 = x_0 - x_2$ and $u_1 = x_1 - x_2$, which allows us to write the above equations as

$$f(x_0) - f(x_2) = a\,u_0^2 + b\,u_0 \tag{B.6}$$
$$f(x_1) - f(x_2) = a\,u_1^2 + b\,u_1 \tag{B.7}$$

The system of Eqs. (B.6) and (B.7) can be solved by any standard method and one obtains

$$a = \frac{u_1\left[f(x_0) - f(x_2)\right] - u_0\left[f(x_1) - f(x_2)\right]}{(x_0 - x_2)(x_1 - x_2)(x_0 - x_1)}$$

$$b = \frac{u_0^2\left[f(x_1) - f(x_2)\right] - u_1^2\left[f(x_0) - f(x_2)\right]}{(x_0 - x_2)(x_1 - x_2)(x_0 - x_1)}$$

With the above information at hand and using Eq. (B.5), one can determine roots of

$$y(x) = 0$$

in order to generate a new approximation of the root of $f(x)$. One obtains

$$x - x_2 = \frac{-b \pm \sqrt{b^2 - 4ac}}{2a} \tag{B.8}$$

Out of the above two roots we choose the one which is closest to x_2. To avoid round-off errors due to subtraction of nearly equal numbers, one multiplies numerator and denominator of

(B.8) by $-b \mp \sqrt{b^2 - 4ac}$ and obtains (we call new solution x by x_3)

$$x_3 - x_2 = \frac{\left(-b \pm \sqrt{b^2 - 4ac}\right)\left(-b \mp \sqrt{b^2 - 4ac}\right)}{2a\left(-b \mp \sqrt{b^2 - 4ac}\right)}$$

$$= -\frac{2c}{-b \pm \sqrt{b^2 - 4ac}} \qquad (B.9)$$

Once x_3 is determined, the 'oldest' point x_0 is ignored by setting $x_0 = x_1, x_1 = x_2, x_2 = x_3$ and the process is repeated. MATLAB implementation of Muller's method using the above algorithm is shown below.

```
% File name: muller.m
function out = muller(f, x0, x1, x2 , epsilon, max)
% Finds zeroes using Muller's method, good for complex roots.
% Variable description:
% out       - result of search
% x1,x2,x3  - previous guesses
% epsilon   - tolerance
% max       - max number of iterations
%
y0 = f(x0);
y1 = f(x1);
y2 = f(x2);
iter = 0;
while (iter <= max)
    iter = iter + 1;
    a =( (x1 - x2)*(y0 - y2) - (x0 - x2)*(y1 - y2)) / ...
        ( (x0 - x2)*(x1 - x2)*(x0 - x1) );
    %
    b = ( ( x0 - x2 )^2 *( y1 - y2 ) - ( x1 - x2 )^2 *( y0 - y2 )) /
        ( (x0 - x2)*(x1 - x2)*(x0 - x1) );
    %
    c = y2;
    %
    if (a ~= 0)
        disc = b*b - 4*a*c;
        q1 = b + sqrt(disc);
        q2 = b - sqrt(disc);
        if (abs(q1) < abs(q2))
            dx = - 2*c/q2;
        else
            dx = - 2*c/q1;
        end
    elseif (b ~= 0)
        dx = - c/b;
    end
```

```
            x3 = x2 + dx;
            x0 = x1;
            x1 = x2;
            x2 = x3;
            %
            y0 = y1;
            y1 = y2;
            y2 = f(x2);
            %
            if (abs(dx) < epsilon)
                out = x2; break;
            end
    end
end
```

Driver for the Muller method is shown below:

```
% File name: dmuller.m
% Driver to test Muller's method
clear all
format short
max = 100;
epsilon = 1e-6;
% starting points
x1 = 0.5;
x2 = 1.5;
x3 = 1.0;
% call to Muller's method
out = muller(@fmuller, x1, x2, x3 , epsilon, max)
```

Test functions for the Muller method are shown below. One can make appropriate choices.

```
% File name: fmuller.m
function out = fmuller(x)
% Test functions for Muller's method
out = x.^6 - 2;
%out = x.^10 - 0.5;
%out = x - x.^3/3;
%out = x.^3 - 5*x.^2 + 4*x;
```

B.2.1 Tests of Muller's method

Using the above functions, we tested Muller's algorithm on two functions:

1. calculations of $\sqrt[6]{2}$, after Fausett [9]. We define function $f(x) = x^6 - 2$, and assume starting values to be: $x_0 = 0.5, x_1 = 1.5, x_2 = 1.0$. Obtained solution is $x = 1.1225$.

2. finding root of $f(x) = e^x + 1 = 0$. The analytical method gives $x = i\pi$. Assuming starting values $x_0 = 1, x_1 = 0, x_2 = -1$, one obtains from Muller's method $x = -0.0000 + 3.1416i$.

B.3 Numerical differentiation

Very often scientific and engineering problems are formulated in terms of differential equations. To solve them numerically, we need to establish schemes to replace derivatives by finite differences.

A typical scientific or engineering problem is described by the second-order differential equation. Examples include Newton's second law $m\ddot{x} = F$, or Laplace's equation $V''(x) = 0$. General second-order differential equation is written as

$$\frac{d^2 y(x)}{dx^2} + a(x)\frac{dy(x)}{dx} = b(x)$$

with $a(x)$ and $b(x)$ known functions. The above can be rewritten as two first-order equations:

$$\frac{dy(x)}{dx} = z(x)$$
$$\frac{dz(x)}{dx} = b(x) - a(x)z(x)$$

where $z(x)$ is a new variable. The process can be generalized for a differential equation of arbitrary order, say $n-th$ order, and when applied, results in n first-order differential equations, as

$$\frac{dy_i(x)}{dx} = f_i(x, y_1, y_2, \ldots y_n), \quad i = 1, 2, \ldots n$$

where the functions f_i on the right-hand side are known.

In order to have a well-defined mathematical problem, the above set of equations must be supplemented by boundary (or initial) conditions. Boundary conditions are algebraic conditions on the value of the functions y_i to be satisfied at discrete specified points, but not between those points.

The (first) derivative of a function $f(x)$ is defined as

$$\left.\frac{df(x)}{dx}\right|_{x_0} = f'(x_0) = \lim_{x \to x_0} \frac{f(x) - f(x_0)}{x - x_0}$$

For a finite (and nonzero) value of $\Delta x = x - x_0$, the above derivative can be approximated as

$$f'(x_0) \approx \frac{f(x) - f(x_0)}{\Delta x}$$

In Table B.2 we summarize common finite-difference formulas.

Table B.2 Summary of common finite-difference formulas.

Type of approximation	Formula	Truncation error
Forward differences	$f'_i = \dfrac{f_{i+1} - f_i}{h}$	$O(h)$
	$f''_i = \dfrac{f_{i+2} - 2f_{i+1} + f_i}{h^2}$	
Backward differences	$f'_i = \dfrac{f_i - f_{i-1}}{h}$	$O(h)$
	$f''_i = \dfrac{f_i - 2f_{i-1} + f_{i-2}}{h^2}$	
Central differences	$f'_i = \dfrac{f_{i+1} - f_{i-1}}{2h}$	$O(h^2)$
	$f''_i = \dfrac{f_{i+1} - 2f_i + f_{i-1}}{h^2}$	

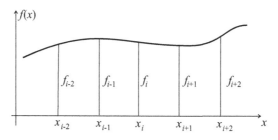

Fig. B.2 Illustration used in determining derivatives.

B.3.1 Numerical differentiation using Taylor's series expansion

We want to determine a derivative of function $f(x)$ at point x_i and also its second derivative. Using notation from Fig. B.2, we write Taylor expansions

$$f_{i+1} = f_i + hf'_i + \frac{1}{2}h^2 f''_i + \frac{1}{6}h^3 f'''_i + O(h^4) \tag{B.10}$$

$$f_{i-1} = f_i - hf'_i + \frac{1}{2}h^2 f''_i - \frac{1}{6}h^3 f'''_i + O(h^4) \tag{B.11}$$

where the following notation has been used:

$$f_n = f(x_n), \quad x_n = n \cdot h, \quad n = 0, \pm 1, \pm 2, \ldots$$

All numerical schemes defining the first derivative are derived from the above equations. From Eq. (B.10) we obtain the *forward* formula

$$f'_i = \frac{f_{i+1} - f_i}{h} + O(h)$$

From Eq. (B.11) we obtain the *backward* formula

$$f'_i = \frac{f_i - f_{i-1}}{h} + O(h)$$

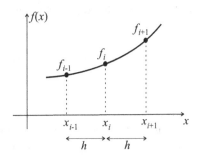

Fig. B.3 Illustration of interpolation process.

Subtracting both, we obtain the *central* formula

$$f'_i = \frac{f_{i+1} - f_{i-1}}{2h}$$

Taylor expansion for next points ($i \pm 2$) gives

$$f_{i+2} = f_i + 2hf'_i + \frac{1}{2}(2h)^2 f''_i + \frac{1}{6}(2h)^3 f'''_i + O(h^4) \qquad \text{(B.12)}$$

and

$$f_{i-2} = f_i - 2hf'_i + \frac{1}{2}(2h)^2 f''_i - \frac{1}{6}(2h)^3 f'''_i + O(h^4) \qquad \text{(B.13)}$$

Multiplying Eq. (B.10) by 2 and adding to Eq. (B.12), we obtain

$$\begin{aligned} f_{i+2} - 2f_{i+1} &= f_i + 2hf'_i + 2h^2 f''_i - 2f_i - 2hf'_i - h^2 f''_i + O(h^3) \\ &= -f_i + h^2 f''_i + O(h^3) \end{aligned}$$

or

$$f''_i = \frac{f_{i+2} - 2f_{i+1} + f_i}{h^2}$$

Other schemes can be obtained in a similar way.

B.3.2 Numerical differentiation using interpolating polynomials

To illustrate a basic concept, consider a second-order polynomial $f(x)$ passing through three data points, Fig. B.3:

$$(x_i, f_i), (x_{i+1}, f_{i+1}), (x_{i+2}, f_{i+2})$$

The analytical form of that polynomial is

$$f(x) = a_0 + a_1 x + a_2 x^2 \qquad \text{(B.14)}$$

Function $f(x)$ has to pass through all three points defined previously. This requirement creates three equations:

$$f_i = a_0 + a_1 x_i + a_2 x_i^2$$
$$f_{i+1} = a_0 + a_1(x_i + h) + a_2(x_i + h)^2$$
$$f_{i+2} = a_0 + a_1(x_i + 2h) + a_2(x_i + 2h)^2$$

The above equations will now be solved to find coefficients a_0, a_1, a_2. The scheme must be independent of the choice of point x_i, so we can choose $x_i = 0$, which significantly simplifies algebra. One obtains

$$f_i = a_0$$
$$f_{i+1} = a_0 + a_1 h + a_2 h^2$$
$$f_{i+2} = a_0 + 2a_1 h + 4a_2 h^2$$

The solutions are

$$a_0 = f_i$$
$$a_1 = \frac{-f_{i+2} + 4f_{i+1} - 3f_i}{2h} \quad \text{(B.15)}$$
$$a_2 = \frac{f_{i+2} - 2f_{i+1} + f_i}{2h^2}$$

Differentiate Eq. (B.14) and use Eqs. (B.15) to obtain

$$f_i' = f'(x_i = 0) = a_1 = \frac{-f_{i+2} + 4f_{i+1} - 3f_i}{2h}$$
$$f_i'' = f''(x_i = 0) = 2a_2 = \frac{f_{i+2} - 2f_{i+1} + f_i}{h^2}$$

The above are forward-difference approximations of the order $O(h^2)$.

Given a vector $x = \{\ldots x_{k-1}, x_k, x_{k+1} \ldots\}$ containing points x_k, one can derive approximations to the first derivative $f'(x) = \frac{dy}{dx}$ of the function $y = f(x)$. Three basic possibilities are [1]:

$$\text{Backward difference,} \quad f'(x_k) \approx \frac{f(x_k) - f(x_{k-1})}{x_k - x_{k-1}} \quad \text{(B.16)}$$

$$\text{Forward difference,} \quad f'(x_k) \approx \frac{f(x_{k+1}) - f(x_k)}{x_{k+1} - x_k} \quad \text{(B.17)}$$

$$\text{Central difference,} \quad f'(x_k) \approx \frac{f(x_{k+1}) - f(x_{k-1})}{x_{k+1} - x_{k-1}} \quad \text{(B.18)}$$

To compute first derivative we use MATLAB function diff(x), which determines the difference between adjacent values of the vector x. The returned array of differences contains one less element than the original array. One should observe that trimming the last value of x produces a forward difference and trimming the first value gives a backward difference.

To obtain the second derivative, we apply the above algorithm a second time. Using the backward difference, one has

$$f''(x_k) \approx \frac{f'(x_k) - f'(x_{k-1})}{x_k - x_{k-1}} \tag{B.19}$$

B.3.3 Crank-Nicolson method

We illustrate the method on the diffusion equation, which we write in the form

$$\frac{\partial u}{\partial t} = \frac{\partial^2 u}{\partial x^2} \tag{B.20}$$

The Crank-Nicolson (CN) method is the finite-difference discretization of the above equation. It was invented to overcome stability limitations and to improve the rate of convergence of the solution of diffusion equation. It has $O(\Delta t)^2$ convergence rate compared to $O(\Delta t)$ of the implicit and explicit methods.

The CN method is essentially an average of the implicit and explicit methods. It is created as follows. Applying a forward difference approximation to the time derivative in the diffusion equation gives the explicit scheme

$$\frac{u_i^{n+1} - u_i^n}{\Delta t} + O(\Delta t) = \frac{u_{i+1}^n - 2u_i^n + u_{i-1}^n}{(\Delta x)^2} + O(\Delta x)^2 \tag{B.21}$$

When using backward difference to the same equation, one creates the implicit scheme

$$\frac{u_i^{n+1} - u_i^n}{\Delta t} + O(\Delta t) = \frac{u_{i+1}^{n+1} - 2u_i^{n+1} + u_{i-1}^{n+1}}{(\Delta x)^2} + O(\Delta x)^2 \tag{B.22}$$

The algebraic average of the two schemes gives

$$\frac{u_i^{n+1} - u_i^n}{\Delta t} + O(\Delta t) = \frac{1}{2}\left\{\frac{u_{i+1}^n - 2u_i^n + u_{i-1}^n}{(\Delta x)^2} + \frac{u_{i+1}^{n+1} - 2u_i^{n+1} + u_{i-1}^{n+1}}{(\Delta x)^2}\right\} + O(\Delta x)^2 \tag{B.23}$$

From the above equation one finally obtains the Crank-Nicolson scheme

$$u_i^{n+1} - \frac{1}{2}\alpha\left(u_{i+1}^{n+1} - 2u_i^{n+1} + u_{i-1}^{n+1}\right) = u_i^n + \frac{1}{2}\alpha\left(u_{i+1}^n - 2u_i^n + u_{i-1}^n\right) \tag{B.24}$$

where $\alpha = \frac{\Delta t}{(\Delta x)^2}$. One can show (see Problems) that the scheme is accurate to $O(\Delta t)^2$ rather than $O(\Delta t)$. Note that in CN scheme $u_{i+1}^{n+1}, u_i^{n+1}, u_{i-1}^{n+1}$ are determined implicitly in terms of $u_{i+1}^n, u_i^n, u_{i-1}^n$.

B.3.4 Simple methods of numerical differentiation

Euler method

The simplest method is the Euler method, which is

$$y_{n+1} = y_n + hf(x_n, y_n)$$

which advances a solution from x_n to $x_{n+1} = x_n + h$. It is not a practical method, because it is not very accurate and also not very stable (see discussion below).

Let us illustrate several methods which originate from the Euler method. We will use definition of an acceleration in one-dimensional motion as an illustration. It is

$$\frac{d^2x(t)}{dt^2} = a(x, v)$$

In general, acceleration can depend on position and velocity. First, write the above equation as a system of two first-order differential equations:

$$\frac{dv}{dt} = a(x, v) \quad \text{(B.25)}$$

$$\frac{dx}{dt} = v$$

Introduce τ as a time step. The right-derivative formula for a general function $g(t)$ is

$$\frac{dg(t)}{dt} = \frac{g(t+\tau) - g(t)}{\tau} + O(\tau)$$

Application of that formula to a system (B.25) gives

$$\frac{v(t+\tau) - v(t)}{\tau} + O(\tau) = a(x(t), v(t))$$

$$\frac{x(t+\tau) - x(t)}{\tau} + O(\tau) = v(t)$$

We have explicitly indicated the dependence of all quantities involved on the specific time. From the above

$$v(t+\tau) = v(t) + \tau \cdot a(x(t), v(t)) + O(\tau^2)$$
$$x(t+\tau) = x(t) + \tau \cdot v(t) + O(\tau^2)$$

For the future use, let us introduce the following notation:

$$g_n \equiv g((n-1)\tau), \quad n = 1, 2, 3, \ldots \quad \text{(B.26)}$$

and

$$g_1 \equiv g(t = 0)$$

In the above, n refers to time steps. In our notation, the Euler method for the acceleration problem reads

$$v_{n+1} = v_n + \tau \cdot a_n$$
$$x_{n+1} = x_n + \tau \cdot v_n$$

The algorithm based on Euler method goes as follows:

1. specify initial conditions (i.e. values corresponding to $t = 0$) x_1 and v_1
2. choose time step τ
3. calculate acceleration for current values of x_n and v_n

4. use Euler method to compute new x and v
5. loop through step 3, until all time steps are done.

An implementation of the Euler method is left as a problem.

Euler-Cromer (E-C) and mid-point methods

Both of these methods are simple modifications of the Euler method just discussed. E-C method consists in replacing v_n by v_{n+1}, the value of velocity at the next time step (it must be calculated first, so the calculations must be done in a proper sequence). Explicitly, the E-C method is

$$v_{n+1} = v_n + \tau \cdot a_n$$
$$x_{n+1} = x_n + \tau \cdot v_{n+1}$$

Another possible modification (and in a similar spirit) results in a mid-point method:

$$v_{n+1} = v_n + \tau \cdot a_n$$
$$x_{n+1} = x_n + \tau \cdot \frac{1}{2}(v_n + v_{n+1})$$

where the last term represents algebraic average of velocities at present and the next time steps.

Leap-frog method

The development of this method starts with Eqs. (B.25). Use the central derivative approach to write (note that central time value is different for both equations)

$$\frac{v(t+\tau) - v(t-\tau)}{2\tau} + O(\tau^2) = a(x(t))$$
$$\frac{x(t+2\tau) - x(t)}{2\tau} + O(\tau^2) = v(t+\tau)$$

Applying notation (B.26), the above system reads

$$\frac{v_{n+1} - v_{n-1}}{2\tau} = a(x_n)$$
$$\frac{x_{n+2} - x_n}{2\tau} = v_{n+1}$$

or

$$v_{n+1} = v_{n-1} + 2\tau \cdot a(x_n)$$
$$x_{n+2} = x_n + 2\tau \cdot v_{n+1}$$

In order that this scheme works, we need to know values at proper initial points. Note that position and velocity are evaluated at different points, as

position is evaluated at points: $x_1, x_3, x_5, \ldots$

velocity is evaluated at points: $v_2, v_4, v_6, \ldots$

Verlet method

To develop the Verlet method, instead of using Eqs. (B.25), we will write them as [3]

$$\frac{dx}{dt} = v$$

$$\frac{d^2x}{dt^2} = a$$

Use central difference formulas for first and second derivatives and obtain

$$\frac{x_{n+1} - x_{n-1}}{2\tau} + O(\tau^2) = v_n$$

$$\frac{x_{n+1} + x_{n-1} - 2x_n}{\tau^2} + O(\tau^2) = a_n$$

From above, we obtain formulas which define Verlet method:

$$v_n = \frac{1}{2\tau}(x_{n+1} - x_{n-1})$$

$$x_{n+1} = 2x_n - x_{n-1} + \tau^2 a_n$$

B.4 Runge-Kutta (RK) methods

B.4.1 Second-order Runge-Kutta

Our goal is to solve linear system of linear differential equations

$$\frac{d\vec{x}(t)}{dt} = \vec{f}(\vec{x}(t), t)$$

with vector $\vec{x}$ expressed as $\vec{x} = [x_1 x_2 \ldots x_n]$. In order to establish second-order RK method, let us start with the simple Euler approximation to the first derivative (which is also Taylor expansion to first order):

$$\vec{x}(t+\tau) = \vec{x}(t) + \tau \cdot \vec{f}(\vec{x}(\xi), \xi)$$

where τ is some arbitrary time step and $t < \xi < t + \tau$. We have freedom to choose ξ to have the most convenient value for us. The second-order RK method originates when $\xi = t + \tau/2$ and also

$$\vec{x}(t+\tau) = \vec{x}(t) + \tau \cdot \vec{f}(\vec{x}^*(t+\tau/2), t+\tau/2)$$

where

$$\vec{x}^*(t+\tau/2) = \vec{x}(t) + 1/2 \cdot \tau \cdot \vec{f}(\vec{x}(t), t)$$

B.4.2 Fourth-order Runge-Kutta

This is one of the best schemes available and is commonly used in practice. It is defined as

$$\vec{x}(t+\tau) = \vec{x}(t) + 1/6 \cdot \tau \cdot (\vec{F}_1 + 2\vec{F}_2 + 2\vec{F}_3 + \vec{F}_4)$$

where

$$\vec{F}_1 = \vec{f}(\vec{x}, t)$$
$$\vec{F}_2 = \vec{f}(\vec{x} + \tau/2 \cdot \vec{F}_1, t + \tau/2)$$
$$\vec{F}_3 = \vec{f}(\vec{x} + \tau/2 \cdot \vec{F}_2, t + \tau/2)$$
$$\vec{F}_4 = \vec{f}(\vec{x} + \tau \cdot \vec{F}_3, t + \tau)$$

MATLAB has some built-in routines implementing RK methods.

B.5 Solving differential equations

Here we provide some practical methods of solving ordinary differential equations (ODE), both single and systems, using MATLAB function *ode45.m*.

B.5.1 Single differential equation

To start with, consider an equation with an initial condition

$$\frac{dx(t)}{dt} = x(t)\left(\frac{2}{t} - 1\right), \quad x(0) = 0.009 \qquad (B.27)$$

We want to solve it and plot the solution in the interval [0.1, 10]. MATLAB code is divided into two m-files: one (the driver) contains all needed parameters and calls the appropriate m-function which defines the function.

```
% File name: ode_single.m
% Illustrates application of R-K method to a single diff. equation
% Results are compared with an exact solution
%
clear all
tspan = [0.1 10];                    % time interval
x0 = 0.009;                          % initial value
%
[t,x] = ode45('func_ode_single',tspan,x0);
t_ex = linspace(0.1, 10, 100);
x_ex = t_ex.^2.*exp(-t_ex);          % Exact solution
plot(t,x,t_ex,x_ex,'.','LineWidth',1.5)
```

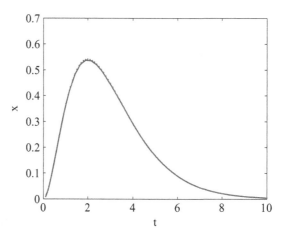

Fig. B.4 Comparison of the solution of a single ordinary differential equation using numerical and analytical methods.

```
xlabel('t','FontSize',14)
ylabel('x','FontSize',14)
set(gca,'FontSize',14);          % size of tick marks on both axes
pause
close all
```

The function itself is

```
function xdot = func_ode_single(t,x)
% Function called by ode_single.m
xdot = x*(2/t -1);
```

The result of solving the above problem is plotted in Fig. B.4 along with an exact solution, which is $x(t) = t^2 e^{-t}$.

B.5.2 System of differential equations

To illustrate application of *ode45.m* function in solving systems of differential equations, consider the Lorenz model [3] defined by the following system:

$$\begin{aligned}
\frac{dx_1(t)}{dt} &= \sigma(x_2 - x_1) \\
\frac{dx_2(t)}{dt} &= rx_1 - x_2 - x_1 x_3 \\
\frac{dx_3(t)}{dt} &= x_1 x_2 - b x_3
\end{aligned} \qquad (B.28)$$

where σ, r, b are positive constants. This system was originally developed to describe buoyant convection in a fluid. It shows chaotic behaviour. The following initial conditions are assumed: $x_1(t) = 1, x_2(t) = 1, x_3(t) = 10.04$. The analysis is conducted in the time interval [0, 10]. That system of equations is coded as

```
function xdot = func_ode_sys(t,x)
% Function called by ode_sys.m
%
sigma = 10; b = 7/3; r = 25;   % parameters
%
xdot(1) = sigma*(x(2) - x(1));
xdot(2) = r*x(1) - x(2) - x(1)*x(3);
xdot(3) = x(1)*x(2) - b*x(3);
xdot = xdot';
```

The MATLAB code of the driver of the above system is

```
% File name: ode_sys.m
% Illustrates application of R-K method to system of diff. equations
%
clear all
tspan = [0 10];    % time interval
x0 = [1, 1, 10.04];                   % initial value
%
[t,x] = ode45('func_ode_sys',tspan,x0);
plot(t,x(:,1), t,x(:,2), t,x(:,3),'LineWidth',1.5)
xlabel('time','FontSize',14)
set(gca,'FontSize',14);                 % size of tick marks on both axes
pause
close all
```

The results of two runs using the above functions are shown in Fig. B.5.

B.6 Numerical integration

The problem is to calculate a numerically definite integral

$$I = \int_a^b f(x)dx$$

The interval $[a, b]$ can be split into equidistant subintervals with total number $N = \frac{b-a}{h}$, with h separation between neighbours. The above integral is thus replaced by

$$I = \int_a^b f(x)dx = \int_a^{x_i} f(x)dx + \int_{x_i}^{x_2} f(x)dx + \cdots = \sum_i A_i$$

where A_i is the area of an elementary element. In the following, we therefore only consider an integral within a single subinterval:

$$A_i = \int_{x_{i-1}}^{x_i} f(x)dx \tag{B.29}$$

Different schemes will originate depending on how we evaluate the above integral.

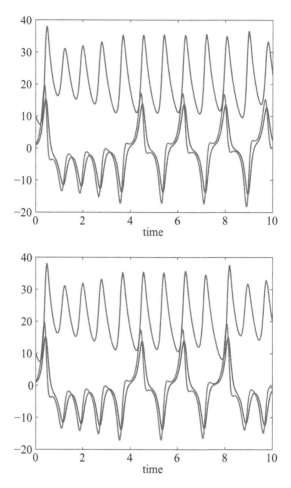

Fig. B.5 Solution of a system of ordinary differential equations showing chaos for two different values of initial conditions; [1,1,10.04] (top) and [1,1,10.02] (bottom).

B.6.1 Euler's rule

Subarea is defined within an interval $x_{i-1} < x_i < x_{i+1}$. The integral (B.29) thus takes the form

$$A_i = \int_{x_{i-1}}^{x_i} f(x)dx \approx h \cdot f_i$$

with $f_i = f(x_i)$ taken at the right end of subinterval. The value of the function can also be taken at the left point, or in fact at any point within the subinterval $x_{i-1} < x_i < x_{i+1}$. Total area is thus

$$I \approx \sum_i A_i$$
$$= h \cdot (f_1 + f_2 + \cdots f_n)$$

B.6.2 Trapezoidal rule

In this method the integral is approximated by a series of trapezoids. The area of each subinterval is evaluated as

$$A_i = \int_{x_{i-1}}^{x_i} f(x)dx \approx \frac{1}{2}(f_{i-1} + f_i) \cdot (x_i - x_{i-1})$$

with $h = x_i - x_{i-1}$. Taking h as a constant, the integral is

$$I = \int_a^b f(x)dx \approx \sum_i \frac{1}{2}(f_{i-1} + f_i) \cdot h$$
$$= \frac{1}{2}h\{(f_0 + f_1) + (f_1 + f_2) + (f_2 + f_3) + \cdots + (f_{n-2} + f_{n-1}) + (f_{n-1} + f_n)\}$$
$$= \frac{1}{2}h\{f_0 + 2f_1 + 2f_2 + \cdots 2f_{n-1} + f_n\}$$

In a condensed form, the integral is

$$I = h\sum_{i=1}^{n-1} f_i + \frac{1}{2}h(f_0 + f_n)$$

B.6.3 Simpson's rule

Simpson's method gives more accurate results than trapezoidal rule. Trapezoidal rule connects consecutive points by a straight line; Simpson's rule connects three points. Normally, a second-order polynomial (a parabola) is used to do so. Consider one:

$$p(x) = c_2 x^2 + c_1 x + c_0$$

Without loss of generality, we can take $x_i = 0$. Then, $x_{i-1} = -h$ and $x_{i+1} = h$. Evaluating the above polynomial gives

$$p(-h) = c_2 h^2 - c_1 h + c_0 = f_{i-1}$$
$$p(0) = c_0 = f_i$$
$$p(h) = c_2 h^2 + c_1 h + c_0 = f_{i+1}$$

Solving the above system gives

$$c_0 = f_i$$
$$c_1 = \frac{1}{2h}(f_{i+1} - f_{i-1})$$
$$c_2 = \frac{1}{2h^2}(f_{i+1} + f_{i-1} - 2f_i)$$

Evaluating elementary integral for an area A_i gives

$$\begin{aligned} A_i &= \int_{-h}^{h} p(x)dx \\ &= \int_{-h}^{h} \left\{ \frac{1}{2h^2}(f_{i+1}+f_{i-1}-2f_i)x^2 + \frac{1}{2h}(f_{i+1}-f_{i-1})x + f_i \right\} dx \\ &= \frac{1}{3}h(f_{i-1}+4f_i+f_{i+1}) \end{aligned}$$

The complete integral is

$$\begin{aligned} I &= \int_a^b f(x)dx = \sum_i A_i \\ &= \frac{1}{3}h\{f_0+4f_1+2f_2+4f_3+\cdots+2f_{n-2}+4f_{n-1}+f_n\} \end{aligned}$$

B.7 Symbolic integration in MATLAB

MATLAB allows us to perform symbolic manipulations. It uses Maple, a powerful computer algebra system, to manipulate and solve symbolic expressions. To use symbolic mathematics, we must use the *syms* operator to tell MATLAB that we are using a symbolic variable, and that it does not have a specific value.

In this book we only use symbolic manipulations to evaluate indefinite integral. We illustrate symbolic concepts of MATLAB with the example below.

Example Evaluate integral of the function $f(x) = x^2$.

The shortest method is to type on a prompt: $>> int(x2)$. The answer is $ans = x3/3$.

Alternatively, we can define x symbolically first, and then remove the single quotes in the *int* statement.

```
>> syms x
>> int(x^2)

ans =

x^3/3
```

B.8 Fourier series

According to Fourier, any function can be expressed in the form known as the Fourier series:

$$f(t) = \frac{1}{2}a_0 + \sum_{n=1}^{\infty} a_n \cos nt + \sum_{n=1}^{\infty} b_n \sin nt \qquad (B.30)$$

Fourier series

Fourier series representation is useful in describing functions over a limited region, say $[0, T]$, or on the infinite interval $(-\infty, +\infty)$ if the function is periodic. We assume that the Fourier series converges for any problem of interest, see Arfken [15]. Cosine and sine functions form a complete set and are orthonormal on any 2π interval. Therefore they may be used to describe any function. The orthogonality relations are

$$\int_0^{2\pi} \sin mt \, \sin nt \, dt = \begin{cases} \pi \delta_{m,n}, & m \neq 0 \\ 0, & m = 0 \end{cases} \tag{B.31}$$

$$\int_0^{2\pi} \cos mt \, \cos nt \, dt = \begin{cases} \pi \delta_{m,n}, & m \neq 0 \\ 2\pi, & m = n = 0 \end{cases} \tag{B.32}$$

$$\int_0^{2\pi} \sin mt \, \cos nt \, dt = 0, \quad \text{all integral } m \text{ and } n \tag{B.33}$$

One also has the relation

$$\frac{1}{2\pi} \int_0^{2\pi} \left(e^{imt}\right)^* e^{int} \, dt = \delta_{m,n} \tag{B.34}$$

Using the above orthogonality relations, the coefficients of expansion in Eq. (B.30) are

$$a_n = \frac{1}{\pi} \int_0^{2\pi} f(t) \, \cos nt \, dt \tag{B.35}$$

and

$$b_n = \frac{1}{\pi} \int_0^{2\pi} f(t) \, \sin nt \, dt \tag{B.36}$$

Also, the exponential form of the Fourier series is often used where the function is expanded as

$$f(t) = \sum_{n=-\infty}^{n=+\infty} c_n \, e^{int} \tag{B.37}$$

Comparison of the above relations gives formulas for coefficients c_n as

$$\begin{aligned} c_0 &= \tfrac{1}{2} a_0 & n &= 0 \\ c_n &= \tfrac{1}{2} (a_n - i b_n) & n &> 0 \\ c_{-n} &= \tfrac{1}{2} \left(a_{|n|} + i b_{|n|}\right) & n &< 0 \end{aligned} \tag{B.38}$$

B.8.1 Change of interval

In the above formulation, the discussion has been restricted to an interval $[0, 2\pi]$. If the function $f(t)$ is periodic with the period $2T$, we may write

$$f(x) = \frac{1}{2} a_0 + \sum_{n=1}^{\infty} a_n \cos nt + \sum_{n=1}^{\infty} b_n \sin nt \tag{B.39}$$

where

$$a_n = \frac{1}{T} \int_{-T}^{T} f(t') \cos \frac{n\pi t'}{T} \, dt', \quad n = 0, 1, 2, 3, \ldots \tag{B.40}$$

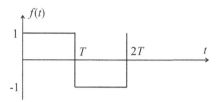

Fig. B.6 Square wave.

and

$$b_n = \frac{1}{T}\int_{-T}^{T} f(t') \sin\frac{n\pi t'}{T}\, dt', \quad n = 1, 2, 3, \ldots \tag{B.41}$$

Expressions (B.30) and (B.39) are related by obvious change of variables, namely $t = \frac{\pi t'}{T}$. The choice of the symmetric interval $[-T, T]$ is not essential; in fact, for periodic function with a period $2T$ it is possible to choose any interval $[t_0, t_0 + 2T]$.

B.8.2 Example

Use the Fourier series to approximate a square wave function $f(t)$ as shown in Fig. B.6. Using Heaviside step function $H(t)$, the function $f(t)$ can be written as

$$f(t) = 2\left[H\left(\frac{t}{T}\right) - H\left(\frac{t}{T} - 1\right)\right] - 1$$

where

$$H(t) = \begin{cases} 0 & t < 0 \\ 1 & t > 0 \end{cases}$$

Determine coefficients a_n and b_n in the Fourier series expansion. Write a MATLAB program which plots the above approximation for various number of terms in the series.

From Fig. B.6 one observes that $f(t) = f(2T - t)$; i.e. function $f(t)$ is odd. Therefore coefficients a_n vanish. Coefficients b_n are evaluated as follows:

$$b_n = \frac{1}{T}\int_0^{2T} f(t)\sin\left(\frac{n\pi t}{T}\right)dt = \frac{1}{T}\int_0^T \sin\left(\frac{n\pi t}{T}\right)dt - \frac{1}{T}\int_T^{2T}\sin\left(\frac{n\pi t}{T}\right)dt =$$

$$= \frac{1}{n\pi}\int_0^{n\pi}\sin x\, dx - \frac{1}{n\pi}\int_{n\pi}^{2n\pi}\sin x\, dx$$

$$= \begin{cases} \frac{4}{n\pi} & n \text{ odd} \\ 0 & n \text{ even} \end{cases}$$

The Fourier series is

$$f(t) = \frac{4}{\pi}\sum_{n=1,3,5,\ldots}^{\infty} \frac{1}{n}\sin\left(\frac{n\pi t}{T}\right)$$

The MATLAB program which evaluates the above is shown overleaf. The result is shown in Fig. B.7.

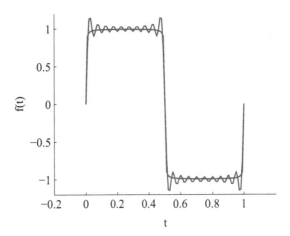

Fig. B.7 Fourier series approximation to the square wave. Line with oscillations corresponds to $N = 20$ terms; the other one was obtained using 100 terms.

```
% File name: square_wave.m
clear all
L = 0.5;
t = [0:100]/100;                    % arguments of independent variable
hold on
for N = [20 100];
    f = zeros(1,101);               % function initialization
    for n = 1:2:N
        pref = 4/(pi*n);            % evaluate prefactor
        f = f + pref*sin(n*pi*t/L); % summation of function
    end
    plot(t,f,'LineWidth',1.5)
end
axis([-0.2 1.2 -1.2 1.2])
ylabel('f(t)','FontSize',14)
xlabel('t','FontSize',14)
set(gca,'FontSize',14);             % size of tick marks on both axes
pause
close all
```

B.9 Fourier transform

Based on the results from previous section, we now introduce Fourier transform (FT). Consider function $f(t)$ to be nonperiodic on the infinite interval. FT is introduced by considering first a Fourier series representation (B.30) of the function which is periodic on the interval $[0, T]$, where T is the period. First, make substitution

$$t \to \frac{\pi}{T}t$$

in Eq. (B.37) and obtain

$$f(t) = \sum_{n=-\infty}^{n=+\infty} c_n e^{in\frac{\pi t}{T}}, \quad c_n = \frac{1}{T} \int f(t) e^{-in\frac{\pi t}{T}} dt \qquad (B.42)$$

Next, introduce discrete angular frequencies

$$\omega_n = n\frac{\pi}{T}$$

The difference between two successive frequencies $\Delta\omega$ is evaluated as

$$\Delta\omega = \omega_{n+1} - \omega_n = (n+1)\frac{\pi}{T} - n\frac{\pi}{T} = \frac{\pi}{T}$$

Using the above definition, the original series given by Eq. (B.42) can be written as

$$f(t) = \sum_{n=-\infty}^{n=+\infty} c_n e^{int\Delta\omega}, \quad c_n = \frac{1}{T} \int_{-T}^{+T} f(t) e^{-int\Delta\omega} dt \qquad (B.43)$$

Introducing new function g by the relation

$$c_n = \frac{\Delta\omega}{\sqrt{2\pi}} g(n \cdot \Delta\omega) \qquad (B.44)$$

one obtains

$$g(n\Delta\omega) = \frac{1}{\sqrt{2\pi}} \int_{-T}^{+T} f(t) e^{-int\Delta\omega} dt \qquad (B.45)$$

and

$$f(t) = \frac{1}{\sqrt{2\pi}} \sum_{n=-\infty}^{n=+\infty} \Delta\omega g(n\Delta\omega) e^{int\Delta\omega} \qquad (B.46)$$

Taking the limit $T \to \infty$, summation becomes an integral and $n\Delta\omega$ becomes continuous variable ω. One finds

$$g(\omega) = \frac{1}{\sqrt{2\pi}} \int_{-\infty}^{+\infty} f(t) e^{-i\omega t} dt \qquad (B.47)$$

and

$$f(t) = \frac{1}{\sqrt{2\pi}} \int_{-\infty}^{+\infty} g(\omega) e^{i\omega t} dt \qquad (B.48)$$

The above relations are known as the Fourier transform and inverse Fourier transform. There is a freedom in choosing prefactor in Eq. (B.44). By making different choices, one can introduce different Fourier transforms.

B.10 FFT in MATLAB

Fast Fourier transform (FFT) in MATLAB is defined as

$$X(k) = \sum_{n=1}^{N} x(n) e^{-j2\pi(k-1)(n-1)/N}, \quad 1 \le k \le N \quad (B.49)$$

which corresponds to the analytical definition (B.47) but without $\frac{1}{\sqrt{2\pi}}$ factor. FFT in MATLAB is performed by the function *fft.m*.

The inverse FFT (IFFT) which is performed by MATLAB function *ifft.m* is defined in MATLAB as

$$x(n) = \frac{1}{N} \sum_{k=1}^{N} X(k) e^{j2\pi(k-1)(n-1)/N}, \quad 1 \le k \le N \quad (B.50)$$

and it corresponds to the analytical definition (B.48). Again, the $\frac{1}{\sqrt{2\pi}}$ factor is absent. Let us illustrate a simple application of FT in MATLAB.

Example 1 As the simplest illustration of *fft.m* and *ifft.m* functions, we provide below MATLAB code to conduct FT and its inverse for a Gaussian pulse. In the end, one recovers original Gaussian pulse.

```
% File name: FT_example_1.m
% Doing preliminary Fourier transform in Matlab
clear all
N = 320;
T_0 - 10;
T_domain = 200;
Delta_t = T_domain/N;
T = Delta_t*(-N/2:1:(N/2)-1);
A_in = exp(-(T/T_0).^2);      % input gaussian pulse
plot(A_in)                    % plot of input gaussian pulse
pause
A_freq = fft(A_in);
plot(A_freq)                  % not a correct pulse
pause                         % (I explain it in the next example)
orig = ifft(A_freq);
plot(orig)                    % plots original input gaussian pulse
pause
close all
```

Example 2 As the previous example shows, Fourier transform in MATLAB must be done carefully. In this example we illustrate the workings of FFT in MATLAB. MATLAB code

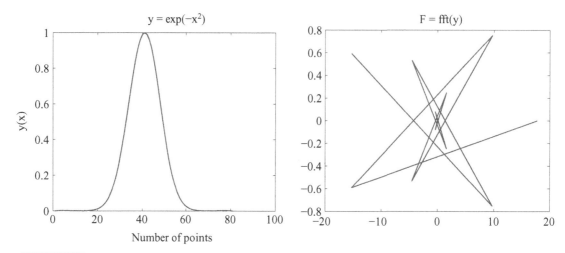

Fig. B.8 Original Gaussian pulse (left) and and its FT (right).

is shown in the following. First, a Gaussian pulse is created and plotted against a number of points along the x-axis. When needed, one can also plot it against x values. At the top of each figure (as a title) we provide an expression for the plotted function.

As the second figure, we plotted normalized Fourier transform using MATLAB function $fft()$. The plot does not resemble the expected result because the Fourier transform produces a complex function of the frequency. Vector F contains complex numbers and when you plot it in MATLAB, it plots the real part against the imaginary part.

To correct for the above deficiency, in the third figure we plotted the above transform plus the MATLAB *abs()* function. Altogether, both functions produce the desired result, i.e. magnitude of the Fourier transform of the Gaussian pulse.

In the remaining two figures we performed the inverse Fourier transform of the created Fourier transform.

Observe the use of function *fftshift.m* which puts the negative frequencies before the positive frequencies. The reason for using it is that when MATLAB function *fft.m* computes FFT, it outputs the positive frequency components first and then outputs the negative frequency components.

Below we show full MATLAB code for this exercise and all the figures created starting with an initial Gaussian pulse. In Figs. B.8–B.10, we show Gaussian pulse in time and its Fourier transform. It was obtained with the following MATLAB program.

```
% File name: FT_example_2.m
% Testing Fourier transform in Matlab
clear all;
x=-4:0.1:4;                  % Interval of independent variable
y=exp(-x.^2);                % Define Gaussian function
plot(y,'LineWidth',1.5);     % Plot Gaussian (Fig.1)
xlabel('Number of points','FontSize',14); ylabel('y(x)','FontSize',14);
set(gca,'FontSize',14);      % size of tick marks on both axes
```

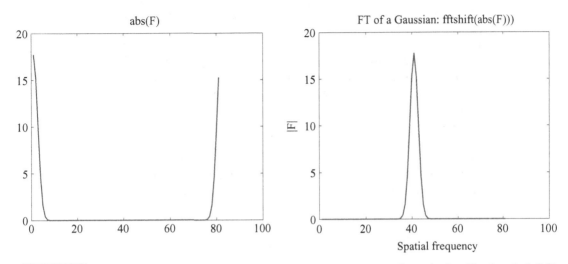

Fig. B.9 The next steps in the application of Fourier transform to the Gaussian pulse. After application of function $abs()$ (left), and $fftshift()$ (right).

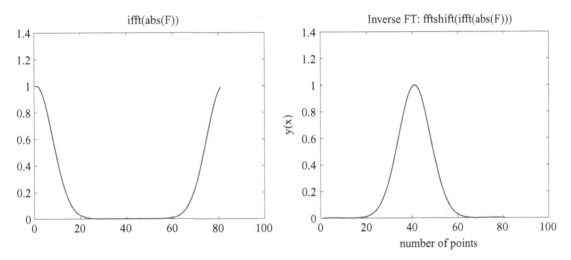

Fig. B.10 Steps in producing inverse Fourier transform of an original Gaussian pulse. After application of $ifft()$ (left), and after application of fftshift(ifft()) (right).

```
title('y=exp(-x^2)');
pause
%
F=fft(y);                        % Calculate FFT
plot(F,'LineWidth',1.5);         % Plot of  FT (Fig.2)
title('F=fft(y)');
pause
%
plot(abs(F),'LineWidth',1.5);    % Plot of abs value (Fig.3)
```

```
title('abs(F)');                    % Take absolute value and plot
pause
%
plot(fftshift(abs(F)),'LineWidth',1.5);    % Use Matlab fftshift function
xlabel( 'Spatial frequency'); ylabel('|F|'); % (Fig.4)
title('FT of a Gaussian: fftshift(abs(F)))');
pause
%---------------------------------------------------------
% Take inverse Fourier transform
%---------------------------------------------------------
plot(ifft(abs(F)),'LineWidth',1.5);   % Use Matlab function ifft (Fig.5)
title('ifft(abs(F))');
pause
%
plot(fftshift(ifft(abs(F))),'LineWidth',1.5); % Use functions fftshift
title('Inverse FT: fftshift(ifft(abs(F)))');  % and ifft (Fig.6)
xlabel('number of points'); ylabel('y(x)');
pause
close all
```

B.11 Problems

1. Use Newton's method to find roots of the following function: $e^x = 1 - cos3x$. Plot it first and use the graph to determine the starting point used by Newton's method.
2. Verify orthogonality relations, Eqs. (B.35) and (B.36) (where the limits of integration are $[0, 2\pi]$) for another limit, i.e. $[-\pi, \pi]$.
3. Derive expressions for coefficients a_n, b_n, c_n using new limits.
4. Show that the Crank-Nicolson scheme is accurate to $O(\Delta t)^2$.
5. Use the 2D version of Euler method to analyse projectile motion [3].
6. Analyse the effect of scaling within Fourier transform in MATLAB.

References

[1] S. E. Koonin. *Computational Physics*. The Benjamin/Cummings, Menlo Park, CA, 1986.
[2] P. L. DeVries. *A First Course in Computational Physics*. John Wiley & Sons, New York, 1994.
[3] A. L. Garcia. *Numerical Methods for Physics. Second Edition*. Prentice Hall, Upper Saddle River, NJ, 2000.

[4] C. F. Garald and P. O. Wheatley. *Applied Numerical Analysis*. Addison-Wesley, Reading, MA, 1999.

[5] S. S. Rao. *Applied Numerical Methods for Engineers and Scientists*. Prentice Hall, Upper Saddle River, NJ, 2002.

[6] M. T. Heath. *Scientific Computing. An Introductory Survey. Second Edition*. McGraw-Hill, Boston, MA, 2002.

[7] G. W. Recktenwald. *Numerical Methods with Matlab. Implementations and Applications*. Prentice Hall, Upper Saddle River, NJ, 2000.

[8] W. H. Press, S. A. Teukolsky, W. T. Vetterling, and B. P. Flannery. *Numerical Recipes in Fortran 90. The Art of Scientific Computing. Second Edition*. Cambridge University Press, Cambridge, 1992.

[9] L. V. Fausett. *Applied Numerical Analysis Using MATLAB*. Prentice Hall, Upper Saddle River, NJ, 1999.

[10] C. F. van Loan. *Introduction to Scientific Computing. Second Edition*. Prentice-Hall, Upper Saddle River, NJ, 2000.

[11] T. L. Harman, J. Dabney, and N. Richert. *Advanced Engineering Mathematics with MATLAB*. Brooks/Cole, Pacific Grove, 2000.

[12] B. Bradie. *A Friendly Introduction to Numerical Analysis*. Pearson Prentice Hall, Upper Saddle River, NJ, 2006.

[13] I. Barrodale and K. B. Wilson. A Fortran program for solving a nonlinear equation by Muller's method. *Journal of Computational and Applied Mathematics*, **4**:159–66, 1978.

[14] K. E. Atkinson. *An Introduction to Numerical Analysis. Second Edition*. Wiley, New York, 1989.

[15] G. Arfken. *Mathematical Methods for Physicists*. Academic Press, Orlando, FL, 1985.

Index

absorption
  coefficient, 181
  in a two-level system, 168, 169
  infrared, 109
  of power in photodetectors, 242
  spectrum, 243
  ultraviolet, 109
acceptance angle, *see* critical angle
active region, 173, 176
  in a VCSEL, 173
air mass, 369
amplifier gain, 206
amplifier, erbium-doped fibre (EDFA), 209–13
  typical characteristics, 213, 215
amplifier, semiconductor (SOA)
  Fabry-Perot (FPA), 223
  travelling-wave (TWA), 223, 227
antireflection (AR) coating, 50–1, 205
  half-wave layer, 53
  quarter-wave layer, 53
array waveguide grating (AWG), 318
asymmetry parameter, 67
attenuation, *see* loss

band gap, 243
  in solar cells, 376
bandwidth
  3-dB, 207
  of a photodetector, 247
  of an NRZ signal, 144
  of an optical amplifier, 207
  of an RZ signal, 143
  of gain, 206, 224
Bessel functions
  modified, 113, 114, 129, 132, 133
  ordinary, 113, 129, 131, 132
bit error rate (BER), 240, 252–7, 259, 336
bit-pattern independency, 242
bit-rate transparency, 241
boundary conditions
  absorbing (ABC) in 1D, 272–5
  absorbing (ABC) in 2D, 277–9
  for a magnetic field, 37–8
  for an electric field, 36–7
  Mur's first order, 272–3

Bragg mirror, 53, 54, 62
  reflectivity spectrum, 57, 62
Brewster's angle, 49–50

C-band, *see* transmission bands
carrier
  generation rate, 182
  leakage rate, 183
  lifetime, 184
  recombination rate, 183
chirping, 192–3
cloaking, 398–9
coefficient of finesse, 29
coordinate transformation, 398
Courant-Friedrichs-Levy (CFL) stability condition, 270
Crank-Nicolson scheme, 301, 302
critical angle, 18, 30, 64, 106
cross-gain modulation (XGM), 234–5
cross-phase modulation (XPM), 233, 235
current
  bias, 144
  dark, 246
  forward diode, 372
  injection, 143
  leakage, 182
  photocurrent, 242, 245, 371, 374
  reverse saturation, 374
  short-circuit, 369
  threshold, 143
current density, 35
cutoff wavelength
  in optical fibres, 122–3, 128
  in photodiodes, 245

Debye medium, 280
depletion region, 370
detection
  coherent, 240
  incoherent (direct), 240
diffraction, 396
dispersion
  equation of the waveguide, 66
  group velocity (GVD), 106, 124
  in free space, 24
  material, 124–8, 136, 280

modal, 124
multipath, 108
numerical, 269, 270, 283
of a 1D wave, 21
of a Debye medium, 280
of a Lorentz medium, 280
of a pulse in optical fibre, 127–8
waveguide, 124–127
distributed Bragg reflector (DBR), 173, 174
Drude relation, 389
dynamic range, 241

E-band, *see* transmission bands
effective index method, 89–92
effective medium, 391, 394
effective thickness of a slab waveguide, 77
EH modes, 116, 117, 119
electric
  boundary conditions, 36–37
  field intensity, 35
  flux, 35
electron, *see* carrier
emission
  amplified spontaneous (ASE), 215, 248, 255
  spontaneous, 168, 169, 208
  stimulated, 168, 169
equivalent circuit, 193
  for bulk laser, 194–6
  for PIN photodetector, 245, 250
  for solar cell, 373–6, 380
error function, 231
Euler differentiation method, 429–31
Euler's rule (integration method), 436
Euler-Cromer differentiation method, 431
evanescent waves, 396
excited state, 168
external modulator, 6, 193
eye diagram, 337

Fabry-Perot (FP)
  interferometer, 29–30
  resonance conditions, 171
  resonant modes, 171, 172
  resonator, 170
Faraday rotator, 323
Fast Fourier transform (FFT), 443–4
Fermi golden rule, 180
Fermi level, 176
fibre Bragg grating, 318
filter function, 258
finite difference method, 425–9
  backward differences, 426
  central differences, 426
  forward differences, 426
finite-difference (FD) approximation, 297–9

focal
  length, 19, 20
  plane, 19, 20
  point, 19
four-wave mixing (FWM), 233, 352
Fourier series, 438–9
  change of interval, 439–40
  of a square wave, 440–1
Fourier transform, 441–2
  of a chirped Gaussian pulse, 142
  of a Gaussian pulse, 141
  of a rectangular pulse, 139, 140, 158
  of an electric field, 146
Fresnel
  coefficients and phases for TE polarization, 44–7
  coefficients and phases for TM polarization, 47–8
  reflection, 45, 47
Fresnel approximation, 290

gain
  approximate formula, 206
  differential, 178
  in an EDFA, 213
  in semiconductors, 177–81
  peak, 178, 180
  ripple, 227, 237
  saturation, 190, 207, 216, 234
  spectrum of semiconductor laser, 172
Gaussian probability density function (pdf), 255
generation of carriers
  in laser diodes, 183
  in PIN photodiodes, 244
  in solar cells, 371
geometrical optics, 17–21
Goos-Hänchen (G-H) shift, 58–9, 397
graded index, 20, 107
grid, *see* mesh
GRIN system, 20–1
group
  delay, 124–5, 148
  velocity, 32
guides modes
  in optical fibres, 114
  in slab waveguides, 65–6

HE modes, 116, 117, 119
Helmholtz equation, 288, 289, 292, 299
hole, *see* carrier
hybrid modes, 116, 119

impedance, 41
interference
  in dielectric films, 26–7
  intersymbol (ISI), 255
  multiple, 27–9
invisibility cloak, *see* cloaking

L-band, *see* transmission bands
large-signal analysis (in laser diodes), 192
laser diode
   distributed feedback laser (DFB), 173, 174
   in-plane laser, 172
   vertical cavity surface-emitting laser (VCSEL), 172, 173
leap-frog differentiation method, 431
left-handed materials (LHM), *see* negative index materials (NIM)
lens
   perfect, 396
   thin, 19
linear system, 338
linewidth enhancement factor (Henry factor), 228
Lorentz medium, 280
loss
   extrinsic, 110
   in optical fibres, 108, 129
   intrinsic, 109–10
LP modes, 119–20

Mach-Zehnder interferometer (MZI), 235
magic time step, 270
magnetic
   boundary conditions, 37–8
   field intensity, 35
   flux, 35
Maxwell's equations, 262
   differential form, 35
   in cylindrical coordinates, 111–12
   integral form, 36
   source-free, 69, 388
mesh
   1D generation algorithm, 102
   staggered grid, 266
   Yee grid in 1D, 266, 267
   Yee grid in 2D, 276
mid-point differentiation method, 431
mode number
   azimuthal, 116
   radial, 116
modulation format
   non-return-to-zero (NRZ), 143–5
   return-to-zero (RZ), 143–5
modulation of semiconductor lasers, 143, 144
modulation response function, 188, 190–2
Muller's method, 100, 422–5

negative index materials (NIM), 385, 386
Newton's method, 420–1
noise
   Gaussian, 255–7
   in optical amplifiers, 208–9
   in photodetectors, 248–9
   shot, 249
   thermal (Johnson), 249–50

nonlinear Schrödinger equation, 353–7
normalized guide index, 67, 94
Numerical aperture (NA), 65, 106–7

O-band, *see* transmission bands
optical black holes, 399–400
optical cavity, 168
   Fabry-Perot, 170
optical communication system, 331–3
optical coupler, 319–22
optical fibre
   bit-rate, 108
   extrinsic loss, 110
   intrinsic loss, 109–10
   modes, 116–17
   single-mode, 122–3
optical isolator, 322–3
optical splitter, 319–20

p-n junction
   double heterostructure, 176–7
   homogeneous (homojunction), 175–6
   multijunction, 376–7
paraxial approximation, 288–92
permeability, 35
   negative, 386, 391–5
permittivity, 35
   negative, 387, 389–91, 395
   plasmon-like, 386
photocurrent, 245
   in photodiodes, 242
   in solar cells, 370, 371, 374
photodiode
   avalanche (APD), 242, 248, 258
   metal-semiconductor-metal (MSM), 246
   PIN, 242, 244
   power, 252
   sensitivity, 241, 257
photon
   density, 183
   lifetime, 183, 184, 187
Planck's radiation law, 169
plane wave, 40–1
plasma, 390
   frequency, 390, 391
Poisson distribution, 249, 252, 253
polarization
   circular, 42–3
   elliptical, 42–3
   linear, 42
   transverse electric (TE), 44
   transverse electric (TM), 47
polarization of dielectric medium, 148
population inversion, 167, 204, 208
power budget, 333–4
Poynting vector, 59–60

propagation constant, 41, 42
   in slab waveguides, 66, 95
pulse broadening, 12, 151
   in optical fibres, 106
pulse half-width, 142
pulse type
   chirped Gaussian, 141–2, 159
   Gaussian, 139–40, 159, 163, 164, 281–6
   rectangular, 138–9, 157, 340–2, 344
   Super-Gaussian, 140–1, 159, 162
pulse wave, *see* waveform
pumping, 167, 168

quantum efficiency, 247
quantum well, 173, 232
quasi-Fermi level, 175, 176

rate equations in EDFA, 211
rate equations in laser diodes
   for an electric field, 184–7
   for carriers, 182–3
   for photons, 183
   parameters, 184
ray optics, 17
   in metamaterials, 389
   in slab waveguides, 64–9
Rayleigh scattering, 110
receiver, 331, 343
recombination of carriers, 183
reflection
   at a plane interface, 17, 18
   coefficient, 17, 45, 46, 48
   external, 48, 49
   internal, 48
   of TE polarized waves, 44–7, 61
   of TM polarized waves, 48, 61
   total internal, *see* critical angle
refractive index, 17
   in GRIN structures, 20
   negative, 384
   numerical values for popular materials, 18
   relative difference, 65
relaxation-oscillation frequency, 189
Resonant cavity, *see* optical cavity
rise time, 144
rise time budget, 333–6
Runge-Kutta method
   fourth order, 433
   second order, 432

S-band, *see* transmission bands
Sellmeier equation, 31, 125–6, 135
signal-to-noise ratio (SNR), 208, 248, 336
Simpson's rule (integration method), 437–8
slowly varying envelope approximation (SVEA), 150, 296–7

small-signal analysis (in laser diodes)
   with linear gain, 188–9
   with non-linear gain, 189–92
Snell's law, 18
soliton
   interactions, 363
   period, 362
spectral
   intensity of solar energy, 368, 369
   responsivity, 247
split-ring resonator (SRR), 391–4
split-step Fourier method, 357–60
steady-state analysis
   in a laser diode, 187
   in an EDFA, 211–12
step index, 107
Stokes relations, 24–6
susceptibility, 186

TE modes
   in optical fibres, 116–18
   in slab waveguides, 71–2
three-level system, 210
threshold
   carrier density, 178, 182
   current, 143
time constant in photodetectors, 246, 247
time division multiplexing (TDM), 11
time-harmonic field, 39–41
TM modes
   in optical fibres, 116–18
   in slab waveguides, 71–2
train of pulses, *see* waveform
transatlantic telecommunications cable (TAT), 6
transfer function, 339
transfer matrix approach
   for antireflection (AR) coatings, 51–3
   for Bragg mirrors, 54–7
   for slab waveguides, 79–85
transitions
   in a two-level system, 169–70
   in semiconductors, 174–5
transmission bands, 10
transmittance of Fabry-Perot interferometer, 29, 33
transmitter, 331, 342
transparency density, 178
transparent boundary conditions, 304–6
transverse resonance condition, 66–7
   normalized form, 67–9
trapezoidal rule (integration method), 437
two-level system (TLS), 167, 169

U-band, *see* transmission bands

velocity
   group, 23–4, 124, 126
   phase, 21–3, 32
Verlet differentiation method, 432

wave equation, 38–9
  for TE modes, 72
  for TM modes, 72
  in cylindrical coordinates, 112
  in metamaterials, 388
waveform, 145, 160
waveguide
  2D, 88–92
  asymmetric slab (planar), 75–9, 95
  cylindrical (optical fibre), 110–23
  lossy, 86
  symmetric slab (planar), 72–5
wavevector, *see* propagation constant
weakly guiding approximation (wga), 118–19
wire medium, 395

Y-junction, 14, 325, 327
Yee algorithm
  lossless in 1D, 266–8, 282
  lossless in 2D, 275–7, 285
  lossy in 1D, 271–2